The Biophysical Characterisation of the Cell Surface

The Biophysical Characterisation of the Cell Surface

G. V. SHERBET

Department of Clinical Biochemistry
Cancer Research Unit
The University
Newcastle-upon-Tyne

1978

ACADEMIC PRESS
LONDON · NEW YORK · SAN FRANCISCO
A Subsidiary of Harcourt Brace Jovanovich, Publishers

ACADEMIC PRESS INC. (LONDON) LTD
24–28 Oval Road
London NW1

U.S. Edition published by
ACADEMIC PRESS INC.
111 Fifth Avenue
New York, New York 10003

Library of Congress Catalog Card Number: 77-15334
ISBN: 0-12-642050-5

PRINTED IN GREAT BRITAIN BY
J. W. ARROWSMITH LTD, BRISTOL

Preface

"கற்றது கை மண்ணளவு
கல்லாதது உலகளவு"

ஔவையார்

What we have learnt is a fistful of earth
What we are yet to learn is as vast as the Earth*

St Avvaiyyar
(Tamil Poetess 11th Century, India)

A major part of the research activity in the field of membrane biology has been concerned with the cell surface and has involved the characterisation of the components of the membrane surface, their organisation and topographical distribution. The cell surface owes this privileged position to its ubiquitous participation, mediated by the macromolecular components, in diverse biological events such as cell division, growth, differentiation, morphogenesis, neoplasia, cell recognition, antigenicity, and in the communication of environmental information to the cell. Therefore little need be said in justification of a book which aspires to review the field of cell surface biology, notwithstanding the limitations imposed on it by the state of my knowledge and interpretation of the events. This branch of membrane biology is so vast that I have approached it with a sense of humility which has been impressed on me by the quoted verse written by the 11th Century Tamil poetess St Avvaiyyar,* and which also serves to emphasise the finite state of our current knowledge of the complexities of biological phenomena.

The growth of science and the advance of scientific thought has, as the history of science would show, been generally non-uniform, with bursts

* Translation by courtesy of Dr M. S. Lakshmi.

of scientific activity interspersed with relatively more quiet periods. The peaks of activity have nearly always accompanied the invention and development of new technology. Bernal's* description of science as "ordered technique" aptly describes this association. Therefore science as a whole or any of its branches can be treated in two different ways, namely as unfolded by technical innovation and advance or as a compilation of observation and discussion. In this book I have taken the former course, and have discussed the theoretical aspects of some biophysical methods and have examined their application in the characterisation of the cell surface. I have then attempted a collation and integration of the different kinds of data relating to the cell surface in its normal state and as affected by some disease processes. I hope I have succeeded in giving the book a sense of cohesion rather than let it appear as a mixture of methods and results. In the main the book is about the cell surface as visualised by a number of bioelectric and electrokinetic techniques.

The scope of this book is wide simply by virtue of the subject being treated. Although the book would appear somewhat specialised in the sense that it deals only with the cell surface, I expect that it would prove its relevance in several fields of study such as cell differentiation, embryology, cancer research, cell biology, immunology and virology. It has been intended for use at the research level but I feel confident that it would prove useful also at the undergraduate level. If indeed it did, I would consider the time taken to write it well spent.

I am grateful to my many friends and colleagues who read through parts or whole of the manuscript and offered valuable criticism. I would like especially to acknowledge the help I received from the late Professor J. A. V. Butler, FRS; Professor David Kessel; Dr M. S. Lakshmi; Professor J. S. Mitchell, FRS; Professor K. R. Rees and Dr P. A. Riley. I am thankful to the large number of fellow scientists and publishers who most graciously allowed me to reproduce their published data and figures. Most of the work in my laboratory while at the University College Hospital Medical School and at the Chester Beatty Research Institute was done in collaboration with Dr Lakshmi without whose help perhaps there would have been little research of my own, and without whose constant encouragement I could not have written this book. I received financial support for my research from The Beit Memorial Fellowship, The Damon Runyon Memorial Fund, The Lord Dowding Fund for Humane Research, The Medical Research Council, The Peel Medical Research Trust, The Tenovus and The Williams Fellowship and

*Bernal, J. D. (1959) "Science in History". Penguin Books, England, p. 3.

the Central Research Fund of London University, to whom I am deeply indebted. Finally, I would like to thank Academic Press for the most cordial treatment accorded to me and to my book. It has been a considerable pleasure to work with them.

February 1978

G. V. SHERBET
Department of Clinical Biochemistry
Cancer Research Unit

Contents

4. Cell Electrophoresis

5. Isoelectric Equilibrium Studies of Cell Surfaces

6. Partition of Cells in Aqueous Two Phase Systems

Abbreviations

ADP	Adenosine diphosphate
ALS	Antilymphocyte serum
APF	Aggregation promoting factor
CCD	Counter current distribution
Con A	Concanavalin A (lectin from *Canavalia ensiformis*)
c.p.m.	counts per minute
CPDS	6,6′-Dithiodinicotinic acid, carboxypyridine disulphide
DAB	Dimethylaminoazobenzene
DFP	Diisopropylfluorophosphate
DMSO	Dimethylsulphoxide
EB virus	Epstein-Barr virus
EDTA	Ethylene diaminetetraacetic acid
EF	Encephalitogenic factor
EI	Ethyleneimine
EKZ	Electrokinetic zone
EM	Electron microscope
EO	Ethylene oxide
EPM	Electrophoretic mobility
e.s.u.	Electrostatic units
FDNB	Fluorodinitrobenzene
5-HT	5-Hydroxytryptamine
IEF	Isoelectric focusing
IEZ	Isoelectric zone
LPS	Bacterial lipopolysaccharide
LVD	Low viscosity dextran
MEM	Macrophage electrophoretic mobility
MSF	Macrophage slowing factor
MW	Molecular weight
NANA	*N*-Acetylneuraminic acid (sialic acid)
NANase	Neuraminidase (RDE, receptor destroying enzyme)
4-OHA	4-Hydroxyanisole
PAGE	Polyacrylamide gel electrophoresis

PAS	Periodic acid-Schiff reagent
PEG	Polyethylene glycol
PFU	Plaque forming unit
pH	Hydrogen ion concentration
pI	Isoelectric point
pIE	Isoelectrophoretic point
pII	Isoionic point
pK	Ionisation constant
PO	Propylene oxide
PPD	Protein derivative of tubercle bacillus (used as antigen in macrophage electrophoretic mobility test)
PPHE	Post-pH equilibrium
PTSC	Paratoluenyl sulphonyl chloride
PVS	Polyvinyl sulphate
Py	Polyoma virus
Py3T3	Polyoma virus-transformed 3T3 mouse fibroblasts
RDE	Receptor destroying enzyme (neuraminidase)
RNA	Ribonucleic acid
RNAase	Ribonuclease
RSPD	Receptor saturation pI differential value
RSV	Rous sarcoma virus
SDS	Sodium dodecyl sulphate
SL	Stationary level (phase) in electrophoretic cell
SV-40	Simian virus-40
SV-CHK	Simian virus-40-transformed Chinese hamster kidney cells
SV-TRK	Simian virus-40-transformed rabbit kidney cells
SV-3T3	Simian virus-40-transformed 3T3 mouse fibroblasts
TU	Tiselius unit for EPM ($=10^{-5}$ cm sec^{-1}V^{-1}cm)
WGA	Wheat germ agglutinin

Symbols

A	Hamaker constant, area
Å	Ångström (1 Å = 10^{-8} cm)
d	Thickness of electrical double layer
D	Dielectric constant of water (78·54 at 25°C)
D	Diffusion constant
e	Electronic charge $4{\cdot}8 \times 10^{-10}$ e.s.u.
η	Viscosity of solvent
E	Potential gradient in V cm^{-1}
E_m	Membrane potential
f	pH compensation factor for calculating EPM from isoelectric data
F	The Faraday 96 500 coulombs mol^{-1}
H	Distance between particles
i	Current in amperes
I	Ionic strength
k	Boltzmann constant ($1{\cdot}3803 \times 10^{-23}$ J°K^{-1})
K	Specific conductance
K	Debye–Hückel function; partition coefficient
M	Molarity of solution, gram mole
N	Normality of solution
N	Avogadro's number ($6{\cdot}023 \times 10^{23}$ mol^{-1})
P	Potential of the surface of particle
ψ	Potential at the interface
Q	Net surface charge
r	Radius of curvature of particle
R	Molar gas constant (8·3144 J mol^{-1} °K^{-1})
R	Resistance in ohms
S	Svedberg unit
σ	Electrical charge density
t	Time
T	Absolute temperature (absolute zero = −273·15°C)
v	Electrophoretic mobility
V	Velocity; volume
V	Volt

X	Field strength
ζ	Zeta potential
z	Valency of ion

This book is dedicated to the memory of
Professor John Alfred Valentine Butler, FRS
and Sir Alexander Haddow, FRS

1. Membrane Structure and Organisation

INTRODUCTION

Cellular membranes perform several functions essential to the life of the cell, and account for 80% of the dry weight of a cell (O'Brien, 1967). Membranes may be subdivided into three groups, namely the plasma membrane, the cytoplasmic and the organelle membranes. The plasma membrane forms the interface between the cell and its environment and maintains the structural integrity of the cell as a stable but dynamic unit and acts as a complex control system for the passage of water, electrolyte ions and other materials required for the metabolic activity of the cell. The plasma membrane also serves as a link in the communication of environmental information to the cell and controls cell division, growth and metabolism. In addition, it plays a significant role in differentiation and morphogenesis, and in cell recognition and antigenicity. Most of these functions are mediated by the macromolecular components of the membrane. Characterisation of cell membrane components and elucidation of their topographical distribution and organisation have therefore formed a major part of research in membrane biology. This area of research is so vast and the growth of the literature so rapid that it would be too ambitious to attempt to survey the whole field. This book is therefore restricted to the discussion of biophysical data, especially bioelectric and electrokinetic, relating to the cell membrane. The purpose of this chapter is to provide a brief description of the salient features of the structure and organisation of the membrane, in order to put the discussions in subsequent chapters in proper perspective. An exhaustive and complete discussion of this subject may be found in the recent reviews by Nicolson (1974a,b, 1975, 1976a,b).

MEMBRANE STRUCTURE

LIPID BILAYER STRUCTURE

Overton (1895) first suggested that membranes were composed of lipids. This was based on the readiness with which lipid-soluble substances penetrated the plasma membrane of the cell. In 1927 Gorter and Grendel extracted lipids from erythrocyte membranes. When these lipids were spread as a monolayer at an air–water interface, they covered an area twice as much as the erythrocyte surface area. This observation led to the postulation of the lipid bilayer. But the surface tension of the cell membrane is much lower than if the membrane had consisted of the lipid bilayer alone. Thus from considerations of surface tension, permeability characteristics and electrical conductivity measurements, Danielli and Davson (1935) deduced that the lipid bilayer is coated on both sides by proteins. The structure of unimolecular films of phospholipids and cholesterol at air–water interface indicated that these lipids were orientated in such a way that their polar groups projected into the aqueous phase. Therefore the lipid bilayer was visualised as a bimolecular leaflet with its non-polar fatty acyl chains orientated inwards perpendicular to the membrane surface. The polar groups of phospholipids were postulated to occur at the external surface, coated in addition by proteins and polysaccharides. This was the early concept of membrane structure generally accepted as the "sandwich" or "unit" model (Robertson, 1959; Davson and Danielli, 1952).

MICELLAR STRUCTURE

Alternative proposals for membrane structure include the globular or hexagonal micelle structure. Electron microscopy has revealed globular or hexagonal micelles in some membrane systems (Sjorstrand, 1963a,b,c; Lucy and Glauert, 1964). Sjorstrand described globular components of approximately 50 Å diameter in membranes from mouse kidney cells, and proposed that membranes may be composed of these globular units with protein molecules between them. This possibility was supported by the earlier finding of Fernandez-Moran (1957) and by subsequent work of Gent *et al.* (1964), Robertson (1963) and Blasie *et al.* (1965). Lucy and Glauert (1964) suggested, on the basis of their work on artificial lipid mixtures, that penta- or hexagonal micelles of lecithin and cholesterol occurred in plasma membranes. Pores existed between the lipid micelles. Proteins, of course, were postulated to occur as a layer on the surface.

FLUID MOSAIC MODELS

The lipid bilayer concept, although an attractive one and accounts for several properties of the cell membrane, does nonetheless present a static picture of membrane structure. It appears from recent work that many components of the cell surface are liable to and capable of rapid redistribution (Fig. 1). Surface immunoglobulins and antigens show aggregation and patching (Taylor, 1971; De Petris and Raff, 1972, 1973; Davis, 1972; Edidin and Weiss, 1972; Kourilsky *et al.*, 1972) leading to the endocytosis of some of the immunoglobulin complexed with the surface (De Petris and Raff, 1974). In heterokaryons produced by cell fusion, an intermixing of the surface antigens occurs in 30–40 minutes at 37°C (Frye and Edidin, 1970; Edidin and Weiss, 1972). Cell surface receptors of lectins are also known to move laterally and aggregate (Comoglio and Filigamo, 1973; Inbar *et al.*, 1973b; Inbar and Sachs, 1973; Nicolson, 1972a, 1973, 1974b; Bretton *et al.*, 1972; Rosenblith *et al.*, 1972; Garrido *et al.*, 1974).

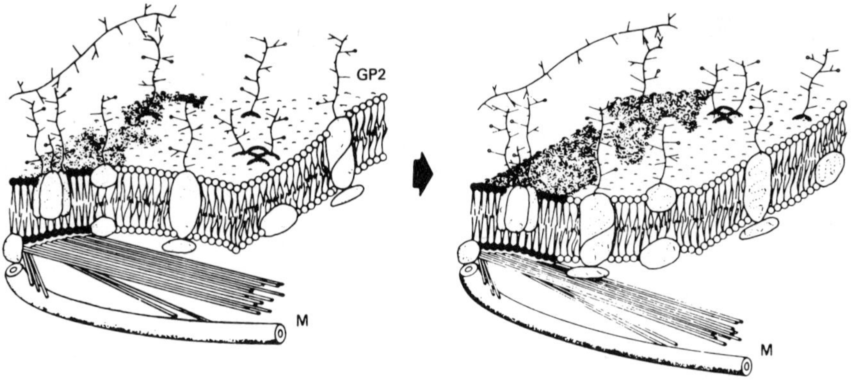

Fig. 1. Fluid mosaic model for the cell membrane showing lateral diffusion of glycoprotein complex GP2. M is membrane-associated cytoskeletal elements (from Nicolson, 1976).

The lipid bilayer concept has since been refined and restated by Singer and Nicolson and described as the fluid mosaic model (Singer, 1972; Singer and Nicolson, 1972; see also Vanderkooi, 1973; Vanderkooi and Green, 1970). This model takes into consideration the motional properties of lipid molecules arranged in a bimolecular leaflet. These dynamic properties include (a) rapid internal motion within each lipid molecule, (b) a lateral diffusion of molecules in the plane of the bilayer, (c) a rotational motion along the axis of the molecule and (d) a "flip-flop" motion of the molecule from one side of the membrane to the other.

These motional properties confer the fluidity which allows a redistribution of the surface components.

As in the Danielli-Davson model, the fluid mosaic model considers phospholipids as structurally the most important component. The arrangement of the phospholipids as a bilayer is dictated by its thermodynamic stability, with lipid acyl groups entering into hydrophobic interactions to exclude the aqueous phase (Singer, 1971). There may exist some structural asymmetry in the lipid bilayer (Bretscher, 1971, 1972, 1974). The proteins associated with the membrane may be immersed in the lipid bilayer to varying degrees. Some of these proteins may be amphipathic in nature, i.e. asymmetric with regard to the hydrophilic and hydrophobic regions of the molecule. By virtue of this asymmetry, these protein molecules penetrate right through the lipid bilayer, with the result that the hydrophobic region is buried in the hydrophobic lipid zone and the hydrophilic region of the molecule presented to the aqueous phase. Evidence for the occurrence of such transmembrane globular proteins and glycoproteins such as glycophorin (Fig. 2) of erythrocytes has come from the use of freeze fracture technique (Branton, 1966; Pinto da Silva and Branton, 1970; Pinto da Silva and Nicolson, 1974) and by chemical method involving surface labelling using the acylating agent ^{35}S-formylmethionyl sulphone methyl phosphate (Bretscher, 1971b,c). In freeze fracturing the bilayer is cracked along the midplane, replicas of the exposed interior surfaces are prepared and examined under the electron microscope. Freeze fracture has

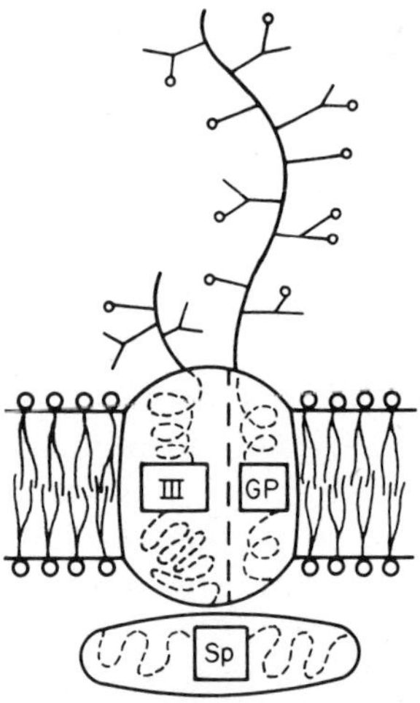

Fig. 2. A postulated transmembrane configuration of glycophorin (GP). III is component III and Sp is spectrin (from Pinto Da Silva and Nicolson, 1974).

revealed the occurrence of intramembranous particles in the hydrophobic matrix. Some of these particles may represent the class of proteins which remain embedded in the lipid bilayer. On the other hand, it has been suggested that these may form the hydrophobic portion of the glycoprotein molecule, with the carbohydrate-rich sections being exposed to the exterior of the membrane (Marchesi *et al.*, 1972; see pp. 67–68). Winzler (1969) showed that the carbohydrate moiety is attached to ten different sites along the peptide backbone to the N-terminal segment of the protein. Winzler also showed that this segment was associated with a hydrophobic core. That the remainder of this molecule, which constitutes its second hydrophilic segment, occurs on the inner surface of the membrane, is demonstrated by the surface labelling experiments of Bretscher (1971b; Segrest *et al.*, 1973). Both hydrophilic segments are labelled by membrane-impermeable reagents. However, only the N-terminal segment can be stained or excised by proteolytic enzyme from the exterior of intact cells or red cell ghosts. Under these conditions the C-terminal segment remains unaffected. Bretscher has described another membrane glycoprotein, the component *a* with a molecular weight of approximately 105 000, which has a transmembrane configuration similar to that described for glycophorin (Bretscher, 1971c,d; Phillips and Morrison, 1971; Hubbard and Cohn, 1972).

Singer and Nicolson (1972) have postulated another class of membrane proteins that can be easily detached by altering ionic conditions or using chelating agents. These proteins have been called the peripheral proteins to distinguish them from the integral components such as glycophorin or glycoprotein *a* of Bretscher. The inner surface of the erythrocyte membrane has been described as possessing a protein called spectrin (Mazia and Ruby, 1968; Marchesi *et al.*, 1969; Clarke, 1971; Nicolson *et al.*, 1971). Nicolson and Painter (1973) believe that spectrin is non-covalently bound to the glycophorin complex. These transmembrane glycoprotein complexes may be linked to one another by cytoskeletal elements, such as microtubules and microfilaments. Such elements have been shown to be associated with membranes (McNutt *et al.*, 1971; Perdue, 1973) and postulated to be responsible for the ligand-induced mobility of surface receptors (Yahara and Edelman, 1973a,b; De Petris, 1974; Loor, 1973; Yin *et al.*, 1972; Ukena *et al.*, 1974; Kram and Tomkins, 1973). This mobility of receptors is an energy-dependent process (Unanue *et al.*, 1972a; 1973; Loor *et al.*, 1972; Yahara and Edelman, 1972; Edelman *et al.*, 1973) and it can be blocked by cytochalasin B which acts on microfilaments and by colchicine and vinblastine which are known to disrupt microtubules (De Petris, 1974; Poste *et al.*, 1975a,b; Ryan *et al.*, 1974a,b).

In summary, the current postulated organisation of the cell membrane may be described as given in Fig. 3. Three levels of organisation may be detected (see Nicolson, 1974), (a) the integral

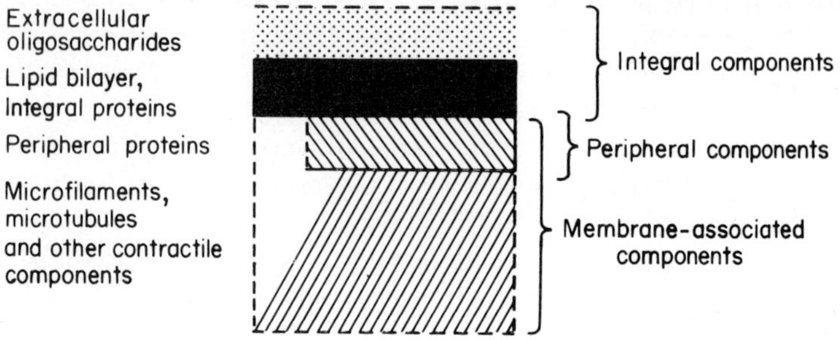

Fig. 3. A representation of the organisation of the cell membrane (from Nicolson, 1974a).

components which comprises the lipid bilayer, the integral proteins and the glycoproteins with oligosaccharide moieties exposed to the aqueous phase, (b) the peripheral protein components and (c) membrane-associated cytoskeletal elements.

2. Electrometric Titration of Cells

Electrometric titration has been extensively used over the past several years for the biophysical and biochemical characterisation of biological macromolecules especially proteins (Linderstrøm-Lang and Nielson, 1959; Jacobsen *et al.*, 1957; Edsall and Wyman, 1958; Tanford, 1961, 1962). To describe in a simple form this method consists of taking a protein solution of known concentration and adding to it a strong base or an acid, beginning with an arbitrary reference pH, and quantitating the amount of base or acid needed to attain a new pH value. When the amount of acid or base bound to the protein in the process of change of pH from the reference to the final pH is plotted against the pH range the titration curve for the protein concerned is obtained (Tanford, 1962).

EARLY EXPERIMENTS

The general principle of electrometric titration was applied to the study of the cell membrane by Kozawa (1914). Kozawa added dilute acid to a suspension of mammalian erythrocytes until the surface of the cells was brought to an isoelectric point. Whilst Kozawa should be admired for this novel idea of applying the titration technique to the cell membrane, he does not seem to have attempted a quantitation of the acid that needed to be bound to the erythrycyte surface to bring it to an isoelectric state.

THE ESTIMATION OF SURFACE CHARGE BY COLLOIDAL TITRATION METHOD

DESCRIPTION OF THE METHOD

In the 1950s a more sophisticated approach to the method of titration of the cell surface was made by Terayama (1952, 1954, 1962). This method has been called the method of colloid titration. It consists of titrating a positive colloid ion against a negative colloid ion. The end point of the titration is determined by the use of metachromatic dyes such as toluidine blue, brilliant cresyl blue or methylene blue, etc., which combine with minute excess of the negative colloid ion following attainment of the equilibrium point between the negative and positive polymer ions, and undergo a change in colour.

Colloid titration is performed over a range of pH (1–12). But the titration curve does not always show similarity to acid–base titration curve, which, as the author suggests, may be due to steric hindrance and ion exchange combinations.

Terayama (1952) has performed colloid titration of several compounds which have included proteins. In fact, it has been stated that any substance which combines stoichiometrically and undissociably with a standard polymer anion could be titrated in this way. The titration is dependent upon the presence of organic and inorganic salt like any electrometric determination. It is also dependent upon the ambient temperature.

After describing the use of this simple titration method for proteins, Terayama suggests the possibility of employing this method for elucidating the surface structure of bacteria, blood cells, yeast, etc. Terayama did in fact follow up these possibilities with publications on the colloid titration of bacterial cells (1954) and rat ascites hepatoma cells (1962).

PROTOCOL OF EXPERIMENTS

The colloid titration experiments with cells are performed as follows. A polyelectrolyte cation (such as clupein) of known concentration is placed in a flask. The cells whose surface is being investigated, are added to the flask in known quantities (determined as mg of nitrogen). The metachromatic dye indicator is now added. This mixture is titrated against a polyelectrolyte anion such as polyvinyl sulphate. The excess cation remaining in the mixture (i.e. the excess amount remaining after the polymer cation has bound the negatively charged groups on the cell surface) combines with the polymer anion which is being gradually added to the titrant mixture. A point of equilibrium is soon reached, when further addition of anion produces a change of colour of the metachromatic dye indicator. A parallel titration is also performed but without cells. The polyelectrolyte binding capacity is then calculated as follows.

Polyelectrolyte binding capacity

$$= \frac{\underset{\text{(cation)}}{\text{Concentration of titrant}} \times \underset{\text{anion}}{\text{Concentration of}} \times 1000}{\text{mg nitrogen}}.$$

COLLOID TITRATION OF BACTERIA

Terayama has studied several bacteria with the colloid titration method using macramine (*N*-polymethylated chitosan) which is a cationic polyelectrolyte. Colloid titration curves were obtained for bacteria which were subjected to heat treatment at 60°C for 30 minutes. The experiments showed that the bacterial surface was negatively charged. None of the bacteria that were examined reacted with polyvinyl sulphate (PVS). The surface pI of bacteria were found to be in the range pH 1–4. At pH 5–6 the macramine-combining ability of Gram negative bacteria underwent an abrupt change. The macramine-binding ability of Gram positive bacteria appears to be less than that of Gram negative organisms. In other words, the former have less number of ionisable groups contributing to the overall net negative charge than the latter (the Gram negative organisms). It may be recalled in this connexion that Gram staining of bacteria is in part attributed to the occurrence of surface RNAs in the Gram positive bacteria (Bartholomew and Umbreit, 1944). These two observations do not seem to be compatible unless it is assumed that the charged phosphate groups of the Gram positive organisms are not actually expressed on the surface or that they are shielded in some way from interacting with macramine molecules.

From Fig. 1 of Terayama's paper (1954) it seems that the pI of *Escherichia coli* var. communicor is between 2 and 4·5. The pI of *E coli* C which was determined by isoelectric focusing (pp. 185–188) was 5·78. The difference is indeed considerable, and may be attributable to the different factors which affect the pI in the two methods. This will be discussed critically in Chapter 5.

It may be stated that in general the colloid titration curves show resemblances with those of mixtures of lecithin, proteins and mucoproteins. The investigations are not extensive and as stated by Terayama himself the titration curves do not allow us to infer the nature of the ionisable groups. But Terayama did show that at least one of the acidic components of the surface of Gram positive bacteria, can be removed by treating them with bovine pancreatic ribonuclease (RNAase), and demonstrated that the titration curve changed rather dramatically after such a treatment. RNAase treatment was performed on *Streptococcus lactis* and *Nocardia mesenterica.* The untreated *Streptococcus* cells show no macramine binding at pH 2. At slightly higher pH (approximately 2·5) macramine binding is noticed. Following this there occurs a plateau until pH 6 is reached, when there occurs a second dramatic increase in macramine binding. In RNAase-treated cells no macramine bound below pH 5. The changes in the titration curve following RNAase treatment of *Nocardia* (which are Gram negative) are not as remarkable, but there is seen a distinct predominance of basic groups. It is difficult to interpret these results, especially to attribute them directly to the removal of ribonucleic acids from the surface of the organisms despite the fact that Terayama demonstrated a concomitant loss of Gram staining. It is possible that the changes in the titration curves are brought about merely by the adsorption of RNAase on the surface.

At this point the following criticism may be made of these titration studies of bacteria. (a) The treatment of bacteria at 60°C for 30 minutes may have resulted in a denaturation of the surface protein, and as a result the position of the ionogenic groups may have been completely altered. (b) Terayama does not seem to have taken into account the electrostatic interactions between the surface charges, nor applied the appropriate corrections in respect of the ionic strength of the medium in which the cells are suspended for the purposes of titration.

COLLOID TITRATION OF RAT ASCITES HEPATOMA CELLS

More recently Terayama has applied the colloid titration method to a number of ascites hepatoma cell lines, and has studied the binding of clupein by a variety of ascites hepatoma cells, with toluidine blue to monitor the end point of titration. The different cell lines showed characteristic protamine-binding pattern over the pH range 2–10·5. The majority of the types examined showed an isoelectric point at approximately pH 3·5. Ehrlich ascites carcinoma cells were also found to have a pI of 3·5 below which these cells showed PVS binding. The pI given for Ehrlich ascites cells does in fact correspond closely with the pI determined by means of electrophoretic mobility measurements of these cells (Cook *et al.*, 1962).

A GENERAL ASSESSMENT OF COLLOID TITRATION METHOD

The colloid titration method is liable to criticism on several counts.

(a) A minor point of criticism is that the end point of titration indicated by metachromasia of the dye may not be judged accurately, especially with the presence of cells in the titrant. Terayama has overcome this by ascertaining that the end point has in fact been reached by removing the cells from the titrant by centrifugation.

(b) A comparatively more serious point has been mentioned early in connexion with bacterial titrations, namely that the electrostatic interactions and ionic strength of the titrant are not considered, or appropriate corrections applied in this respect (see Debye–Hückel parameter discussed on pp. 45–47).

(c) It appears that while the method might give an estimate of the net charge present on the surface of the cells, it does not provide any information as regards the nature of ionogenic groups that are responsible for creating a net negative electric potential at the surface.

(d) The method, unfortunately, provides no information about the disposition of the groups that bind the polyelectrolyte ion. In fact, the most critical point in this method is the actual binding of the polyelectrolyte by the active groups on the cell surface. The colloidal cation-binding ability of a cell would make sense only if it were assumed that the distribution of the positive and negative charges on the surface of the cells and the polyelectrolyte ion is similar, especially as regards their spacing. If the distribution of the charges is not uniform and similar, which is very likely, the binding molecule may buckle or hump thereby shielding some groups and not result in the binding of all the active groups. In other words, the binding capacity is likely to give an erroneous estimation of the negative charge present on the cell surface.

(e) In assessing the state of net electric charge from their binding ability, it should be borne in mind that some of the polyelectrolyte cations might actually enter the cell. For example, using tritium-labelled calf thymus histones it has been shown that these proteins with molecular weights in the region of about 35 000 are able to cross the cell membrane and become distributed in the cytoplasm and nuclei of chick embryo cells (Sherbet, 1966). Blazsek and Gyergyay (1966) have shown that histones labelled with fluorescent dyes enter Ehrlich ascites carcinoma cells via the endoplasmic reticulum and become localised in the nuclear membrane as well as in the nucleolus. These authors have particularly drawn attention to the fact that the cell membrane is continuous with the nuclear membrane through intermediary ramifications of the endoplasmic reticulum (Mirsky and Osawa, 1961; Palay, 1960; Feldherr, 1962).

THE ESTIMATION OF SURFACE CHARGE BY CATION EXCHANGE METHOD

The cation exchange method is one of the analytical methods used in the investigation of the properties of soil. Schollenberger (1927) and Dreibelbis (1930) have estimated the exchangeable cations present in soil using a 2 M solution of ammonium acetate at pH 7. Essentially the method consists of repeatedly extracting the soil sample with ammonium acetate. As a result of this procedure the cations of the salt are exchanged for a variety of cations present in the soil. The exchanged cations are then individually estimated by chemical means. The total exchangeable cations present in the soil can be estimated by quantitating the ammonia adsorbed by the soil.

In a very preliminary study Majhi *et al.* (1967) have used this method to estimate the surface charge of the neural tissue of chick embryos from 8 to 20 days of development. These investigators repeatedly leached different sections of the central nervous system, namely forebrain, midbrain, hindbrain and the spinal cord of the embryos, with ammonium acetate at pH 7. The ammonia bound by the various regions of the central nervous system was then distilled. They estimated that the charge associated with the tissue on the basis that 1 ml of N NH_4OH contained one milliequivalent of charge i.e. 96·5 coulombs.

The charge associated with neural tissue as estimated by Majhi *et al.* (1967) is given in Table 1. These results show a certain degree of correlation between the differentiation of the nervous system in time and the charge associated with the tissue. Although these results are interesting, they can be accepted only with considerable reservations.

Apart from some experimental inadequacies the paper published by Majhi *et al.* lacks many important experimental details. Therefore an evaluation of these results at this stage may be inappropriate.

Table 1 *Electric charge of nervous tissue estimated by cation exchange method*

Age of embryo (days)	Charge in coulomb g^{-1} of nervous tissue			
	Forebrain	Midbrain	Hindbrain	Spinal cord
8	6·72	13·58	8·95	9·36
12	7·12	5·34	5·81	6·77
15	9·24	7·04	5·34	6·43
20	6·54	5·68	11·90	14·80

From Majhi *et al.* (1967).

It should be conceded that this approach to the estimation of charge associated with tissue is new and should be explored further. The most important question which cannot be answered is what the figures given in Table 1 might actually represent. From the title of their paper and the short introduction it appears that Majhi *et al.* consider that their estimations pertain to the surface charge of the tissue. They do not seem to have dissociated the tissue before extracting it and one wonders if the estimations would represent only a part of the total charge associated with a gram of the tissue. It appears to us perfectly possible that a certain amount of the ammonia may have been bound intracellularly. In our opinion therefore these estimations, at least at this stage, may not be regarded as pertaining to the surface charge.

3. Bioelectric Potential and the Cell Surface

The electrophysiological properties of the cell membrane have been used for the characterisation of the cell surface. In the past twenty years the bioelectric potentials across the cell membrane have been studied intensively and attempts have been made to relate it to the surface charge characteristics and the electrokinetic properties of cells with a view to achieving a characterisation of the cell surface.

ESTIMATION OF SURFACE CHARGE BY DONNAN DILUTION POTENTIAL

THE DONNAN MEMBRANE EQUILIBRIUM

Let us suppose that a chamber is divided into two compartments by a semipermeable membrane. Let us further suppose that one of the two compartments contains a sodium salt NaS which cannot diffuse into the opposite chamber in an undissociated form, nor can the ion S^- diffuse. Let us also suppose that the other chamber contains NaCl. Let the concentrations of NaS and NaCl be *a* and *b* respectively. At the initial stage the composition of the ions in the two compartments can be represented as in Fig. 4.

Since the membrane is permeable to Na^+ and Cl^- ions there will be a movement of Na^+ and Cl^- from compartment 2 into compartment 1, and Na^+ ions from compartment 1 to compartment 2.

When a state of equilibrium is reached between the two compartments if *x* amounts of NaCl have diffused into compartment 1 from compartment 2 the ionic concentrations at equilibrium would be as shown in Fig. 5.

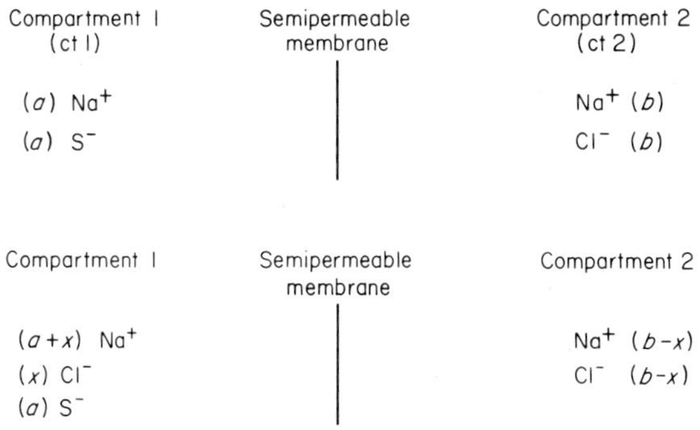

Figs 4 and 5. Schematic representation of a chamber divided into two compartments by a semipermeable membrane and containing NaS and NaCl respectively. Figure 4 gives the composition of ions at the initial stage. If it is supposed that NaS cannot diffuse in its undissociated form into compartment 2 nor can S^- ions diffuse, Fig. 5 represents the distribution of the ions at a state of equilibrium.

Since in the state of equilibrium the rate of diffusion of ions from compartment 2 to 1 is the same as that of diffusion from compartment 1 to 2, we have:

$$(a+x)x = (b-x)^2$$

$$\frac{(a+x)}{(b-x)} = \frac{(b-x)}{x}$$

i.e.

$$\frac{[Na^+]ct1}{[Na^+]ct2} = \frac{[Cl^-]ct2}{[Cl^-]ct1}. \quad (1)$$

This is the fundamental Donnan equation.

It is also clear from Fig. 5 and the Donnan equation that the concentration of Na^+ in compartment 1 is greater than the concentration of Na^+ in compartment 2—whereas the concentration of Cl^- in compartment 2 is greater than that of Cl^- in compartment 1. The position will of course be reversed if the non-diffusible ions were positive.

As a result of this unequal distribution of diffusible ions an electric potential difference will occur across the semipermeable membrane. The animal cell membrane forms a boundary between an insoluble colloidal phase and an aqueous electrolyte phase. In these two phases also therefore an asymmetric distribution of diffusible ions occurs and therefore a bioelectric potential occurs across the cell membrane. This potential difference has been used in determining the titration curves of cell surfaces.

MEASUREMENT OF NORMAL DONNAN POTENTIAL

Let us assume that a colloidal phase with a characteristic colloidal charge density is separated by a boundary from an electrolyte phase composed of sodium chloride. At a state of equilibrium between the two phases, the normal Donnan potential across the interphase, as derived by Joseph and Engel (1959) is given by the equation:

$$E_D = \frac{2{\cdot}303RT}{96\,500}\log_{10} r \tag{2}$$

where E_D is the normal Donnan potential, r is the Donnan ratio which is equal to:

$$\frac{[\mathrm{Na^+}]\ \text{colloidal phase}}{[\mathrm{Na^+}]\ \text{electrolyte phase}} = \frac{[\mathrm{Cl^-}]\ \text{electrolyte phase}}{[\mathrm{Cl^-}]\ \text{colloidal phase}}$$

96 500 is the value in coulombs of Faraday, R the molar gas constant and T the absolute temperature. It should be noted that in eqn (2) the transport of cations and anions does not appear because the rate of their diffusion is constant at equilibrium and therefore the ratio of concentrations of cations and anions in the two phases are reciprocals but when the phases are not in equilibrium (see eqn 3) the Donnan potential does depend upon the fraction of current transported by the cations and the anions.

MEASUREMENT OF THE DISPLACED DONNAN POTENTIAL

This is essentially the difference in the Donnan potential produced as a result of altering the ionic concentration in the electrolyte (or solution) phase. Joseph and Engel (1959) measure this by equilibrating the colloidal boundary with a standard isotonic (0·15 M) sodium chloride solution. The standard saline is then removed and replaced by 0·015 M sodium chloride. As a result of decimating the ionic concentration in the electrolyte phase, the Donnan potential across the boundary is altered. Joseph and Engel denote the Donnan potential measured against isotonic saline as E_D and the potential difference obtained against 0·1 isotonic sodium chloride as $E_{D'}$. The latter is also described as the displaced Donnan potential. The difference between E_D and $E_{D'}$ is denoted as E_d. E_d is described by Joseph and Engel as the dilution potential.

Now, in the displaced state the ratios of cations and anions in the two phases could be given as:

$$\frac{[\mathrm{Na}^+]\ \text{colloidal phase}}{[\mathrm{Na}^+]\ \text{electrolyte phase}} = 10r = \frac{[\mathrm{Cl}^-]\ \text{electrolyte phase}}{[\mathrm{Cl}^-]\ \text{colloidal phase}}$$

where r is the Donnan ratio.

In such a situation Joseph and Engel (1959) represent the displaced Donnan potential as follows:

$$E_{D'} = \frac{2{\cdot}303RT}{96\,500}[(n_1' - n_2') + \log_{10} r] \qquad (3)$$

n_1' and n_2' being the fractions of current transported by cations and anions respectively.

The dilution potential $E_d (= E_{D'} - E_D)$:

$$E_d = \frac{2{\cdot}303RT}{96\,500}(n_1' - n_2') \qquad (4)$$

and the general formula for the dilution potential:

$$E_d = \frac{2{\cdot}303\,RT}{96\,500}(n_1' - n_2')\log_{10}\frac{0{\cdot}15}{C} \qquad (5)$$

where C is the ionic concentration of the substituted electrolyte phase.

For a full derivation of the equation for the dilution potential the readers are referred to the original paper by Joseph and Engel (1959) and Joseph *et al.* (1959).

The dilution potential E_d is related to the density of the colloidal negative charge (x). The value of x can be estimated at different pH values at a given temperature and titration curves could be plotted. These are considered equivalent to acid–base titration curves (Engel *et al.*, 1968).

The displaced Donnan potential method of titration has been used in the study of many biological surfaces. These include human epidermis (Joseph and Engel, 1959), mammalian connective tissues (Joseph *et al.*, 1959), skeletal muscle (Engel *et al.*, 1960) and cultured mammalian cells (Engel *et al.*, 1968).

EXPERIMENTAL METHODS

The measurements of the Donnan potential and the displaced Donnan potential are made in the following circuit

$$\text{Hg} \mid \text{Hg}_2\text{Cl}_2 \mid \text{KCl} \mid \text{Solution I, II or III} \mid \text{Surface A, Surface B} \mid \text{Solution I}' \mid \text{KCl} \mid \text{Hg}_2\text{Cl}_2 \mid \text{Hg}$$

as represented in Fig. 6.

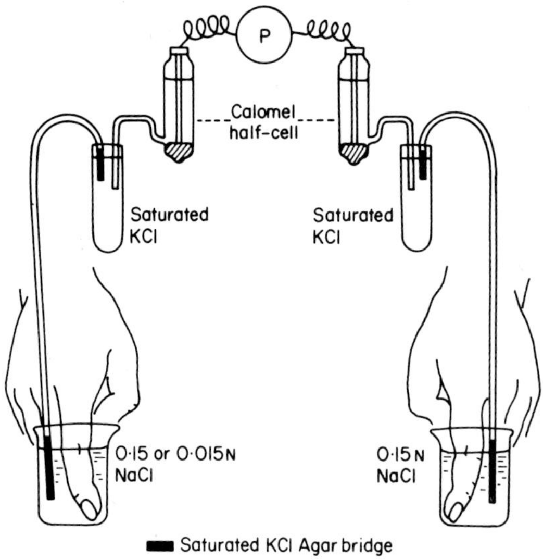

Fig. 6. Diagrammatic representation of the arrangement of circuitry for measurement of Donnan potential and the displaced Donnan potential (from Joseph and Engel, 1959).

The potentiometer P is connected to saturated KCl-calomel half-cells, liquid junctions with the electrolyte solutions being made with saturated KCl-agar in a saturated KCl bridge.

For examining the potential in human epidermis, the surface A in the circuit is provided by the forefinger of the right hand and surface B by the left forefinger. Solution I is isotonic NaCl, solution I′ is also isotonic NaCl. When the surfaces have equilibrated the potential is measured (E_1). Now solution I is replaced by solution II which is 0·015 M NaCl and the potential (E_2) is measured. Finally solution II is replaced by solution III which is a N KCl, and potential E_3 is determined. The difference between E_2 and E_1 is the dilution potential E_d; while $E_1 - E_3$ is the normal Donnan potential. The pH of solutions I and II (the latter a one-tenth dilution of the former) can be changed and the normal and displaced Donnan potentials measured.

The titration values obtained by Joseph and Engel (1959) for human epidermis (which represent average values obtained for five subjects) are given in Table 2.

Table 2 *Donnan dilution potential titration of human epidermis*

pH	E_D	r	Colloidal charge equivalent (x) Kg H_2O^{-1}	Dilution potential (E_d) = ($E_2 - E_1$) mV
7·4	12·8 ± 1·0	1·63	−0·153	18·6 ± 1·5
6.0	8·1 ± 0·8	1·37	−0·097	10·8 ± 1·9
4·6	0·7 ± 0·4	1·03	−0·009	−10·2 ± 1·0
3·7	−2·4 ± 0·9	0·91	0·028	−18·4 ± 1·0
2·9	−7·2 ± 0·9	0·76	0·083	−22·8 ± 1·2
2·2	−12·7 ± 1·2	0·61	0·153	−30·5 ± 1·7

E_D is the normal Donnan potential, r the Donnan ratio calculated from $E_D = 60 \times \log_{10} r$.
From Joseph and Engel (1959).

DONNAN DILUTION POTENTIAL TITRATION CURVES OF DERMIS

The titration curves obtained by Joseph *et al.* (1959) for dermis of rat and rabbit are remarkably similar at all pH values from approximately 2·2 which has been found to be their isoelectric point (pI) to pH 7·5. These authors have also examined the effects of the presence of positively charged substances such as ammonium ions, lysine, protamine, etc. This was done simply by measuring the displaced potential with a solution II (one-tenth isotonic) containing 0·0001 to 0·001 M of the basic substances. The effect was, as expected, a reduction in the colloidal charge of the tissue, but there was no difference in the pI of ammonium and glucosamine titration curves. The curves obtained in the presence of ammonium ions, glucosamine and lysine ran exactly similar courses, but at pH 3 the binding of lysine was much greater than that of ammonium ions or glucosamine, and roughly similar to the level of binding shown by protamine.

The maximum binding was shown by protamine. The titration curve after lysine binding showed a pI of 2·4 which is 0·3 units greater than the control titration curve. The titration curve obtained after protamine binding was 2·7. This upward shift in lysine- and protamine-bound colloidal surface may be related to the dissociation constants of the molecules binding the colloid surface. The dissociation constant of aqueous ammonia at 25°C is 4·75 (value taken from Hodgman *et al.*,

1958). The binding of the weak basic group can hardly be expected to affect the overall pK of the colloid membrane. In contradistinction the ε-NH_3^+ group of the lysine side chain has an observed pK of 10·8 (see Tanford, 1962). The protamines contain a very high proportion of arginine (Phillips, 1962), and the guanidyl group of the arginine side chain has an intrinsic pK of 12·5. It is not surprising therefore that the binding of lysine and protamine should have shifted the pI (see Sherbet *et al.*, 1972 and pp. 182–184).

The effect of anions on the titration curve of cartilage has also been found to be a reduction in the negative colloidal charge. This has been attributed to the fact that the binding of an anion such as picric acid is accompanied by a simultaneous uptake of hydrogen ions thereby resulting in a lowering of the colloidal charge (Joseph *et al.*, 1959).

INTERPRETATION OF DATA OBTAINED USING TISSUES

The displaced Donnan potential titration curve has been used to elicit information on the physiological and physicochemical aspects of metabolic activities of tissues.

The properties of the connective tissue are largely attributable to the nature of the intracellular matrix which is composed of collagen, mucopolysaccharides and mucoproteins. Joseph *et al.* (1959) find a clear correlation between the negative colloidal charge measured with certain tissues and their chemical composition. As may be seen in Table 3 the

Table 3 *Correlation between negative colloid charge and chrondroitin sulphate content of some tissues*

Tissue	Sulphate groups mMkg H_2O^{-1}	Negative colloidal charge (equiv. kg H_2O^{-1}
Tendon	5·0	0·035
Skin	0·8	0·05
Cartilage	120–140	0·16–0·17

From Joseph *et al.* (1959).

chondroitin sulphate content of cartilage is very high and the negative colloid charge is also much greater compared with that of tendon and skin which have a far lower chondroitin sulphate content. According to the authors electrostatic binding between the sulphate and amino groups is favoured because of the predominance of sulphate groups and as a result of this, carboxyl groups are available for titration.

In addition, the shape of the titration curves allows an estimation to be made of the availability of the active groups for titration. In the case of dermis and the tendon, the titration curves are isoelectric around pH.2·2. Between pH 2·2 and 3 the colloidal charge shows a steep increase and the curves then show a plateau. It may be seen from Fig. 7 that approximately 60% of the groups that contribute to the negative colloid charge are titratable between pH 2·2 and 3, a further 30% of the titratable groups occur between pH 3 and 4.

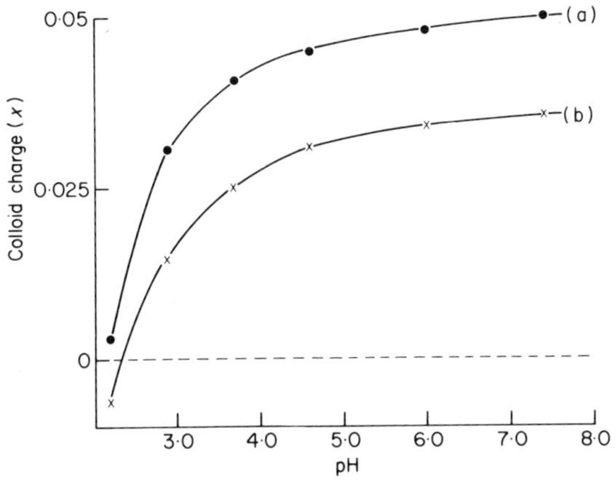

Fig. 7. Titration curves for (a) rat dermis and (b) tendon. Ordinate: colloid charge (x) in equiv. kg^{-1} water and abscissa pH (from Joseph *et al.*, 1959).

The degree or extent of physiological activity occurring in a tissue may be reflected in the negative colloidal charge measureable at the tissue surface. The Donnan dilution potential curve for frog muscle was unaffected by temperatures ranging from 2 to 25°C but treatment of the muscle with metabolic inhibitors such as dinitrophenol, iodoacetate, cyanide, etc. did produce changes in the titration. There was a reduction in the negative colloidal charge. These experiments appear to indicate a redistribution of ions between the intra- and extracellular phases as a result of inhibition of the respiratory processes (Engel *et al.*, 1960).

DONNAN DILUTION POTENTIAL TITRATION CURVE FOR MAMMALIAN CELLS

The Donnan dilution potential measurement has also been carried out with two mammalian cell lines maintained in tissue culture, namely rabbit myocardial cells and lung cells from Swiss white mice. The experimental method followed was as described previously. The cells to be studied were centrifuged in a plastic tube. A pin hole was made in the tube in the region of the cell pellet so that electrical contact may be achieved; yet the hole was small enough to prevent leakage. The electrical contacts with the calomel half-cells were as usual made by saturated KCl-agar junctions (see Fig. 8).

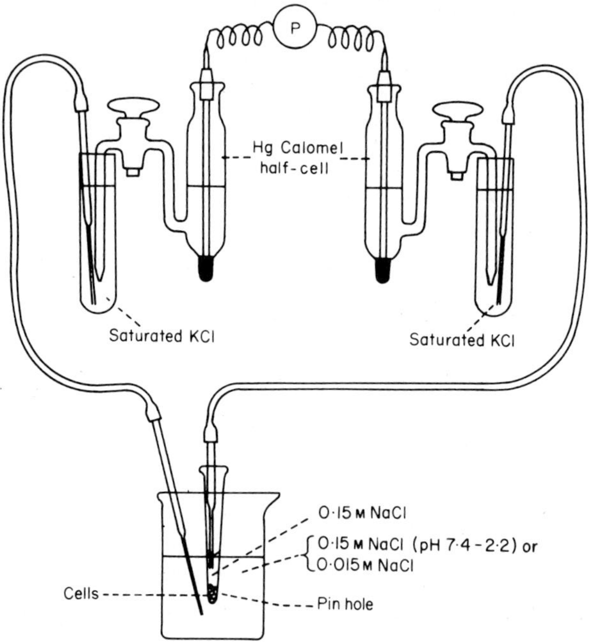

Fig. 8. Electrical circuit for measurement of Donnan dilution potential (from Engel *et al.*, 1968).

From this study, identical titration curves for the two cell types have been illustrated by Engel *et al.* (1968). There was no difference in the titration curves using either 0·15 M NaCl or Ringer solution. A reduction in the colloidal charge occurred in rabbit myocardial cells which had been left in trypsin for 16–17 hours at 37°C. This reduction may be due to the removal of negatively charged sialopeptides from the cell surface. Reduction in electrophoretic mobility of human erythrocytes (Cook *et al.*, 1960) and of murine CL3 cells (Ward and Ambrose, 1969) has been reported as a consequence of treatment of the cells with trypsin. Cook *et al.* (1960) have in fact isolated a sialomucopeptide from human erythrocytes using this enzyme.

Using a similar experimental set up, Catchpole *et al.* (1961) earlier reported on the titration curves of some murine ascites cells. The titration curves of these cells (details of which are not available) are said to be affected by arginine. Unfortunately this report which is in the form of an abstract does not describe the nature of the arginine effect. It would be interesting to know for instance whether the effect was similar to that of protamine on the titration curves of rabbit dermis; especially the effect on the isoelectric point which, it might be recalled, was shifted in the basic direction by protamine (see pp. 59–62).

It should be mentioned here that in most of the tissues studied by the Donnan dilution potential method, the titration curves have run similar courses, as for example in myocardial cells of rabbit, lung cells of Swiss white mice, certain types of connective tissue and even ascites tumour cells. It is even more surprising that in all these varied cell types (and also erythrocytes, see below) the titration curves were isoelectric between pH 2·2 and 2·9.

DONNAN DILUTION POTENTIAL TITRATION STUDIES WITH HUMAN ERYTHROCYTES

Corley and Joseph (1966) have studied human erythrocytes from normal subjects as well as from patients recovering from shock, or after major surgical operations (see Table 4). They find that the colloidal charge of red blood cells of normal human subjects is almost doubled as a consequence of treatment of the cells with low viscosity dextran. These authors have cited Rothman *et al.* (1957) as stating that dextran forms a film at the red cell surface. Dextran is a non-ionic and inert polymer and

Table 4 *Negative colloidal charge of human erythrocytes*

Source of erythryocytes	Negative colloidal charge equiv. kg H_2O^{-1}	
	Untreated	LVD treated
Normal subjects	0·046	0·081
"Shock" patients	0·068	
Patients recovering from major surgery	0·068	0·081

LVD is low viscosity dextran.
From Corley and Joseph (1966).

if it did form a thin sheet over the cell surface the obvious result should have been *status quo* in the negative colloid charge. Low molecular dextrans have been used to inhibit intravascular aggregation of erythrocytes (Gelin, 1956; Gelin and Ingelman, 1961). High molecular weight dextran, on the other hand, causes agglutination of blood cells (Minor and Burnett, 1952, 1953; Wildy and Ridley, 1958). The mechanism by which these opposite effects are produced by dextran are not yet clear—dextran presumably interferes with the electrostatic interactions between the charged groups of the blood cells.

The observations made by Corley and Joseph (1966) that red blood cells from shock patients possess a higher negative charge than do cells from normal subjects is also not compatible with erythrocytes from shock patients showing tendencies towards aggregation. A higher negative charge is more likely to produce a decrease in adhesiveness and not provide conditions conducive to aggregation or agglutination (but see pp. 129–131).

GENERAL COMMENTS

The Donnan dilution potential titration of the cell membrane is a very ingenious application of the principle of Donnan equilibrium. The cell membrane acts as a boundary between the colloidal phase, with the membrane forming a part of the colloidal phase, and the electrolyte phase. Therefore the colloidal charge that is being estimated by Donnan dilution potential measurements probably reflects the electrical charges residing in the cell membrane. We have discussed evidence that binding of tissue surfaces with cationic substances, for instance, alters the negative colloidal charge of the surface. Nonetheless, it would be difficult to equate the colloidal charge to the surface potential of cells (see later discussion and also pp. 45–47).

It appears that the method allows reasonable conclusions to be drawn as regards the composition of the membrane and the intercellular matrix, as in the case of cartilage tissue (and epidermis) which are rich in substances such as chondroitin sulphate and which show considerably higher negative colloidal charge values (x) (see Table 5) than tissues which do not contain such highly acidic substances in excess. In addition, the possibility that it could be employed to monitor physiological changes occurring in cells as a result of changes in physiological conditions, experimentally designed changes or in pathological states, would appear to be a very significant use of this technique.

It may be felt, on the other hand, that it is difficult to interpret some of the data obtained from these experiments especially the isoelectric

Table 5 *Negative colloidal charge of some mammalian tissues*

Tissue	Negative colloidal charge at pH 7·4 equiv. kg H_2O^{-1}
Human epidermis	0·15
Human costal cartilage	0·15
Rabbit myocardial cells	0·065
Mouse lung cells	0·06
Murine ascites tumour cells	0·05
Human erythrocytes	0·046
Rabbit dermis	0·046
Rat dermis	0·03

Data obtained from various publications discussed in the preceding sections.

points which as we have pointed out earlier always fall between 2·2 and 2·9. A simple interpretation of these values as has been done with pI values obtained in isoelectric focusing experiments (see pp. 181–190) may not be possible. It should in theory be possible, however, to elicit much information from the pI values by specifically altering the active groups in the membrane using substances with known characteristics, such as the nature of ionogenic groups, their intrinsic pK values, etc. The results that have been obtained with tissues that have been treated with protamine, for instance, are not amenable to simple interpretation. The intrinsic pK values of the predominant active groups are often altered as a consequence of the secondary structure assumed by such macromolecules.

Finally, it is also difficult to assess the sensitivity and reliability of the method. One wonders, for instance, about the incongruity in the experiments on red blood cells, and their ability to aggregate. Looking at Table 5 one would be inclined to disbelieve the values for murine ascites and human erythrocytes when compared with the negative colloid charge for mouse lung or rabbit myocardial cells. These values are incompatible with their respective electrophoretic mobilities (see pp. 55 and 194). Erythrocytes and ascites tumour cells are among cell types which show the highest electrophoretic mobilities. However, electrophoretic mobilities are dependent upon the ζ potential of the cell. It is possible therefore that the colloidal charge is not strictly equivalent to the ζ potential.

MEMBRANE POTENTIAL AND THE CELL SURFACE

MEMBRANE POTENTIALS (E_m) OF CELLS

As discussed in the foregoing section, a Donnan equilibrium exists when a charged particle is restrained by a semipermeable membrane. When the diffusible ions have attained an equilibrium, a potential (E_D) difference could be observed if the two chambers are connected to reversible electrodes via salt bridges.

The transport of ions across a biological membrane can be mediated by an energy-dependent mechanism. Ions can also move in response to electrical stimuli i.e. gradients of electrical potential. From the converse of this, an unequal distribution of ions should result in a potential difference across that membrane. This has been termed the trans-membrane or membrane potential (E_m), to which the rest of ths chapter will be devoted. The Donnan and membrane potentials are manifestations of the electrophysiological properties of the membrane. Another source of potential at the cell membrane is due to the polar groups residing at the cell surface and to their interaction with the electrical double layer surrounding the cells. The potential at any point in this ionic cloud, which is due to the ions in the region and the charges on the surface of the particle, is the surface (ψ) potential. The latter can be measured as an approximation of the potential at the surface of shear, namely the ζ potential. When a charged particle is suspended in an ionic medium, the surface potential is reduced by the ionic cloud of opposite charge that is attracted by the particle. The ζ potential is a resultant potential arrived at by the sum of the potential at the zone of shear and the potential of ions in the vicinity of the particle. It is therefore dependent upon the ionic concentration and composition of the suspending medium and is related in a still unknown manner to the surface potential. The concepts of ψ and ζ potentials in relation to the characterisation of the cell surface are discussed in detail in Chapter 4 (pp. 45–50).

The membrane potential is measured as the potential difference between an intracellular and an extracellular electrode. Fine tipped (<1 μm diameter at tip) micropipettes may be used as electrodes. One electrode is inserted into the cell while the extracellular (indifferent) electrode is brought close to the membrane. The intracellular electrode has a high impedence (5–50 megohms d.c. resistance) and is filled with 2–3 M KCl. The extracellular electrode is filled with balanced salt solution or culture medium. The intracellular electrode is connected to a capacitance-compensated unity gain amplifier. The extracellular electrode is grounded through a separate electrode.

An optical method for the measurement of the membrane potential using dyes is also available and has been described by Cohen *et al.* (1974) and Ross *et al.* (1977). The changes in the optical properties of the dyes, such as extrinsic absorption, fluorescence, dichroism and birefringence are related and could be used to monitor membrane potential. It is believed that alterations in the membrane potential result in changes in the concentration of the dye in solution and that associated with the membrane. These changes in the dye concentration result in a shift in the equilibrium between the amount of the monomer and dimer (or higher aggregates) of the dye bound to the membrane. Such a shift in the equilibrium is believed to be reflected in altered optical properties of the dye (Sims *et al.*, 1974; Ross *et al.*, 1977).

The membrane potentials of a large number of animal cells have been measured. These potentials are polarised with the cell interior being negative relative to the exterior of the cell, although Lundberg (1955) described a positive membrane potential of 5–10 mV in unfertilised eggs of *Psammechinus*. De Laat *et al.* (1975) have reported a positive potential of 37 mV in the uncleaved egg of *Xenopus laevis*. The potential measurements pertaining to some of the cells examined are given in Table 6.

MEMBRANE POTENTIAL AND MITOTIC REGULATION

Cone (1969, 1970, 1971) proposed that the membrane potential plays an important role in the control of cell division. A correlation between mitotic activity of cells and their membrane potential has emerged from a number of observations. For instance, in unfertilised sea urchin eggs the potential is reported to rise from −10 mV to −60 mV on fertilisation (Steinhardt *et al.*, 1971). Parthenogenetic activation of eggs also produces an increase in membrane potential (Steinhardt and Mazia, 1973). In *X. laevis* the membrane potential rises from −10 mV to −20 mV within 6–8 minutes of the onset of cleavage (De Laat *et al.*, 1973, 1974; De Laat and Bluemink, 1974). In *Rana pipiens* the potential has been described to rise to between −20 mV (Woodward, 1968; Takahashi and Ito, 1968) and −50 mV (Morrill and Watson, 1966; Morrill *et al.*, 1971). De Laat *et al.* (1975) observed a positive potential of 37 mV in the uncleaved egg. A reversal of the polarity of the membrane occurs about 15 minutes prior to cell division and at the onset of division a negative potential of 11 mV is reached. The hyperpolarisation of E_m at the cleavage stage of embryonic development has also been described in

Table 6 *Membrane potentials of some cell types*

Cell type	Membrane potential (−mV)	Reference
Sea urchin unfertilised eggs	10	Steinhardt *et al.* (1971)
Triturus and *Xenopus* oocytes	10 to 20	Kanno and Loewenstein (1963)
Amphiuma erythrocytes	16	Pape *et al.* (1975)
Chick spinal nerve	65	Crain (1956)
Neuroblastoma cells	37·5	Nelson *et al.* (1971)
Chick embryo heart	29 to 40	Fingl *et al.* (1952)
Chick embryo heart cells	69·5	DeHaan and Gottlieb (1968)
Skeletal muscle cells	59·1	Fischbach *et al.* (1971)
FL cells	18 to 20	Redmann *et al.* (1974)
Peritoneal leukocytes of rabbit	7 to 8	Redmann *et al.* (1974)
Chinese hamster lung	27	Sachs *et al.* (1974)
L-strain mouse fibroblasts in interphase	10·5	Cone (1969)
MK-2	12.3	Redmann (1971)
KB cells	13	Redmann (1967)
Human ovarian tumour cells in interphase	2·5	Redmann *et al.* (1972)
HeLa (human cervical tumour) cells	11	Borle and Loveday (1968)
Ehrlich ascites cells	13·7	Smith and Adams (1977)

echinoderm (Tupper, 1972; Tupper and Power, 1973) and teleost embryos (Bennett and Trinkaus, 1970), and thus appears to be a general phenomenon. In addition, inhibition of mitosis by blocking agents such as vinblastine appears to reduce the membrane potential (Redmann *et al.*, 1974).

Earlier Cone (1969) had observed that the membrane potentials of L-strain mouse fibroblasts increased from an average value of −10·5 mV to −16 mV at the initiation of prophase, when it is reported to decline to reach an almost complete depolarisation by the time the cell is in the mid-metaphase stage of mitosis. A two-fold increase in the membrane potential (as compared with the interphase potential) has also been reported in human ovarian tumour cells entering mitosis (Redmann *et al.*, 1972). However, from this work it is not clear in which phase of mitosis the cells might have been. The authors have described them as rounding-up cells. If the rounding-up process described by Cone (1969) for L-cells can apply to ovarian tumour cells in culture, it may be assumed that the latter were in the prophase stage of mitosis. An increase in the membrane potential in that case would be consistent with the findings of Cone (1969) of a maximum polarisation in the prophase which is followed by a gradual depolarisation in the course of mitosis.

Recently Sachs *et al.* (1974) have examined the original suggestion of Cone by measuring the membrane potential of synchronous Chinese hamster lung (V79) cells at different stages of the cell cycle. They found that at the G_1 phase the membrane potential was low (around −20 mV). The latter rose sharply at the onset of the S (synthetic) phase. This high level of E_m was maintained during the S and the G_2 phases, but then a decline ensued as the cells entered mitosis. Ebert's group had earlier come to the conclusion that the membrane potential was related to the ability of the cells to synthesise DNA (McDonald *et al.*, 1972a,b). Steinhardt and Mazia (1973) found that parthenogenetic activation of sea urchin eggs not only caused an increase in the membrane potential but also produced a concomitant increase of thymidine uptake and incorporation into DNA. However, there does not appear to be any causal relationship between the observed changes in membrane potential and the cellular events occurring in the cell cycle. For, Tupper (1973) has shown that the hyperpolarisation of the E_m occurring in association with the process of fertilisation could be blocked without affecting the synthesis of DNA, merely by altering the external potassium ion concentration. The hyperpolarisation of the potential of cleavage-stage embryos has been attributed to an increased permeability of the membrane to K^+ ions (see De Latt *et al.*, 1974; Lassen *et al.*, 1976). The membrane which is newly formed during the cytokinesis of *Xenopus* eggs shows a five-fold increase in permeability to K^+ as compared with the original membrane. The intracellular sodium ion activity also shows concomitant increase with hyperpolarisation of the potential (De Latt *et al.*, 1974). In most of this work De Latt *et al.* (1974, 1975) have shown that the observed potential could be described by the constant field equation (Goldman, 1943; Hodgkin and Katz, 1949) which gives the relationship between the membrane potential and the ionic gradients across the cell membrane:

$$E_m = \frac{RT}{F} \ln \frac{P_{Na} a^0_{Na} + P_K a^0_K}{P_{Na} a^i_{Na} + P_K a^i_K} \qquad (6)$$

where R is the molar gas constant, T the absolute temperature, F the Faraday, P_{Na} and P_K the permeabilities of Na^+ and K^+, a^0_{Na} and a^0_K the external activities and a^i_{Na} and a^i_K the intracellular ionic activities of sodium and potassium respectively.

One may conclude therefore that a close association does exist between the membrane potential and the stage of the cell cycle. It may be expected that the selective membrane permeability will alter to suit the metabolic requirements of the cell and such alterations may affect the membrane potential.

However, the only investigations which give credence to the idea of a functional relationship between mitotic activity and membrane potential are those described by Cone (1969). Cone found that blockage of mitosis can be achieved by maintaining cells in mannitol solutions. The degree of mitotic blockage produced was related to the extracellular concentration of mannitol. When this is taken with the exact agreement he found between the E_m level and mitotic activity, the functional relationship between the latter two features of the cell becomes highly suggestive. Nevertheless, the mechanisms by which extracellular mannitol alters the mitotic activity or the potential are not known. Cone suggests that mannitol may block metabolic pathways essential for the phase preparatory to mitosis, and presumably also the increased tonicity results in the maintenance of a sufficiently negative potential by the sodium pump system. Also compatible with this view is the mitotic stimulation achieved in L-cells by pulsed infusion of sodium ions using 0·17 M NaCl. Cone suggests that the infusion treatment accomplishes the natural process involved in the phenomenon of polarisation associated with prophase initiation, namely a significant increase in the intracellular concentration of sodium ions.

MEMBRANE POTENTIAL OF MALIGNANT CELLS

A variety of electrophysiological and biochemical observations may have prompted the investigation of the possible role played by the membrane potential in malignant transformation of cells. Firstly, Loewenstein and his colleagues showed over a decade ago using intracellular electrical techniques, that ionic communication between adjacent cells occurs through sections of membrane which have low electrical resistance. When two such areas are apposed a channel of communication is opened between cells (Lowenstein, 1966). These channels of communication also allow the passage of molecules with molecular weights up to 60 000 into the cells. Whilst such pathways have been detected in embryonic and adult cells and normal cells in culture, no such communication has been observed to exist between tumour cells (Loewenstein and Kanno, 1966, 1967; Loewenstein and Penn, 1967; Jamakosmanović and Lowenstein, 1968). The attractive notion that control substances which might influence the mobility, organisation, division and differentiation may pass from cell to cell, has been mooted by these observations.

Secondly, it was known since the turn of this century that the calcium content of tumour cells was less than that of homologous normal cells (Beebe, 1904; Clowes and Frisbie, 1905; Carruthers and Suntzeff, 1944; Suntzeff and Carruthers, 1944; but see Hickie and Kalant, 1967 who find increased calcium levels in a Morris minimum-deviation hepatoma). Cone (1969) has argued that the variations in membrane potential are caused by modulation in active Na^+ pumping rate and/or Na^+ conductivity of the membrane and that the inability of cancer cells to incorporate Ca^{2+} may affect both these parameters, and produce a consequent alteration in the potential. Cone has also cited in support of this view the observation that lack of Ca^{2+} in nerve renders the cell membrane unable to maintain a high negative potential.

Finally, the correlation between mitotic activity and the membrane potential and also the possible functional relationship between them has contributed significantly to efforts into examining the involvement of the membrane potential in neoplastic phenomena.

Compatible with some of these observations is the reported low membrane potential in tumour cells as compared with homologous normal cells. Jamakosmanović and Loewenstein (1968) examined the membrane potentials of normal thyroid cells of the hamster and the rat and the corresponding tumour cells. They consistently obtained significantly reduced membrane potentials in tumour cells. A similar reduction (of about 40%) was reported by Kanno and Matsui (1968) for neoplastic stomach epithelium of man (see Table 7). Redmann *et al.* (1972) reported potentials as low as $-2{\cdot}5$ mV for interphase cells of human ovarian tumour cells, although unfortunately no values of corresponding normal cells are available. It is possible that a reduced membrane potential is a characteristic feature of the neoplastic cell, without it being causally related to the neoplastic process itself.

On the basis of the experiments in which he achieves inhibition of mitosis using mannitol (which supposedly interferes with sodium pumping) and stimulation of mitosis by infusion of Na^+, Cone (1969) has proposed that a relatively high negative potential blocks cells from entering the phases preparatory to mitosis. Ehrlich ascites cells which show a high rate of proliferation are known also to have an abnormally high permeability to Na^+ (Maizel *et al.*, 1968). From the foregoing discussion it would seem that a cell entering G_1 has relatively lower potential than subsequent phases such as S and G_2 where high levels of

Table 7 *Membrane potentials of normal and tumour cells*

Cell type	Membrane potential (−mV)		Reference
	Normal	Neoplastic	
Rat thyroid	47·2	22·3	Jamakosmanović and Loewenstein (1968)
Hamster thyroid	38·5	22·6	Jamakosmanović and Loewenstein (1968)
Human stomach epithelium	22·4	14·9	Kanno and Matsui (1968)
Human ovarian tumour cells		2·9–10	Redmann *et al.* (1972, 1974)
Muscle cells (source not known)	90	15	Cone (1968)

potential are maintained. Normal muscle cells are said to have a high negative potential (approximately −90 mV) and rarely divide. But the homologous myosarcoma cells have a low potential (−15 mV) (Cone, 1969) and would therefore continue to proliferate. There are undoubtedly significant differences between values of membrane potential of normal and malignant cells. But it should be pointed out that although the measurements have been made on homologous cells, valid comparisons could be made only between normal and tumour cells at the same stage in the cell cycle. On the other hand, these correlations may not be relevant to the process of neoplastic transformation at all but relevant only to their mitotic status. Human embryo cells have been reported to have very high membrane potentials (Redmann, 1971). The situation appears to be analogous to the relationship between surface charge densities of tumour cells from malignant human astrocytomas and non-malignant meningiomas and human embryo cells (Sherbet and Lakshmi, 1974a) (see also pp. 223–225).

RELATIONSHIP BETWEEN MEMBRANE POTENTIAL AND ELECTROKINETIC POTENTIAL

Surface Groups and Membrane Potential

Although a considerable amount of research has gone into the relevance of membrane potentials in the control of cell division and in neoplastic processes, it is yet unclear if membrane potentials have any relevance to the structure of the cell surface.

Some attempts have been made to examine possible relationship between surface components and the electrophysiological properties of the membrane. Weiss and Levinson (1969) and Gomulkiewicz (1973) investigated the effects of enzymic excision of surface groups on ion fluxes. Treatment of Ehrlich ascites cells with neuraminidase (NANase) or ribonuclease (RNAase) produced significant reductions in electrophoretic mobilities of the cells but undirectional K^+ fluxes were largely unaffected. This indicates that neither sialic acids nor RNAase-sensitive surface groups are associated with the regulation of intracellular cation concentration (Weiss and Levinson, 1969). This is further supported by Pape *et al.* (1973) who observed no differences in membrane potentials of *Amphiuma* red cells even after a major proportion of surface sialic acids had been removed. In cultured heart cells removal of sialic acid with NANase increased Ca^{2+} exchangeability but potassium ion fluxes were not affected (Langer *et al.*, 1976). Gomulkiewicz (1973) found that permeability of the membrane to sulphate ions was unaffected by the removal of sialic acid from the erythrocyte membrane. However, binding of divalent ions of calcium has been found to alter the permeability of the squid axon membrane to K^+. In *Amphiuma* red cells, raising the concentration of extracellular Ca^{2+} ions causes a transient hyperpolarisation of the potential (Lassen *et al.*, 1976). This has been attributed to an increase in K^+ permeability of the membrane. This hyperpolarisation can be blocked by antihistamines and chlorpromazine. The effects of these compounds may be due to their membrane-stabilising or local anaesthetic properties (Gardos *et al.*, 1976). Forrester (1965) showed that binding of Ca^{2+} to BHK-21 cells caused a reduction in their surface negative charge. It appears possible therefore that the surface potential of cells may regulate the flux of K^+ which in turn might be reflected in changes in the membrane potential. Although Weiss and Levinson (1969) do not believe that such a regulation might occur, it ought to be pointed out that their experiments did indicate a 9·5–17% reduction in the permeability to K^+ of Ehrlich ascites cell membrane, on removal of sialic acid moieties from the surface. In yeast cells an inhibition of $^{86}Rb^+$ is caused by UO_2^{2+} and La^{3+} and this has been attributed to a reduction of surface potential by the polyvalent cations (Theuvenet and Borst-Pauwels, 1976). Levinson *et al.* (1972) and Smith *et al.* (1972) showed that La^{3+} ions rapidly associated with Ehrlich ascites cell membrane and caused a hyperpolarisation of the membrane potential. But subsequent work of Smith (1976) has shown that La^{3+} association, despite causing a concentration-dependent reduction in the EPM of Ehrlich ascites cells, exerted little influence on the passive movements of potassium and sodium ions.

Electrophoretic Behaviour and Membrane Potential of Cells

On the other hand, Redmann *et al.* (1974) have reported a detailed investigation of the possible relationship between membrane potentials and electrophoretic mobilities of cells. They observed that FL cells subjected to mitotic inhibition showed a marked reduction in membrane potential. A reduction in the ζ potential was also noticed in these cells, but the difference may be described as less remarkable though statistically significant. Redmann *et al.* (1974) also found that subjection of these cells to low temperature produced characteristic and parallel changes in both potentials. Immediately after exposure to 17°C, both membrane and ζ potentials showed a small increase followed by a reduction in the potentials. Over a period of 1–2 hours the ζ and the E_m fell by 23 and 30% respectively, below their corresponding control values (see Table 8). The effect of hyperthermia also appears to be a decrease in both potentials—the higher the temperature the lower are the potentials obtained.

Table 8 *Effect of hypothermia on* ζ *and* E_m *of FL cells*

Temperature (°C) treatment	ζ potential (−mV)	E_m(−mV)
37	11·4	18·6
17 (less than 10 minutes)	12·1	20·8
17 (over 60 minutes)	8·8	13·1

From Redmann *et al.* (1974)

A second interesting correlation which has emerged from these investigations of Redmann and colleagues is between changes in electrophoretic mobility and membrane potentials of cells on entering mitosis (see Table 9). These authors point out not only that both these features show an increase in mitosing cells but also that the magnitude of the increases is roughly comparable. Unfortunately they have not determined the electrophoretic mobilities themselves nor have they compared the membrane potentials and mobilities of homologous normal and tumour cells at interphase and in mitosis. Nonetheless, the assembled data do lend some support to the correlation between ζ potential and E_m shown by Redmann *et al.* (1974). Even the effects of ionising radiation have been claimed to be similar. The only instance where this parallelism does not obtain is in experiments where the extracellular Cl^- ion concentration is manipulated. Increase of the latter

Table 9 *Membrane potentials and electrophoretic mobilities of some cell types in interphase and mitosis*

	E_m(−mV)/EPM(−TU)			
Cell type	Interphase	Mitosis	Increase (%)	Reference
Human ovarian tumour	2·5–9·9 av. 6·2	5·8–12 av. 9·5	52·0	Redmann *et al.* (1974)
FL cells	14·0	17·0	21·0	
L-cells	10·0	16·0	50·0	Cone (1969)
	EPM (−TU)			
HeLa	9·8	12·6	29·0	Brent and Forrester (1967)
RPM	10·0	12·0	20·0	Mayhew (1966)

EPM is the electrophoretic mobility in Tiselius units (TU) where 1 TU = 10^{-5} cm sec^{-1} V^{-1} cm (see Catsimpoolas *et al.*, 1976).

reduces the membrane potential but not the ζ potential of peritoneal exudate leukocytes. Obviously changes in the ion gradients, the status of ionic permeability and the metabolic state of the cell does affect the membrane potential.

Hato *et al.* (1976) have examined the ζ and membrane potentials of the slime mould *Physarum polycephalum* and their alteration in response to a number of ionic species, and other chemical stimuli. They noticed that both the potentials showed parallel reductions in response to increases in concentration of Ca^{2+} and La^{3+} ions above a defined threshold level. Similar changes have also been described using cyclic AMP, ATP and some sugars. There is such close agreement between the changes produced by the various chemical stimuli in the ζ and membrane potentials, that Hato *et al.* have suggested that the expression of the two potentials may be causally related.

Since the ζ potential or the electrophoretic mobilities are related to the structure and composition of the surface, metabolic changes may be expected to affect the electrokinetic behaviour. But in the present state of knowledge no tangible relationship between the two potentials can be firmly established. This is a prerequisite to the employment of E_m characteristics for the characterisation of the cell surface. Correlation of isoelectric characteristics of cells (see pp. 146–235) with membrane potentials may be another avenue which could open up interesting and rewarding possibilities.

4. Cell Electrophoresis

Electrophoresis is the mobility of particles under the influence of an electric current and attributed to the free charges on their surface. Reuss (1809) may be considered to have provided the first demonstration of electrophoretic phenomena, although he may not have realised the nature of the phenomenon he was demonstrating. Reuss used a block of moist clay into which he inserted two glass tubes 14 cm apart. The tubes (9 cm high) were sunk 1 cm into the block. Both these tubes were filled with fine washed sand to 1 cm high and topped up with water. Electrodes were connected to the glass tubes and a current was allowed to pass. Soon after, the clay base of the positive electrode glass tube began to swell and this was followed by a diffusion of clay particles into the positive electrode tube, while the negative glass tube was free of clay particles.

The mobility of a particle under the influence of an electric current appears to depend entirely on the nature of the particle surface, and is independent of the size or shape or the nature of the particle itself. If particles of different mobility characteristics are coated with a film of gelatin, the differences are quickly obliterated. The particles then assume the mobility characteristics of the gelatin film deposited on their surfaces.

Pro- and eukaryotic cells similarly show electrokinetic behaviour by virtue of the net negative electric charge they carry on their surfaces. The mobility characteristics of cells were recognised several decades ago, and both qualitative and quantitative studies have been reported.

THE MEASUREMENT OF ELECTROPHORETIC MOBILITY

THE MOVING BOUNDARY METHOD

Lodge (1886) measured the electrophoretic mobilities of ions in terms of the shift in the boundary of a coloured solution. In 1892 the method was applied to larger particles of colloidal systems by Picton and Linder. Observations of the motion of boundary in coloured solutions or suspensions of large particles was easy. While working with colourless solutions the shift in the boundary could be followed by taking advantage of the Tyndall effect or using fluorescence or u.v. absorption. Although the method was simple the practical use of the moving boundary method is complicated by a number of factors such as the uniformity of ionic concentration in the solution being investigated, the electro-osmotic effects, etc. Abramson (1934) has discussed the advantages and disadvantages of the moving boundary method.

In the early days of the study of electrokinetic behaviour of cells the moving boundary method of electrophoresis was used. Coulter (1920–21) observed using this method that erythrocytes showed no mobility when suspended in buffer at pH 4·6. Howitt (1934) measured the electrophoretic mobilities (EPMs) of erythrocytes from a number of vertebrate species (Table 10, p. 55). Subsequently Anderson and McKie (1939) examined the effects of colloidal silicic and tannic acids on the mobility of sheep erythrocytes.

A simple apparatus for the moving boundary method of measuring EPM was described by Todd (1927) and has been used by many investigators after minor modifications. Essentially the apparatus consists of a U-tube in which the cell suspension is placed. Both the limbs are carefully topped up with saline solution without disturbing the saline–cell suspension boundaries in the limbs. The limbs are then connected to non-polarising electrodes. When a current is applied the level of the boundary in the anodic limb shifts. The degree of shift is given by the difference in the levels of the boundaries between the two limbs of the U-tube, and it is used as a measure of the electrophoretic mobility.

The electrophoretic mobility is calculated using the following equation:

$$v = \frac{h}{tE} \tag{7}$$

where v is the electrophoretic mobility, h the shift in the boundary (cm) and E the potential gradient (volts cm^{-1}).

Now

$$E = i\mathrm{R}$$

where i is the current (amperes) and R is the resistance (ohms).

Since

$$\mathrm{R} = \frac{1}{A\mathrm{K_c}}$$

where A is the area of cross section of the electrophoretic cell and K_c the specific conductance (reciprocal ohms).

$$E = \frac{i}{AK_c}.$$

Substituting E in eqn (7)

$$v = \frac{hAK_c}{ti} \tag{8}$$

v being expressed in cm sec^{-1} V^{-1} cm.

Although the moving boundary method is simple, there are a number of drawbacks in the technique. For instance, a lack of symmetry occurs between the two limbs of the electrophoretic tube especially as regards their ionic concentration and specific conductance. It is essential to obtain a uniform conductivity throughout the experiment (Tiselius, 1930). This condition is never fulfilled. Further, the ascending boundary tends to be sharper and the descending boundary diffuse which makes accurate measurements of shift difficult.

THE MICROSCOPIC METHOD

The moving boundary method is simple in theory and operation but it has been considered unsuitable where the electrophoretic behaviour of large particles is to be studied or when high ionic strengths are required. In such cases the microscopic method is preferred.

The Rectangular Cell

Northrop (1921) described a rectangular cell for cell electrophoresis. The cell consisted of a thin glass slide resting on two strips (0·8 mm thick) of glass cemented to a thick glass slide. Two blocks of glass are fixed on top of the slide at each end of the cell—giving a rectangular cell of dimensions 75 mm × 10 mm × 0·8 mm (Fig. 9), connected through side arms at the ends, to the electrode chambers. Northrop and Kunitz (1925) subsequently described a flat glass capillary cell for the same purpose.

The rectangular cell may be mounted in three different positions, (a) horizontal, (b) horizontal but the cell turned 90° on its axis as compared with position (a), and (c) vertical (Fig. 10). The choice of position largely depends on the size of cells under investigation. Position (a) mounting is used when examining the EPMs of bacteria. With larger particles position (b) may be used. The cells may show sedimentation but they

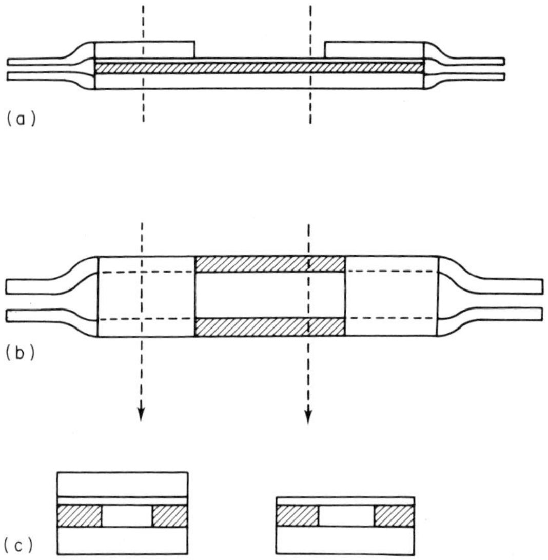

Fig. 9. A rectangular electrophoretic cell constructed by Northrop (1921). (**a**) Side view, (b) top view and (c) cross sections passing through regions indicated.

continue to remain in the stationary level (see pp. 41–45). The EPM of the cells is not affected by sedimentation. If the cell is mounted in position (c) it would be necessary to determine the rate of sedimentation of the particles being studied (Ambrose *et al.*, 1956; Ambrose, 1966).

Sachtleben *et al.* (1961) and Fuhrman and Ruhenstroth–Bauer (1965) have described an improved assembly of the rectangular cell. They employed a rectangular cell of dimensions 32 mm × 14 mm × 0·7 mm as recommended by Northrop and Kunitz (1925). The salient features of the improved cell include a water jacket through which water can be circulated and measurements of EPM made at the required temperature. Also included is a special microscope assembly which could magnify images up to 800-fold. The reversible electrode system is separated by semipermeable membranes so that the measuring cell is completely independent. This enables electrophoretic measurements to be made in a variety of suspending media.

The Cylindrical Cell

Cylindrical electrophoresis chambers have been used by a number of investigators (Mooney, 1924; Mattson, 1933; Smith and Lisse, 1936; Alexander and Saggers, 1948; Rutgers *et al.*, 1950; Bangham *et al.*, 1958).

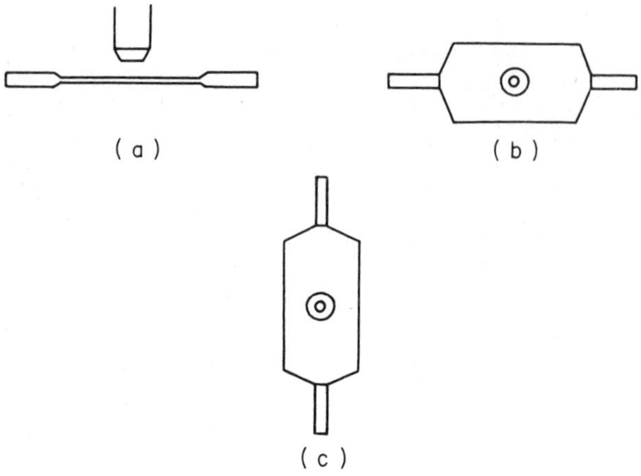

Fig. 10. Mounting positions of the rectangular cell (from Ambrose, 1966).

The most widely used cylindrical electrophoresis apparatus has been described by Bangham *et al.* (1958) which is a modified and improved version of that described by Alexander and Saggers (1948). The micro-electrophoresis tube consists of a cylindrical precision capillary of "veridia" glass, of internal diameter 2·0–3·0 ± 0·01 mm and 150 mm length. The tube is fixed horizontally and permanently in position in a holder which is attached to a calibrated vertical traverse which, in turn, is mounted on a cross bar on a water bath. The electrophoresis tube and the holder are immersed in a water bath, the temperature of which can be controlled. The optical system consists of a microscope tube attached horizontally to the cross bar. An adjustable zero dial test indicator is incorporated to enable a rapid and accurate setting of the objective to be made at the correct level. The microscopic attachment gives a magnification of 400 diameters. Bangham *et al.* consider that for working at such a magnification the tube diameter of 2·0–3·0 mm is most suitable. They suggest that at 1·5 mm diameter the mobility determinations may be subject to large errors even for small errors in the setting of the depth in the tube in the stationary region. In electrophoresis tube of diameter more than 3 mm, thermal convection effects are another source of error.

The electrode system employed by Bangham *et al.* (1958) consisted of electrodes of grey platinum prepared by platinisation as described by Findlay (1949). According to Seaman (1965) platinum electrodes are satisfactory while working with ionic concentrations below 0·1 M. At higher ionic strengths reversible silver/silver chloride or copper/copper sulphate electrode systems may be used.

The velocity of the particles is measured by timing them over a given number of graticule squares in the eyepiece, and at a given potential gradient. The particles can be timed again by reversing the direction of current. The electrophoretic mobility is expressed as μm sec^{-1} V^{-1} cm i.e. as velocity per unit field strength. Catsimpoolas *et al.* (1976) have suggested that EPM be expressed in units named after Tiselius. They proposed that one Tiselius unit (TU) should represent 10^{-5} cm sec^{-1} V^{-1} cm of EPM. In this book the suggestion of Catsimpoolas and colleagues has been adopted and accordingly EPM is expressed in Tiselius units.

Determination of the Stationary Level

In a narrow closed electrophoresis cell containing a suspension of particles in an electrical field, two electrokinetic phenomena occur, namely the electro-osmotic and electrophoretic effects. The electro-osmotic effect is produced by the electrical charges assumed by the capillary walls relative to the liquid, which results in the movement of the liquid relative to the wall of the electrophoretic cell towards the anode and a return flow of the liquid through the centre of the cell. Between these regions of electro-osmotic movement there are regions where no net movement of the liquid occurs, which form the stationary level.

The electrokinetic effect refers to the motion of the particles with respect to the medium, and therefore for measuring the true electrophoretic mobility the stationary level ought to be determined. At levels outside the stationary levels the observed electrophoretic mobility (V_0) of the particle is the sum of the osmotic flow (V_w) and the true mobility of the particle (V_e).

Now, if the velocity of particles, such as iron-free quartz for instance, is determined at various depths in the electrophoretic cell, of cylindrical or rectangular design, a smooth parabola for the velocity depth curve (Fig. 11) is obtained. In this figure SL_1 and SL_2 are levels in the cell where there is no net movement of the liquid; hence the velocity of particles at these levels, will represent their true velocity. It is therefore crucially important to determine the position of the stationary phase, which can be derived as given below (Bangham *et al.*, 1958).

According to Lamb (1888), in a closed cylindrical chamber the velocity of the liquid (V_w) varies radially across the cell. At a distance s from the axis of a tube of radius r, it can be shown that:

$$V_w = \frac{P(s^2 - 0{\cdot}5r^2)}{4\eta} \tag{9}$$

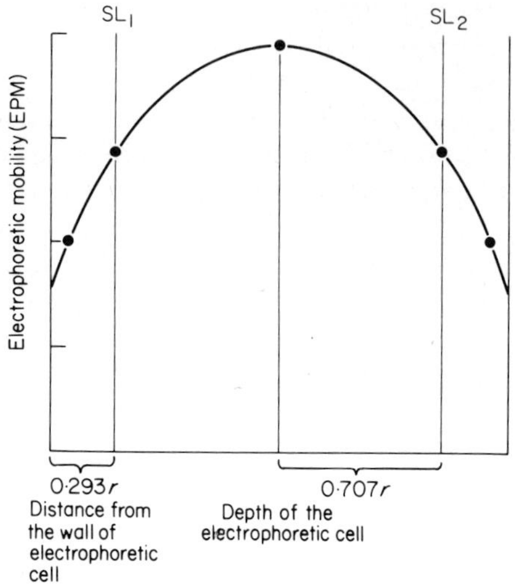

Fig. 11. Electrophoretic mobility (EPM) of particles in relation to depth in the electrophoretic cell. SL_1 and SL_2 indicate stationary levels which occur at $0{\cdot}707r$ from the axis of the cell. Velocities measured in other regions than the stationary levels will result in over or under estimations.

where P is the hydrostatic pressure gradient parallel to the axis, η the coefficient of bulk viscosity.

At the wall of the capillary tube $s = r$ and the velocity of the liquid $(V_s) = Pr^2/8\eta$. Hence eqn (9) can be written as:

$$V_w = V_s(2s^2/r^2 - 1). \tag{10}$$

Since the observed velocity of particle (V_0) is the sum of its true velocity V_e and V_w:

$$V_0 = V_e - V_s + V_s(2s^2/r^2). \tag{11}$$

Now, $V_w = 0$ when $2s^2 = r^2$ or

$$s = r/2^{\frac{1}{2}} = 0{\cdot}707r.$$

In other words, the stationary phase is $0{\cdot}707r$ from the axis of the tube or $0{\cdot}293r$ from the wall of the tube. For a rectangular electrophoretic cell stationary level conditions obtain at distances of $0{\cdot}21T$ and $0{\cdot}79T$ where T is the depth of the cell, provided that the ratio of width to depth is greater than 20. It should be emphasised, however, that there is no exact solution to the hydrodynamic equations of the rectangular cell.

In order to increase the accuracy of measurements of velocity, some modified versions of the cylindrical cell have been designed. It has been argued that despite accurate determinations of the stationary phase the measurement of particle velocities may not be sufficiently reliable, if the particle being timed is not placed symmetrically in the stationary phase (Lukiewicz and Korohoda, 1966). The particle located in the stationary level (SL) is still subjected to the electro-osmotic forces acting in opposite directions (see Fig. 12). The larger the particle size the greater is their liability to be affected by these opposing forces. The major modifications are therefore concerned with diverting the return flow of the liquid down the axis of the electrophoretic cell.

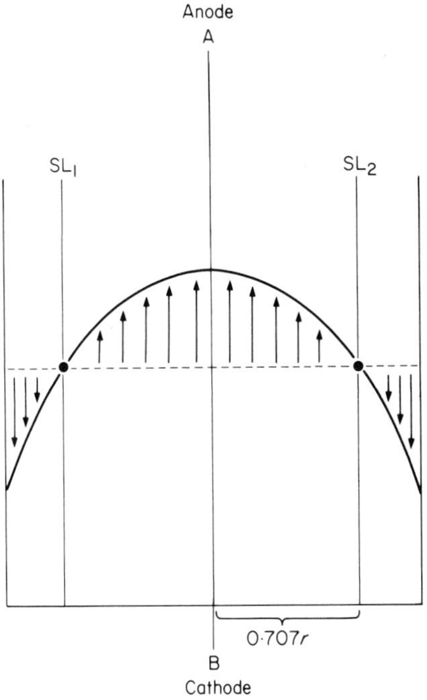

Fig. 12. Velocity of liquid across the conventional electrophoretic cell. SL_1 and SL_2 are the stationary levels.

Smith and Lisse (1936) employed a double-tube electrophoresis cell. The tubes are arranged parallel to each other in a closed system, and the lengths and diameters of the tubes are so arranged that the electro-osmotic return flow of the liquid is diverted into the longer and wider parallel tube (Fig. 13). As a result no fluid movement occurs along the axis of the tube T1. Beniams and Gustavson (1942) employed this principle and constructed a two-path rectangular cell, while Dworkin

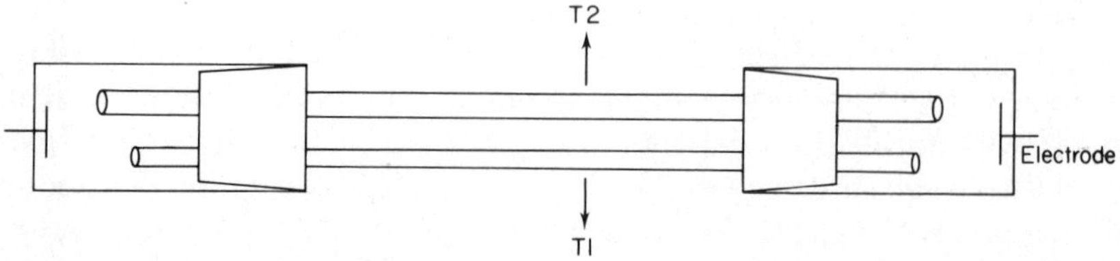

Fig. 13. The Smith and Lisse double tube electrophoretic cell (T1) in which the lengths and diameters of the tubes are so arranged that the electro-osmotic return flow of liquid is diverted into the longer and wider parallel tube T2. The stationary level occurs along the axis of T1.

(1958) substituted a flat rectangular cell in the place of the tube T1. In these double-tube versions there is only one stationary level along the axis of tube T1 of the Smith–Lisse cell or in the middle of the rectangular flat cells of the Beniams–Gustavson and the Dworkin versions of the two-path cell.

The advantages offered by these modifications are two-fold. Firstly, the particles being investigated are not subjected to electro-osmotic forces in opposite directions because in the actual measuring cell (tube T1 or its counter part) the liquid moves only in one direction (Fig. 14). Secondly, velocity of the medium across the measuring cell is less steep as compared with the corresponding velocities in a conventional cylin-

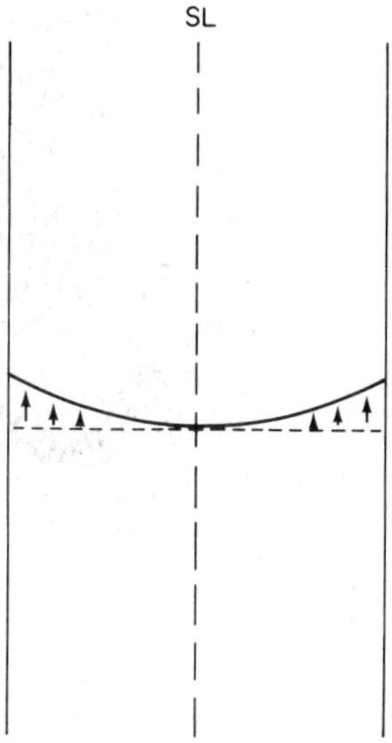

Fig. 14. Velocity of liquid across the measuring (Tl) electrophoretic cell illustrated in Fig. 13. SL occurs along the axis of this tube.

drical cell (Fig. 12). A comparison of Figs 12 and 14 also shows that a small deviation of the particle from the SL in the conventional cylindrical cell will bring in its wake larger errors in the measured velocities than in a two-path cell.

THE CONCEPT OF ζ POTENTIAL

The electrophoretic mobility of a particle is related to the number of net charges present on the particle and a general expression for the relationship between the two quantities can be written as:

$$v = Q\mathrm{D}/kT \tag{12}$$

where Q is the number of net charges, D its diffusion constant, k Boltzmann's constant and T the absolute temperature. This equation will hold good when the particle velocity is proportional to the force exerted, and in the absence of ions in the fluid medium.

When the particle is suspended in an ionic medium, however, the particle will attract ions of unlike sign from the environment and this results in an asymmetrical distribution of ions in the atmosphere. Since the fluid phase in the vicinity of the particle will be charged, the sign of charge being opposite to that borne by the particle, this layer of fluid will tend to migrate in the opposite direction of the particle itself, and therefore the particle tends to be retarded.

For the quantitation of net charges on the surface of the particle from its velocity, this retardation effect must be taken into consideration. For this, the concept of ζ potential has been a most useful notion.

It was mentioned earlier that a charged particle suspended in an ionic medium attracts an ionic cloud of opposite charge in the vicinity of the particle. This has the effect of reducing the potential on the surface of the particle. The resultant potential i.e. the sum of the potential on the surface of the particle and the potentials of the ions in the neighbourhood of the particle, is the ζ potential.

If the particle bears Q net negative charges on its surface, over a period of time the ionic cloud will contain as many positive charges. It could be postulated, albeit artificially, that these positive charges are situated on a shell lying outside the plane of shear at a constant distance. One could thus visualise the formation of a rigid electrical double layer. This notion of an electrical double layer was first conceived by Quincke (1859). Helmolz (1879) provided a mathematical expression for this notion and also put forward the concept of ζ potential.

The potential at any point in the ionic cloud surrounding the particle, due to the surface charge of the particle and the ions in the region, may be regarded as the ψ potential. When the ionic strength of the medium is raised a drop in the ψ potential occurs. The sharpness or magnitude of this reduction in potential is related to the extent of the change introduced in the ionic strength. The ζ potential at the surface of shear is in fact the ψ potential at this level. It is therefore inevitable that any expression of the ζ potential must take into consideration the dimensions of the electrical double layer which in turn necessitates a mathematical expression of the electrical double layer itself in relation to the ionic strength of the medium. This was achieved by Guoy (1910), Chapman (1913) and Debye and Hückel (1923), who considered the electrical double layer as a diffuse layer rather than a rigid one as postulated by Quincke and Helmholz.

Gouy pointed out that Helmholz's rigid layer concept was artificial because the electrical double layer could not be at a constant distance from the surface on account of the electrical forces operating therein and the osmotic forces that tend to maintain a homogeneity. The operation of these forces results in an equilibrium distribution of ions in the diffuse layer. Chapman (1913) developed the idea of the diffuse layer by introducing Poisson's equation to determine the equilibrium distribution of ions. The Gouy–Chapman model is simple in conception and assumes a uniform distribution of charges on the surface, and considers the ionic atmosphere as being made up of point-like ions, and that the medium in which the particle is suspended affects the double layer only by its dielectric constant.

Stern (1924) suggested a model for the double layer which achieved a synthesis of the Helmholz rigid layer and the Gouy–Chapman diffuse layer. Stern suggested that the charges in the ionic atmosphere associated with a particle, are made up of adsorbed ions and those in the diffuse layer. He supposed that the counterions nearest the surface are sufficiently large as to render their nuclei unable to touch the surface of the particle and consequently there occurs between the particle surface and the double layer a region where no ions are present.

These various theories about the structure of the double electrical layer may be graphically represented as in Fig. 15.

An electrostatic potential is measured by the work required to bring an unit charge from infinity to that point. In the diffuse double layer the intensity of the electrical field falls rapidly with increasing distance from the interface (Fig. 15). Consequently in the substructure of the double layer three levels of potential relevant to the electrokinetic behaviour of particles can be discerned. The potential at the interface is the Nernst potential which represents the total potential of the double layer. The Gouy–Chapman potential is the potential between the Stern layer and

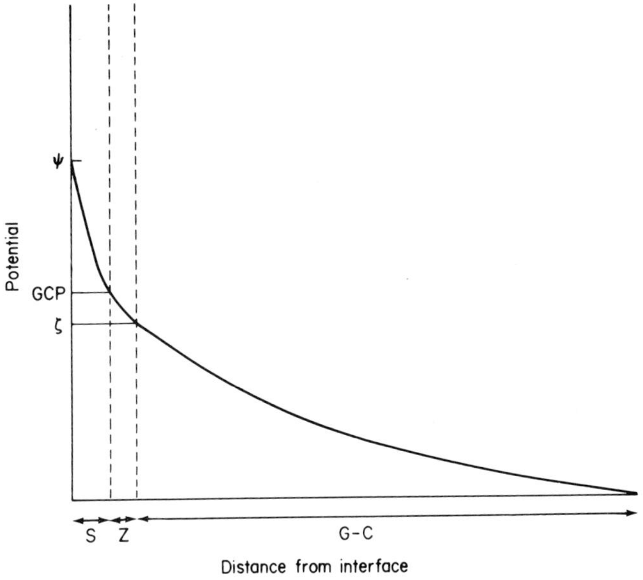

Fig. 15. A graphic illustration of the electrical double layer. S is the Stern layer, Z the zone of shear, G–C the Gouy–Chapman diffuse layer. Abcissa: distance from interface into the ionic atmosphere. Ordinate gives the potentials. ψ is the Nernst potential at the interface, GCP the Gouy–Chapman potential and ζ the zeta potential.

the diffuse layer. A large potential drop (between the Nernst level and the Gouy–Chapman potential) occurs across the Stern layer whose dimensions may be regarded as considerable as compared with the thickness of the diffuse layer. The potential at the plane of shear is the ζ potential. The zone of shear is only a few angstroms thick and therefore the Gouy–Chapman potential and the ζ potential are almost identical.

As derived by Debye and Hückel the equation for the relationship between the thickness of the double layer, $d = 1/K$ and the concentration of ions in the medium, may be given as follows:

$$K = \left[\frac{4\pi Ne^2}{1000\, DkT}\right]^{\frac{1}{2}} \times \left[\sum c_i z_i^2\right]^{\frac{1}{2}} \tag{13}$$

where N is Avogadro's number, e the electronic charge, D the dielectric constant of water, k Boltzmann's constant, T the absolute temperature, c_i the number of ions of ith kind mol l^{-1} and z the valency of the corresponding ion. At 25°C in water $K = 0{\cdot}327 \times 10^8\, I^{\frac{1}{2}}$ cm where I is the ionic strength as defined by Lewis and Randall (1921).

EQUATION FOR A LARGE SPHERICAL PARTICLE

The generalised equation for electrophoretic mobility, v, can be written as:

$$v = \frac{\zeta D}{4\pi\eta} \tag{14}$$

where η is the viscosity of the medium. The Debye–Hückel equation for the thickness of the electrical double layer can be incorporated into this equation as follows. Equation (14) is the Helmholz–Smoluchowski equation for electro-osmosis, which Smoluchowski (1921) showed must also apply to electrophoretic mobility of particles. In electro-osmosis one is dealing with motion of the liquid along a fixed solid surface, while electrophoresis is the motion of particle relative to the solvent.

Consider a spherical particle of radius r bearing Q net negative charges on its surface, suspended in a medium of viscosity, η (see Fig. 16).

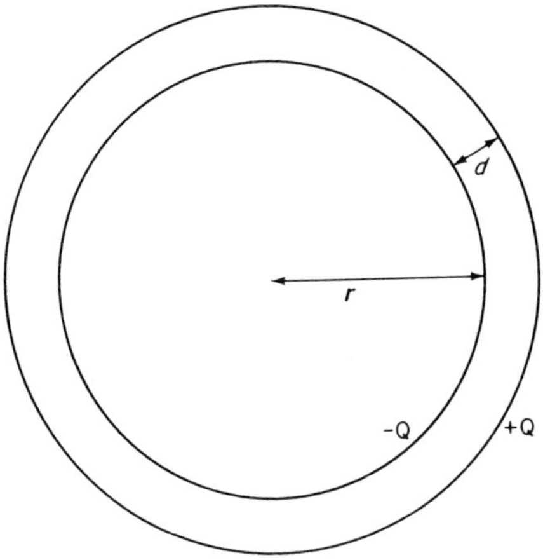

Fig. 16. A schematic diagram of a spherical particle of radius r and negative charge Q, with a highly exaggerated electrical double layer of thickness d visualised as concentric spheres.

The potential, ζ, on the surface of the sphere, if the medium were free from ions, will be:

$$\zeta = -\frac{Q}{Dr}. \tag{15}$$

When this particle is in an ionic medium it will be bounded by an electrical double layer of $+Q$. The potential due to this electrical shell will be

$$\zeta = \frac{Q}{D(r+d)}. \tag{16}$$

The net potential on the surface of the particle is therefore:

$$\zeta = \frac{Q}{D(r+d)} - \frac{Q}{Dr}$$

$$= -\frac{Q}{Dr} \times \frac{d}{r+d}. \tag{17}$$

Since d is very much greater than r:

$$\zeta = \frac{-Qd}{Dr^2}. \tag{18}$$

Substituting eqn (18) in expression for v (eqn 14):

$$v = \frac{Qd}{4\pi r^2 \eta}$$

$$v = \frac{\sigma}{K\eta} \tag{19}$$

where σ is the net negative charge density and $d = 1/K$ where K is the Debye–Hückel function.

It is normal practice to measure electrophoretic mobility at various pH values in a monovalent $I:I$ electrolyte. The surface charge is calculated on the Gouy–Chapman theory for values $\zeta < 25$ mV, using eqn (20):

$$\sigma = 3{\cdot}52 \times 10^4 I^{\frac{1}{2}} \sinh(\zeta/51{\cdot}3) \tag{20}$$

where σ is the charge density, I the ionic strength. σ is related to the surface potential (ψ):

$$\sigma = \frac{DK}{4\pi}\psi \tag{21}$$

where D is the dielectric constant of the solvent and K the Debye–Hückel function (see eqn 13).

The equation for electrophoretic mobility (eqn 19) is applicable to particles of regular shape, the mobility being dependent only on the nature of the surface and independent of shape or size. However, the equation relates only to large particles i.e. for all cases where the radius of the particle is very much greater than the thickness of the electrical double layer. At physiological ionic strength (0·145 M) the layer is approximately 8·0 Å thick. Therefore the equations are applicable to most animal cells.

EQUATION FOR A SMALL SPHERE

The equations described above for large particles of regular shape and small curvature are based on the requirement that the distorted electric field is always parallel to the surface of shear, throughout the electrical double layer. In the case of small particles, however, the lines of force around the particle are practically undisturbed throughout the double layer, and therefore the relative motion within the double layer cannot be regarded as parallel to the applied field.

Alternative equations for a small sphere may be derived as follows.

For a small spherical particle (large enough to obey Stokes' law), the electrical force acting on the particle may be equated to the viscous retarding force:

$$QX = 6\pi\eta r V \tag{22}$$

or

$$v = \frac{Q}{6\pi\eta r} \tag{23}$$

since $v = V/X$.

This equation applies when the solvent is free of ions. When ions are present in the medium, the ζ potential in the case of a large spherical particle is obtained from eqn (17). This equation can also be written:

$$\zeta = \frac{Q}{Dr} \times \frac{1}{1+Kr}. \tag{24}$$

Introducing eqn (24) into the Helmholz–Smoluchowski eqn (14) for the electrophoretic mobility of a large particle:

$$v = \frac{Q}{4\pi\eta r} \times \frac{1}{1+Kr}. \tag{25}$$

When the small particle is suspended in an ionic atmosphere, by an analogous Debye–Hückel treatment eqn (25) may be rewritten for the electrophoretic mobility of the small particle:

$$v = \frac{Q}{6\pi\eta r} \times \frac{1}{1+Kr}. \tag{26}$$

Hückel, who derived eqn (26), believed that it was valid for particles of any size and not only for small particles with a small radius as compared with the thickness of the electrical double layer. Hückel also doubted the applicability of eqn (25) to particles of all sizes. Hückel's equation however is based on the assumption that the applied electrical field is not affected by the presence of the particle, and is therefore valid for small particles only or for particles of any size provided that the surface conductance is the same as that of the medium. Since this is not usually the case, Hückel's equation has no general applicability.

Henry (1931) resolved this question and gave the following generalised equation for a non-conducting spherical particle, after taking into consideration the distortion of the applied field by the electrical double layer and hydrodynamic equations for fluid motion:

$$v = \frac{\zeta D}{6\pi\eta} f(Kr). \tag{27}$$

In the case of small spheres the factor $f(Kr) = 1$ and eqn (26) is valid. For a large sphere $f(Kr) = 1{\cdot}5$ and therefore the Helmholz–Smoluchowski eqn (14) will apply. For particles of intermediate size Gorin (1941) has provided a table of values for the Henry factor.

THE RELAXATION EFFECT ON MOBILITY

Henry (1948) argued that the Helmholz–Smoluchowski eqn (14) rested on the assumption that the applied electrical field and the field of electrical double layer were superimposed without mutual distortion. But if such a distortion did occur the mobility of the particle would be subject to a "relaxation" effect i.e. the mobility would be reduced on account of the distortion of the spherical symmetry of the electrical double layer. Also, the applied field would be modified in the vicinity of the particle by the electrical conductivity of the double layer.

Henry (1948) introduced a correction for the surface conductance:

$$\zeta_{corr} = \frac{\zeta(K_0 + K_s/r)}{K_0} \tag{28}$$

where K_0 is the conductivity of the suspension medium, K_s the surface conductance of the particle evaluated using equations due to Street (1956) and r the radius of the particle.

The relaxation effect may be neglected when (a) the values for ζ potential are far below 25 mV and (b) values for Kr are small (less than 1) or when Kr is very much greater than 100 (Overbeek, 1952a). In the electrophoretic investigations of cells the value of Kr is usually very much greater than 100 and therefore the relaxation effect may be safely neglected, provided that the ζ potential is below 25 mV. However, characterisation of cell surface often involves the chemical modification of the inogenic groups. When the mobilities of these modified cells are measured at certain bulk pH values, the ζ potential may require to be corrected for surface conductance.

MEASUREMENT OF EPM BY ELECTRO-OSMOSIS

In an open capillary tube in an electrical field, the electrical charges assumed by the inner wall of the capillary relative to the liquid result in the movement of the liquid relative to the capillary wall. The liquid flow will occur towards the anode if the wall of the capillary is negatively charged. This is the electro-osmotic movement of the liquid. The higher the surface potential of the capillary wall, the higher will be the electro-osmotic flow, under defined conditions of ionic strength and electrical field. This relationship is given by the Helmholz–Smoluchowski equation v electro-osmotic$=\dfrac{\zeta D}{4\pi\eta}=\dfrac{V}{X}$ where D is the dielectric constant of the liquid, η its viscosity, V the velocity of the liquid flow, and X the field strength. By measuring V in a known electrical field, therefore, the potential of the capillary wall may be determined. By coating the inner surface of the capillary wall with a known protein film and using particles covered with a film of the same proteinaceous material, it has been shown that the electrophoretic mobility and the electro-osmotic mobility are identical (see Abramson *et al.*, 1942).

If the inner wall of the capillary has an attached monolayer of cells the rate of flow of liquid through the capillary would be dependent upon the potential of the cellular layer. Fike and Van Oss (1976) have determined the electro-osmotic mobilities due to monolayers of a number of cell types and have found the values consistent with their true electrophoretic mobilities. They have used glutaraldehyde-fixed erythrocytes as indicator particles. The electro-osmotic mobility due to the cellular layer is given by the difference between the observed velocity of the indicator particles and their true electrophoretic mobility. Van Oss *et al.* (1974) coat the inside of the capillary with agarose and also plug the ends of the capillary by the same substance thus converting the capillary into a closed cell and use it for measuring the true electrophoretic mobility. In the absence of electro-osmotic movement of liquid, the EPM can be measured at any level.

The principal advantage of using electro-osmotic mobility to determine ζ potential is that it obviates the need to detach the cells from the substratum, which is usually achieved by employing chelating agents, enzymes or by mechanical means. It is known that the cell surface undergoes significant changes as a result of the methods employed in harvesting the cells. The electro-osmotic method enables a determination of the ζ potential of the surface in its native state.

The methods described by Fike and Van Oss (1976) are extremely simple. The main problem may be growing the cells in a complete monolayer in the capillaries. A discontinuous monolayer would obviously produce erroneous values of electro-osmotic mobility. Despite the simplicity of procedures, the scope of application of the technique appears somewhat limited.

THE ELECTROPHORETIC INVESTIGATION OF CELL SURFACES

It was mentioned on p. 36 that the electrophoretic mobility of a particle appears to depend entirely on the nature of the particle surface. It was also mentioned that if several particles differing in electrophoretic behaviour were coated with gelatin they would all assume uniform electrophoretic characteristics that can be attributed to the gelatin film deposited on their surfaces. Electrophoresis has therefore provided a simple tool for the physicochemical characterisation of the constituents of the cell surface. The remainder of this chapter will be devoted to a discussion of the electrophoretic data available on a variety of cell types and the interpretation of such data in terms of the physicochemical make up of the cell surface, and in the context of the variety of biological functions that the cell surface subserves.

THE ELECTROPHORETIC ZONE

Before undertaking a discussion of surface characterisation by cell electrophoresis, it should be mentioned that only the outermost region of the membrane, whose dimensions depend upon the ionic concentration of the bulk medium, is amenable to investigation by the electrophoretic method. This region will be referred to as the elektrokinetic zone (EKZ). It was mentioned on pp. 49–50 that at the physiological ionic strength of 0·145 M, the thickness of the electrical double layer is about 8·0 Å. Now if it is assumed that the particle has a smooth surface, and that the boundary of shear is very close to the surface of the cell membrane, it may be expected that only the ionisable groups present in the 8 Å zone will influence the electrokinetic behaviour of the particle. If the particle surface is crenated or rugged with "hills" and "valleys" the ionisable groups situated in the "valleys" will be unable to exert their electrophoretic effect, while the groups on the "hills" will be detected. At low ionic strength e.g. $\frac{1}{3}$ the physiological strength ($I = 0{\cdot}048$ M), the thickness of the electrical double layer will be 13·9 Å, so that groups situated deeper in the membrane are liable to be detected. In other words, at lower ionic strengths the particle will behave as if it had a smooth surface. Thus alteration of the ionic strength will not only detect groups situated at different depths but also for the same reason alter the iso-electrophoretic point (pIE) as well as the isoelectric point (pI) (see discussion on pp. 234–235).

THE ELECTROPHORESIS OF BLOOD CELLS

The formed elements of blood show distinct differences in their biological behaviour. For example, blood cells are usually in a dispersed state. They show differences in adhesiveness to glass. Erythrocytes, for instance, do not adhere to glass like leukocytes. Platelets aggregate easily in the presence of calcium ions but erythrocytes do not. It is also known that erythrocytes do not adsorb protein non-specifically. As will become apparent from later discussion these differences in surface properties may reflect differences in the nature of the surfaces of these cells.

Erythrocytes

Erythrocytes are the most extensively investigated cell type as regards electrophoretic behaviour. Investigations of the erythrocyte surface began as early as 1860 (see Abramson, 1934 for references). Abramson (1929) made a series of electrophoretic measurements on erythrocytes derived from a variety of animal species. The electrophoretic mobility differed between the species but was fairly reproducible within the same species. In human cells the EPM is independent of the age, sex, race and blood group of the donor (Abramson *et al.*, 1942; Seaman, 1958; Ruhenstroth-Bauer and Sachtleben, 1959). The mobility data are summarised in Table 11.

The interpretation of electrophoretic mobilities in terms of the nature of the erythrocyte surface was also attempted early. Abramson (1930) and Furchgott and Ponder (1941) examined the EPM of erythrocytes at various bulk pH ranges, and observed that these cells are negatively

Table 10 *Electrophoretic mobilities of erythrocytes from vertebrate species using moving boundary and horizontal cell electrophoresis*

Donor	Electrophoretic mobility (−TU) Moving boundary method	Horizontal microelectrophoresis
Rat	14·5	14·5
Rabbit	6·0	5·5
Mouse	14·2	13·5
Guinea pig	11·6	11·6

From Howitt (1934) and Abramson (1929).
TU is Tiselius unit = 10^{-5} cm sec^{-1} V^{-1} cm.

Table 11 *Electrophoretic mobility of erythrocytes*

Donor species	EPM[a] (−TU)	σ (Electronic charges cm^{-2}) $\times 10^{-13}$	Reference
Human	11·6	0·73	Eylar *et al.* (1962)
Human	10·8	0·77	Cook *et al.* (1961)
Human	9·7	0·6	Lipman *et al.* (1966)
Human	11·3	0·8	Abramson *et al.* (1942)
Pig	8·8	0·75	Seaman and Uhlenbruck (1962)
Guinea pig	12·1	0·56	Abramson *et al.* (1942)
Toad	8·1	0·5	Lipman *et al.* (1966)
Dog	12·8	1·1	Seaman and Uhlenbruck (1962)
Horse	11·6	0·99	Seaman and Uhlenbruck (1962)
Sheep	14·4	1·24	Seaman and Uhlenbruck (1962)
Chick	8·2	0·69	Seaman and Uhlenbruck (1962)
Chick	9·0	0·57	Eylar *et al.* (1962)
Calf	10·2	0·64	Eylar *et al.* (1962)

[a] EPM measured at 25°C in 0·145 M buffer.

charged until the bulk pH was reduced to 3·6. They also found that the cells showed no mobility at bulk pH of 1·7 (at $I = 0{\cdot}172$ M; 25°C) i.e. this was the isoelectrophoretic point (pIE) (the point of zero mobility). They further noticed that the lipid components of the membrane were isoelectric at pH 2·6, while the stroma protein was isoelectric at pH 4·7. Furchgott and Ponder therefore suggested that the red cell surface was not made up of the stroma proteins, and that the surface did contain strongly acidic groups. The phosphate groups of a phospholipid, possibly cephalin, were suggested as probable candidates. Winkler and Bungenberg de Jong (1940–41) also came to a similar conclusion using a different method.

Winkler and Bungenberg de Jong (1940–41) and Bangham *et al.* (1958) examined the charged groups on the erythrocytes of pig and sheep making use of the principle described by Kruyt (1948) that the surface groups can be characterised by investigating the binding of a series of cations which differed in their effectiveness to reverse the charge of the particle as assessed by EPMs. The charge reversal spectrum was then compared with spectra for model systems with known charged groups. The charge reversal spectrum for erythrocytes showed that uranous oxide (UO_2^{2+}) ions can reverse the charge at a low concentration. According to Kruyt (1948) a low charge reversal concentration of UO_2^{2+} ions is characteristic of phosphatide and phosphate groups. Low charge reversal concentrations of UO_2^{2-} were also seen in the egg-lecithin charge reversal spectrum. These investigations indicated the involvement of phosphate groups in the electrokinetic behaviour of erythrocytes and also supported the earlier observations of Furchgott and Ponder (1941).

In a recent study on erythrocytes of the giant salamander *Amphiuma means*, Pape *et al.* (1975) reduced their EPM by about 16% by increasing the extracellular concentration of Ca^{2+}. This is consistent with the binding of surface phosphate groups by calcium ions. Calcium produced similar reductions in the EPM of cells which had been stripped of the sialic acid residues by neuraminidase (NANase). This indicates that the erythrocytes of *Amphiuma* contained Ca^{2+}-binding groups other than the sialic acids. The pIE of the cells was found to be 2·3 which lies between the pK of the sialic acid carboxyl groups (pK 2·6) and the phosphate groups (pK about 2·1). According to Forrester (1965) increase of extracellular calcium does not produce reductions in EPM of erythrocytes, the origin of which he omitted to mention. Presumably Forrester had either human or hamster erythrocytes in mind. In any case, it is well documented that following removal of sialic acids, the pIE of erythrocytes from humans, calf, chicken and lamb, actually shows an increase (Table 12) which strongly suggests the absence of phosphate groups on the surface of these cells. Pape *et al.* (1975) have unfortunately not examined the effects of NANase on the pIE of the cells, which would have been interesting to compare with the increase in pIE described for several other species. In retrospect, Forrester (1965) may have been justified in making a generalised statement about the lack of effect of

Table 12 *Effect of trypsin and neuraminidase treatment on isoelectrophoretic point of erythrocytes*

Erythrocyte source	Treatment	pIE	Reference
Sheep	none	2·3	Bangham *et al.* (1958)
Human	none	2·2	Seaman and Heard (1960)[a]
	Inactivated DFP-trypsin	2·2	Seaman and Heard (1960)[a]
	Trypsin	3·7	Seaman and Heard (1960)[a]
	NANase	3·5	Seaman and Cook (1965)
Human	NANase	5·5	Eylar *et al.* (1962)[a]
Calf	NANase	4·5	Eylar *et al.* (1962)[a]
Chicken	NANase	5·0	Eylar *et al.* (1962)[a]
Lamb	NANase	4·6	Eylar *et al.* (1962)[a]

[a] Seaman and Heard (1960) measured EPMs at 0·145 M ionic strength while Eylar *et al.* (1962) used 0·072 M ionic strength. The higher pIE values obtained by Eylar *et al.* may be a direct result of low ionic strength buffers employed by them in making the EPM measurements.

extracellular Ca^{2+} on erythrocyte mobility, for *Amphiuma* cells could be only an exception.

Moyer (1940) in fact proposed that the polar groups of the lipids were exposed on the surface thereby conferring a high charge density and a low isoelectric point on the cells.

Despite these interesting investigations, evidence was accumulating which pointed towards the possibility of mucosubstances being involved in the electrokinetic behaviour of erythrocytes. Hirst (1942a,b; 1948) found that erythrocytes adsorbed influenza virus which caused them to agglutinate. After a while the cells came apart and could be re-agglutinated by the virus. Hirst suggested that the erythrocyte surface possessed receptors which bound the virus. But the cells came apart because an enzyme produced by the adsorbed virus destroyed the surface receptors. Such a receptor-destroying enzyme (RDE) was also found to occur in preparations of *Vibrio cholerae* (Stone, 1947). Hirst (1948) suggested from virus adsorption and elution experiments that the receptor might be a mucoprotein. Hanig (1948) showed that when erythrocytes were adsorbed with PR 8 influenza virus with concomitant agglutination and when this was followed by elution of the virus, a decrease in their EPM occurred. This may be due to release of surface material by the action of viral enzymes. Gottschalk (1956, 1957) showed later that an enzyme derived from *V. cholerae* was a glycosidase which he termed neuraminidase because it releases acylated neuraminic acids by cleaving α-glycosidic linkages (see Blix, 1958). At this time it was also demonstrated that this enzyme released *N*-acetylneuraminic acid (NANA) from erythrocyte stroma (Klenk and Uhlenbruck, 1958; Klenk and Lempfrid, 1957) and it appeared that the negative charge on the erythrocyte surface may be due to an acylated neuraminic acid (Klenk, 1958). Cook *et al.* (1961) characterised the material released from the erythrocyte surface by treatment with RDE (NANase). They showed by chromatographic methods and from its crystalline form that the material was a *N* or *O*, *N*-acetylated neuraminic acid. The loss of the material also resulted in a decrease of 80–100% in the EPM of the cells after the treatment (Cook *et al.*, 1961; Eylar *et al.*, 1962). Since the action of the enzyme was unlikely to introduce cationic groups (Gottschalk, 1956), the reduction in the EPM could only be attributed to the loss of the terminal carboxylic groups of the neuraminic acids.

Eylar *et al.* (1962) not only examined the EPM of erythrocytes derived from a variety of species but also determined the amount of sialic acids released by NANase. Table 13 summarises their data. The sialic acid content is expressed as number of molecules μm^{-2} of the erythrocyte surface. The data shows clearly that the amount of surface sialic acid determined the degree of mobility: greater the number of sialic acid molecules μm^{-2} the greater is the mobility. A statistical analysis of the correlation between the two variables gives a correlation coefficient $r = 0{\cdot}95$ ($P < 0{\cdot}01$) indicating that the variables are completely dependent on each other.

Table 13 *Relationship between EPM and number of sialic acid molecules on erythrocyte surface*

Erythrocyte donor	No. of sialic acid molecules $\mu m^{-2} \times 10^{-4}$	EPM (−TU) at $I = 0{\cdot}072$ M, pH 6.4
Pig	5·3	11·8
Chicken	5·7	12·8
Calf	10·5	14·4
Lamb	12·3	16·9
Human	14·7	16·4
Horse	28·2	19·8

Data from Eylar *et al.* (1962). A statistical analysis of correlation between the sets of variables gave a correlation coefficient $r = 0{\cdot}94$ ($P < 0{\cdot}01$) i.e. the variables were completely interdependent.

A dissenting view has been expressed by Sachtleben *et al.* (1972) about the events associated with the reduction of EPM of erythrocytes following NANase treatment. Sachtleben *et al.* (1972) have claimed that the EPM of NANase-treated cells increased when they were treated with anti-NANase antibody. They have interpreted these results as suggesting that at least part of the loss of EPM sustained by the cells on NANase treatment was attributable to the adsorption of the enzyme on the surface and masking of some of the ionogenic groups. When NANase-treated cells were subsequently subjected to anti-NANase antibodies the latter may have reacted with the adsorbed enzyme molecules and presumably removed them. However, the authors have made the assumption that an adsorption of the antibodies themselves does not cause an increase of the EPM. This assumption does not seem justifiable (see pp. 204–211 for a discussion of the effects of antisera on the isoelectric points of cells).

When erythrocyte mobilities are measured at various bulk pH ranges, the cells can be seen to move in the anodal direction until the bulk pH is 2–3. In this region the cells do not show electrophoretic mobility. This pH of zero mobility is the isoelectrophoretic point (pIE). The pIE of erythrocytes appears to lie between pH 2 and 3 (Bangham *et al.*, 1958; see also pH-mobility curves for human erythrocytes given by Seaman and Heard, 1960). The pIE provides valuable information about the groups present on the surface. When erythrocytes are treated with trypsin the pIE increased to approximately 3·7 (Seaman and Heard, 1960). Eylar *et al.* (1962) showed that treatment of erythrocytes with NANase not only reduced the EPM but caused a concomitant increase in the pIE (see Table 12).

The pIE of untreated erythrocytes is in good agreement with the pK of *N*-acetylneuraminic acid which has been determined by Svennerholm (1956) to be 2·6. The pIE of the cells on removal of the sialic acid appears to represent the presence of anionic groups with higher intrinsic pK values. Haydon and Seaman (1967) found that these groups had an overall pK value of 3·35, assuming that they were homogeneous. These authors have argued that this pK value is consistent with that of α-carboxylic groups of proteins but is too high for γ-carboxylic groups of glutamic residues and β-carboxylic groups of aspartic acid residues and too high for saccharide carboxylic groups (Saric and Schofield, 1946). When NANase-treated cells are subsequently subjected to treatment with alkali to release *N*-acetylgalactosamine, the γ-carboxylic groups of the glutamic acid residue through which the prosthetic group was linked to the protein becomes free. This can also be achieved by treating the erythrocyte with aldehyde to release the disaccharide, sialyl-*N*-acetylgalactosamine and then treating the cells with alkali. These treatments have the effect of replacing each disaccharide with the terminal sialic carboxylic group with a γ-carboxylic one of glutamic residue. When such a modification of the surface is made the EPM remains unchanged but the overall pK of the surface increases. Haydon and Seaman (1967) showed that overall pK of the cell after these treatments is consistent with the presence of γ-carboxylic and α-carboxylic groups. By titrating the numbers of α-carboxylic groups using methylene blue these authors also showed that the α-carboxylic groups are all situated in the same plane of the cell surface close to within 12 Å from the cell surface.

The pK of the carboxylic species of proteins lies in the range 3–5, the pK value of α-COOH being 3·0–3·2, β-COOH of aspartic acid residue 3·0–4·7 and γ-COOH of the glutamic acid residue approximately 4·5 (see Edsall and Wyman, 1958; Dawes, 1972).

The absence of significant numbers of cationic groups on the surface has also been demonstrated. Heard and Seaman (1960) showed that anionic substances such as thiocyanate, chloride, fluoride, etc. did not change the EPM of erythrocytes. Their data on surface charge densities for erythrocytes suspended in isotonic solutions composed of fluoride, chloride, iodide and thiocyanate indicate equivalence in the degree of adsorption. They argued that since ion adsorption depends on the energy of hydration of the ion, the degree of adsorption should have followed the order CNS>iodide>chloride>fluoride. The equivalence in the degree of adsorption on the erythrocyte membrane therefore suggested a lack of cationogenic groups. Besides, Heard and Seaman (1960) and Seaman and Heard (1960) also showed that acetaldehyde, 2,4 dinitrobenzene, *p*-toluenesulphonyl chloride or formaldehyde exerted no influence on the EPM. The lack of effect of formaldehyde on the EPM was also confirmed by Eylar *et al.* (1962).

Finally, a reduction in the EPM could be achieved by treating the cells with trypsin or *N*-bromosuccinimide but not by tosyl chloride (Seaman and Heard, 1960). This effect of trypsin had been shown earlier by Ponder (1951). Further investigation of trypsin-released material by Cook *et al.* (1960) indicated that it was a sialomucopeptide structurally analogous to the mucoprotein derived from bovine salivary gland (Gottschalk, 1960a).

The above discussion may be summarised. There is considerable evidence which indicates the occurrence of glycoproteins at the erythrocyte surface. The electrokinetic behaviour of the cells appears to be due mainly to the negative electric charges contributed by the carboxylic groups of sialic acids and α-carboxylic groups of proteins, and probably with minor contributions from β,-γ carboxylic groups.

Nature of the Erythrocyte Surface Sialopeptide

It was mentioned earlier that trypsin reduced the EPM of erythrocytes (Ponder, 1951; Seaman and Heard, 1960). Seaman and Heard also pointed out that inactivated trypsin did not cause any loss of EPM and therefore it was not attributable to the adsorption of the enzyme molecules on the cell surface, but to a release of peptides or mucopeptides by the proteolytic action of the enzyme.

According to Mäkela *et al.* (1960) trypsin releases 50% of the sialic acid and 15% of hexosamine and structural sugars from the human erythrocyte. Cook *et al.* (1960) investigated the nature of the substance released by trypsin. They isolated a substance which showed absorption in the u.v. range of 215–300 nm wavelength, with a minor peak at 270mμ. When treated with acid the substance showed increased absorption at 270–290 nm and at 260 nm on treatment with alkali. This spectral behaviour is characteristic of substances containing sialic acids (Gottschalk, 1955). Further, microchemical tests revealed the presence of nitrogen, carbohydrates and sialic acids which occurred in the ratio of 18 to 10 to 1 by weight. The peptide nature of the material is indicated by the nitrogen content and the ninhydrin-positive reaction. It also appears that the sialic acids released into the medium were in a bound form. These are bound to the mucopeptide by α-glycosidic linkage because sialic acids are easily released by the action of NANase which is an α-*O*-glycosidase (see Gottschalk, 1960b). A mild hydrolysis of the substance with 0·5 N hydrochloric acid at 105°C for 15 hours, released no neutral sugar, while a complete acid hydrolysis (6 N HCl at 105°C for 15 hours) yielded galactosamine and eight aminoacids. The sialopeptide thus contains sialic acid, galactosamine or its acetyl derivative and an aliphatic peptide chain (Cook, 1962). Using NANase and mild alkali treatment, Cook (1962) showed that the saccharide moiety is a galactosamine thus providing a structure for the carbohydrate moiety of the sialopeptide as: 6α-D-sialyl-*N*-acetylgalactosamine. This is similar to the carbohydrate moiety of bovine salivary mucoprotein (Gottschalk and Graham, 1959; Gottschalk, 1960a). The carbohydrate–protein linkage probably occurs through a γ-glutamyl residue since only glutamic acid is shown to occur in the erythrocyte sialopeptide. In the salivary mucopeptides the prosthetic groups may be joined through the glutamic or aspartyl residues (Gottschalk, 1960). About 16% of the prosthetic groups may also involve serine and threonine residues (Gottschalk *et al.*, 1962). Similarly, the sialic acid in the erythrocyte surface, are not always bound through *N*-acetylgalactosamine and glutamic acid. Romanowska (1961) demonstrated the occurrence of sialylgalactose, while Cook and Eylar

(1965) showed that a serine residue was involved in the linkage of the prosthetic group to the peptide. Figure 17 shows a segment of ovine submaxillary mucoprotein, with the sites that are attacked by trypsin and NANase are indicated. The sialopeptide released by trypsin from erythrocyte surface is essentially similar in structure. The N-terminal amino acid in the erythrocyte sialopeptide may be alanine.

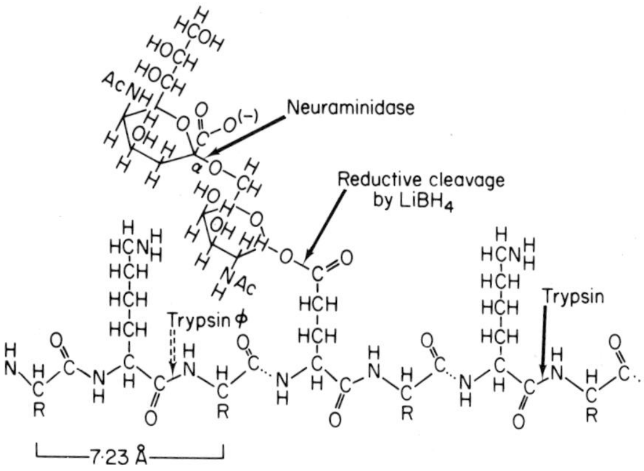

Fig. 17. Structure of a segment of ovine submaxiallary protein proposed by Gottschalk (1960), indicating sites liable to attack by trypsin (site indicated by broken arrow sterically not accessible to the enzyme) and neuraminidase-susceptible glycosidic linkage between prosthetic group and the carboxyl group (γ carboxyl of glutamic acid residue in this case).

These early investigations on the characterisation of erythrocyte surface glycoproteins have been mentioned here only to facilitate discussion and to place them in their historical perspective. More recent work has involved the isolation and characterisation of glycoproteins possessing specific blood group activities from typed erythrocytes. These have been dealt with separately elsewhere in this chapter (pp. 64–68).

Types of Sialic Acids

The sialic acids occurring in erythrocytes derived from different species have been found to vary considerably in their composition. Eylar *et al.* (1962) found that human and chicken erythrocytes contained 95–100% *N*-acetylneuraminic acid. In lamb erythrocytes *N*-acetyl and *N*-glycolyl forms occurred in roughly equal proportions, while in the red cells of calf, pig and horse 70–95% of the sialic acids were of the *N*-glycolyl type. *N,O*-diacetylneuraminic acids described by Blix *et al.* (1956) were not detected (see Table 14).

Table 14 *Types of sialic acids in erythrocyte stroma*

Species	*N*-acetyl (%)	*N*-glycolyl (%)
Human	95–100	0–5
Chicken	95–100	0–5
Lamb	58	42
Calf	40	60
Pig	34	66
Horse	5–10	90–95

From Eylar *et al.* (1962).

Although NANase reduced EPM by 60–90% through a release of sialic acid, the reduction in EPM produced by proteolytic enzymes is considerably smaller than that produced by NANase (Table 15). It has been suggested therefore that a greater part of the sialic acids involved in the electrophoretic behaviour of erythrocytes may not be linked to protein (Weiss, 1967) but to glycolipids. Although this is a possibility there is much contrary evidence. For instance, Yamakawa *et al.* (1960) reported that glycolipids from human, sheep, guinea-pig and rabbit erythrocytes contained no sialic acids. Booth (1963) also reported low lipid-bound sialic acids in human erythrocytes. Further, at least in this erythrocyte species equivalent amounts of sialic acids were released by pronase and NANase (Cook and Eylar, 1965). The reduction in the EPM caused by NANase may still be considered as indicating that most of the sialic acids are bound to protein, in view of the fact that lipid-bound sialic

Table 15 *Effects of eyzymic treatment on EPM of erythrocytes*

Donor species	Treatment	Reduction in EPM (%)	Reference
Human	Trypsin	25	Ponder (1951)
Human	Trypsin	26	Seaman and Heard (1960)
Human	NANase	63	Seaman and Cook (1965)
Human	NANase	71	Zalik *et al.* (1972)
Human	NANase	94	Eylar *et al.* (1962)
Calf	NANase	84	Eylar *et al.* (1962)
Lamb	NANase	74	Eylar *et al.* (1962)
Chicken	NANase	72	
Chicken	NANase	36	Zalik *et al.* (1972)

acids may be resistant to the action of NANase (Faillard, 1956; Klenk, 1958).

Blood Group Substances

No discussion of the erythrocyte surface components may be deemed complete without a discussion of the blood group substances and the recent attempts at isolation and characterisation of specific blood group antigens.

Landsteiner (1900, 1901) discovered the first blood group system, namely the ABO system. Von Dungern and Hirzfeld (1911) showed that groups A and B could be divided into subgroups. A new group system, the MN and P systems, completely independent of the ABO system was discovered by Landsteiner and Levine (1927, 1928) and subsequently Landsteiner and Wiener (1940) announced the discovery of the *Rhesus* (Rh) system which is of considerable clinical significance. At least 14 different blood group systems are now known.

Since the discovery of the ABO system, attempts have been made to isolate the blood group substances from erythrocytes. Early attempts at obtaining aqueous extracts of the substances were not successful. However, some alcohol extracts with blood group activity were prepared. Shortly afterwards extracts of materials with specificities related to A and B substances were prepared from proteolytic digests of mucosal lining of some animal species (see Kabat, 1956).

Much work has been done on the chemistry, molecular structure and immunological specificities of blood group substances. In many of these studies the substances have been obtained from salivary and gastric secretions (Kabat, 1966). Human blood group substances have been isolated in quantity from ovarian cyst fluids (Morgan and Van Heynigen, 1944). Purified preparations have been found to be sialoglycoproteins. The A,B,H and Le^{a} antigens contain five sugars, including *N*-acetylneuraminic acid. The peptide moiety contains 15 residues, two-thirds of which comprise four amino acid residues, namely of threonine, serine, proline and alanine. Sulphur-containing amino acids are virtually absent, while aromatic amino acids occur in small proportions (see Watkins, 1966). On the other hand, A and B blood group activity on red cells also appears to be associated with some glycolipids. These glycolipids contain the same five sugars mentioned earlier (Yamakawa and Suzuki, 1952; Koscielak and Zakrezewski, 1960; Hakamori and Jeanloz, 1961). The glycoprotein nature of the A and B determinants has recently been amply demonstrated (Hamaguchi and Cleve, 1972; Marchesi *et al.*, 1972; Whittemore *et al.*, 1969; Gardas and

Koscielak, 1971). In the active substances obtained from red cells and from secretions, the major serological specificity appears to be associated with *N*-acetylgalactosamine and D-galactose (Watkins *et al.*, 1964). Sialoglycoproteins have also been shown to carry the MN antigenic system (Springer, 1964; Romanowski *et al.*, 1959; Stalder and Springer, 1960; Kathan *et al.*, 1961; Hamaguchi and Cleve, 1972; Kathan and Adamany, 1967; Lisowska, 1969; Cook and Eylar, 1965; Klenk and Uhlenbruck, 1960). The sialoprotein which contains 54–60% carbohydrate has been isolated. It has a rather low subunit molecular weight of 30 000 (Springer *et al.*, 1966; Winzler, 1969; Blumenfeld *et al.*, 1970; Tanner and Boxer, 1972).

It is known that the human erythrocyte membrane contains at least three glycoproteins which can be distinguished by electrophoresis. Most of the membrane sialic acids occur in these glycoproteins (Fairbanks *et al.*, 1971; Winzler, 1969; Tanner and Boxer, 1972). Hamaguchi and Cleve (1972) recovered membrane glycoproteins from the aqueous phase after extraction with a mixture of chloroform–methanol. On SDS-PAGE electrophoresis three major components, namely glycoproteins I, II, and III, and an additional periodic acid-Schiff (PAS)-positive component were recovered (but see p. 67). These components had molecular weights of 58, 37, 24 and 50×10^3 respectively. The authors found that M and N activities were found mainly in glycoprotein I. N activity was also present in glycoprotein III. The latter showed the highest activity of A, I and S antigens.

Since sialic acids make the major contribution to the surface charge of the erythrocyte one may inquire whether the blood group substances are involved in determining the electrokinetic behaviour of these cells. It may be recalled that proteolytic enzymes cause a reduction in the EPM of erythrocytes. Treatment of erythrocytes with proteolytic enzymes not only can destroy blood group activity on the cell surface, but it releases substances possessing blood group activity.

Morton (1962) found that impure preparations of trypsin destroyed M,N,S,Fy^a, Kell, Yt^a and possibly also Lu^a antigens. However a pure preparation of the enzyme destroyed only the M and N antigens. Crystallised chymotrypsin destroyed S, Fy^a and Yt^a activities while papain destroyed M,N,S,Ky^a and Yt^a antigens. When treated with trypsin sialopeptides carrying M and N activity are released from the erythrocyte surface (Cook *et al.*, 1960; Mäkela *et al.*, 1960; Thomas and Winzler, 1971; Jackson and Seaman, 1972).

A number of hereditary MN variants have been shown by several workers (Furuhjelm *et al.*, 1969; Myllyla *et al.*, 1971; Nordling *et al.*, 1969; Sturgeon *et al.*, 1972, 1973a,b) to occur. These variants not only exhibit diminished MN antigenicity and increased saline agglutinability by incomplete antibodies but they also show reduced EPMs and reduced levels of surface sialic acids (Luner *et al.*, 1975; Sturgeon *et al.*, 1972, 1973a,b). Luner *et al.* (1975) also found that the variant behaviour was similar to that of normal cells subjected to light protease treatment.

At low concentrations trypsin is also able to liberate a sialopeptide with I blood group activity (Ohkumma and Ikemoto, 1965) and low concentrations of chymotrypsin cause the release of a sialopeptide with Fy^a group activity (Ohkuma *et al.*, 1966).

A partial characterisation of a MN glycopeptide isolated from human erythrocytes by alkali treatment followed by pronase digestion, has been achieved (Thomas and Winzler, 1971). This glycopeptide, which is one of the four separated by electrophoresis, contained (mol^{-1}) 1 mol of fucose, 1 mol of sialic acid, 3 mol of galactose and mannose each, 4 mol of acetylglucosamine, 1 mol of aspartic acid and small amounts of threonine, serine and glycine. On the basis of sequential hydrolysis with glycosidases a tentative structure has been proposed for this glycoproetin by Thomas and Winzler (1971) (Fig. 18).

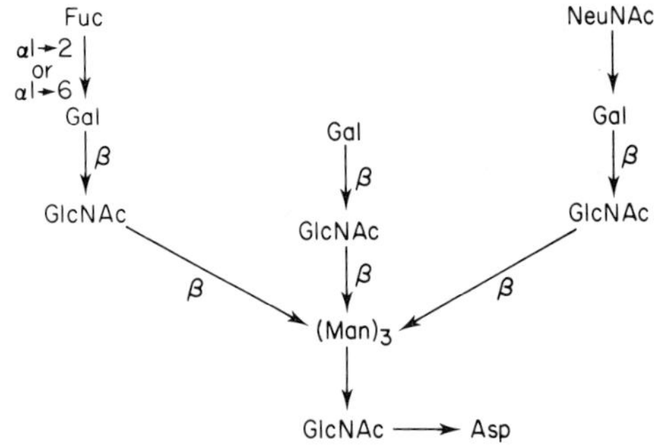

Fig. 18. Proposed structure of MN-glycopeptide of human erythrocyte membrane (Thomas and Winzler, 1971).

According to the proposed structure, three non-reducing branches are attached to a mannose core and these bear terminal fucose, galactose and sialic acids. In branch (i) the penultimate galactose residue is attached in either 2- or 6-position by the terminal fucose, while in branch (iii) the galactose residue is substituted in 2,3,4 or 6-position with *N*-acetylneuraminic acid. In branch (ii) the galactose forms the non-reducing terminal group. The galactose residues in each chain are linked to *N*-acetylglucosamine. It is postulated that the fourth acetylglucosamine residue is bound by an *N*-glycosidic linkage to the amide group of asparagine. This is suggested on the basis that the glycopeptide contains aspartic acid and is alkali stable. The oligosaccharide sequence of sialic acid (or fucose) → galactose → *N*-acetylglucosamine → mannose has also been shown to occur in fetuin, α-acid glycoprotein (Spiro, 1969) and human chorionic gonadotropin (Bahl, 1969) and may turn out to be not an uncommon structural feature of glycoproteins derived from mammalian species.

Much information about the location of these oligosaccharide residues on the glycoprotein molecule, and on the distribution, attachment and orientation of the molecules on the erythrocyte surface has become available in the past few years. Marchesi *et al.* (1972) investigated the chemical structure of the major sialoglycoprotein component, glycophorin, of the human erythrocyte membrane. The glycoprotein isolated using lithium diiodosalicylate is a single polypeptide chain of approximate molecular weight of 50 000. Grefrath and Reynolds (1974) obtained a value of 29 000 for the monomer. This agrees well with the molecular weight calculated from amino acid sequence data. It contains 60% carbohydrate, 40% protein, and carried blood group antigens and receptors for influenza virus and plant agglutinins (Morawiecki, 1964; Uhlenbrook, 1964; Marchesi and Andrews, 1971; Marchesi *et al.*, 1972; Fukuda and Osawa, 1973; Tillack *et al.*, 1972). Tryptic digestion of the glycoprotein gives four unique fragments ($\alpha-1$; $\alpha-2$; $\alpha-3$ and β). By sequential tryptic digestion of the intact membrane and partially digested fragments, the order in which the fragments may occur has been deduced. Cleavage with cyanogen bromide yields five fragments. Of these fragments C5 and C1 contain most of the carbohydrate and these are derived from the N-terminal end of the glycoprotein.

Segrest *et al.* (1973) showed that the primary sequence of glycophorin contains domains of hydrophilic and hydrophobic residues, with the carbohydrate moiety attached to the N-terminal end which is exposed to the exterior of the cell (Winzler, 1969). The glycophorin A molecule is now known to be composed of 131 amino acid residues and 16 oligosaccharide chains, and the sequence of occurrence of the amino acid residues has been determined. The oligosaccharide residues are attached to the amino acid residues of the N-terminal third of the glycophorin molecule. SDS-electrophoresis of human red cell membrane proteins has revealed five polypeptides (Bands 1, 2, 4, 5 and 6) and six sialopeptides (Bands 3, 7 and PAS-1, 2, 3 and 4) stainable by periodic acid-Schiff (PAS) reagent. Of the PAS-stainable sialopeptides, Band 3 and PAS-1 sialopeptides are the most prominent. PAS-1 appears to be a transmembrane and dimeric sialopeptide. Glycophorin A is believed to be the basic subunit of the PAS-1 sialopeptide (Furthmayr *et al.*, 1975; Tomita and Marchesi, 1976).

Winzler *et al.* (1969) had earlier suggested that the hydrophilic N-terminal segment is associated with a hydrophobic core. Marchesi *et al.* (1972) have also suggested that the C-terminal end of the glycoprotein interacts with some membrane component to form intramembranous particles (see Fig. 2) while the carbohydrate-rich N-terminal segment is exposed to the exterior of the membrane.

The occurrence of intramembranous particles has been demonstrated using freeze etching method which involves freeze fracturing of the membrane to reveal its hydrophobic matrix. This is followed by etching which exposes the membrane surface by lowering the ice table (Pinto Da Silva and Branton, 1970; Branton, 1966; Tillack and Marchesi, 1970). The hydrophobic matrix thus revealed can be differentiated into smooth areas and particles (see Weinstein, 1969). These particles have a diameter of approximately 85 Å. Pinto Da Silva and Douglas (1970) found that the A antigen sites were associated with the intramembranous particles.

It is not yet clear, however, if a 1 : 1 relationship exists between the intramembranous particles and the glycoprotein molecules. Marchesi *et al.* (1972) drew attention to the correlation observed between the number of such particles (5×10^5 cell^{-1}) with the number of phytohaemagglutinin receptors ($4{\cdot}5 \times 10^5$ cell^{-1}) and the number of blood group sites (8×10^5 cell^{-1}) (Economidou *et al.*, 1967). A similar correlation between the distribution of intramembranous particles and the distribution of phytohaemagglutinin and influenza receptors has been observed by Tillack *et al.* (1972).

The hydrophilic C-terminal segment of the glycophorin molecule occurs at the inner surface of the plasma membrane (Bretscher, 1971b; Segrest *et al.*, 1973). The sequence analysis made by Tomita and Marchesi (1976) shows that this segment is rich in acidic amino acid residues and therefore this may be conducive to association with cations and peptides that are relatively more cationic in nature. There are indications that the polypeptide Bands 1 and 2 are polymers of spectrin and are in close association with this C-terminal segment (Nicolson and Painter, 1973; Ji and Nicolson, 1974; see also Fig. 2).

Lymphocytes and Platelets

It had been previously recognised that the different cell types in the blood differed in their EPM. Erythrocytes and lymphocytes migrated at different rates to polymorphonuclear leukocytes. It was suggested that this may be because leukocytes may have an adsorbed film of serum protein, and it was also known that erythrocytes do not show non-specific adsorption of protein on their surfaces. Further, blood platelets were found to have EPMs comparable with leukocytes which led to the suggestion that the platelets too may have protein films adsorbed on their surfaces (Abramson *et al.*, 1942).

Interest in the behaviour of lymphocytes and platelets has been revived in the past few years, although the volume of information available is still considerably less than in the case of erythrocytes. For this reason and on account of possible similarities in the nature of their surface, lymphocytes and platelets will be considered together in this section.

Bangham *et al.* (1958) examined the pH-mobility relationship for erythrocytes, granulocytes, lymphocytes and platelets. The behaviour of lymphocytes and platelets was strikingly different from that of erythrocytes and granulocytes, in that the former showed true pIE at pH 2·6 and 3·2. At bulk pH values below the pIE, the cells showed a cathodic mobility. The erythrocytes and granulocytes on the contrary did not exhibit true pIE. This is a clear indication of the presence of positively charged groups on the surface of lymphocytes and platelets. When treated with acetaldehyde platelets showed a 20% increase in EPM at pH 7·0. This confirms the presence of positively charged groups. Further the cationic group blocked by aldehyde appears to have a $pK > 9$ which is indicated by the absence of any increase in the anodic mobility of unmodified cells until the bulk pH is greater than 9 (Seaman and Vassar, 1966).

Mehrishi (1970a, 1973) has estimated that the human lymphocyte and platelets have $8{\cdot}5 \times 10^3$ positive charges per μm^2 area presumably contributed by amino groups.

The pIE of 2·6–3·0 of lymphocytes (Bangham *et al.*, 1958; Mehrishi and Thomson, 1968) are indicative of the presence of carboxylic groups and/or phosphoric acid residues. Mayhew and Weiss (1968) examined the effects of neuraminidase and ribonuclease on the EPM of a number of cell types derived from the peripheral blood. Treatment with NANase reduced the EPM of all cell types. The reduction in the case of erythrocytes was 65% (as discussed on p. 63) while in the case of lymphocytes, platelets and polymorphs, NANase reduced the EPM by 43–66%. In monocytes the reduction in EPM was only about 17%. Thus all the elements of blood share the common feature that their electrophoretic behaviour is largely due to the negative charges contributed by the surface sialic acids (Table 16). Apart from the carboxylic groups of the sialic acids it appears from the investigations of Seaman and Vassar (1966) that an unidentified species of carboxyl group may contribute negative charges to the platelet surface. These acidic groups

Table 16 *Electrophoretic mobilities of human peripheral blood cells after treatment with NANase and ribonuclease*

Cell type	Treatment	EPM (−TU) ($I = 0{\cdot}145$ M, 25°C)	Reduction (%)
Lymphocytes	C	8·74	
	N	5·05	42·2
	R	7·64	12·6
Platelets	C	9·58	
	N	4·22	55·9
	R	9·52	0
Polymorphs	C	7·99	
	N	4·51	43·6
	R	7·27	9·0
Monocytes	C	5·72	
	N	4·65	18·7
	R	5·3	7·3
Erythrocytes	C	10·92	
	N	3·81	65·1
	R	10·74	1·7

C is control EPM measured in Hank's balanced saline; N is NANase-treated cells and R is Ribonuclease-treated cells.
From Mayhew and Weiss (1968).

may have a pK of about 4·0 and may be contributed by the acidic amino acid residues of platelet protein.

The electrophoretic mobility of lymphocytes and polymorphonuclear leukocytes was reduced by 12·6 and 9% respectively by subjecting the cells to RNAase treatment. Mayhew and Weiss have suggested that this is due to the cleavage by the enzyme of ionisable phosphate groups of pyrimidine ribonucleotides. It is also possible that the reduction of EPM after RNAase treatment could be due to conformational changes at the surface, but Mayhew and Weiss (1968) do not consider that this is likely because they found that susceptibility to NANase action remained unaltered even when the cell had been subjected to RNAase treatment prior to treatment with NANase. Further, on the basis of the effects of the enzymes on cellular EPMs measured at lower ionic strengths which have the effect of increasing the thickness of the electrical double layer, they have suggested that the phosphate groups are distributed uniformly in the cell periphery. The blood cells also differ in their susceptibility to RNAase treatment. For instance, the enzyme has only a marginal effect on the EPM of erythrocytes, and none at all on platelets and monocytes. Surface RNAs have also been described by the same authors on other cell lines such as mouse liver cells, sarcoma 37 ascites cells, macrophages and in Ehrlich ascites cells (Mayhew and Weiss, 1958; Weiss and Mayhew, 1966, 1967). Mehrishi (1970b) found that platelets carried groups which were susceptible to alkaline phosphatase. He found that the cells showed reduction of EPM by up to 37%, which corresponds to 5×10^5 to 10^6 electron charges removed from the cell surface (Mehrishi, 1969). Brossmer and Patsche (1968) earlier reported the occurrence of 10^6–10^7 alkaline phosphatase-susceptible groups on the platelet surface.

Since the mechanism of action of the two enzymes is different it seems possible that they act on different components of the cell surface. The possibility also exists that the RNAase molecules which carry a number of net positive charges will produce an overall decrease in the net negative surface charge by adsorbing non-specifically on the surface. In a later publication Weiss and Mayhew (1967) have discounted this possibility.

Bangham *et al.* (1958) have suggested that the phosphate groups detected on the surface of lymphocytes and platelets may be derived from phospholipid. This is indicated by the similarities between the pH-mobility relationships for lymphocytes and erythrocytes and egg lecithin and phosphatidylethanolamine.

It may be concluded that phosphate groups are present on the surface of platelets and lymphocytes, but they do not seem to occur on the erythrocyte surface. This may account for the ease with which platelets show aggregation in the presence of Ca^{2+} and the inability of these cells to stick to glass or most other surfaces in the absence of divalent cations.

The lymphocyte population is made up of two functionally distinct cell types, namely the T-lymphocytes—the thymus-dependent type which are responsible for the cell-mediated immune response and the B-lymphocytes which are dependent on the bursa Fabricius or equivalent lymphoid organ and are responsible for the synthesis of humoral antibodies.

Electrophoresis has been shown to separate lymphocytes into two subpopulations of high and low EPM (Ruhenstroth-Bauer and Lucke-Huhle, 1968; Hannig and Zeiller, 1969; Bert *et al.*, 1971; Zeiller *et al.*, 1972; Zeiller and Pascher, 1973; Nordling *et al.*, 1972a,b; Wiig and Thunold, 1973; Andersson *et al.*, 1973; Durandy *et al.*, 1975; Sabolovic *et al.*, 1974; Donald *et al.*, 1974; Gladstone *et al.*, 1974; Weiss *et al.*, 1976a,b). Apparently the subpopulation with higher mobility are of the T-cell type while those with lower EPM are of the B-cell type (Zeiller *et al.*, 1972; Nordling *et al.*, 1972; Wioland *et al.*, 1972; Donner and

Wioland, 1975; Dumont, 1974a,b, 1975). Additional evidence which supports this view has come from the work of Dumont and Robert (1975) who examined the electrophoretic behaviour of spleen lymphocytes of two H-2 identical strains of mice, Balb/c and DBA/2 (H-2^d). These two strains differ in the responsiveness of their spleen lymphocytes to PHA. Thus, while Balb/c cells respond to PHA, DBA/2 mice cells showed greater response to bacterial lipopolysaccharide (LPS) than did cells from the Balb/c strain. When the cells from these two strains were electrophoresed, Dumon and Robert (1975) noticed not only a bimodal electrophoretic distribution of slow and fast cells, but also significant differences in their relative proportions (Table 17). Of the cells from the

Table 17 *Electrophoretic mobility and composition of thymocyte subpopulations of Balb/c and DAB/2($H-2^d$) mice strains*

Strain	Mitogen response	Slow moving cells (%)	Fast moving cells (%)	Mean EPM of cells (−TU) Slow	Fast
Balb/c	PHA positive	49	51	6·9±0·1	11·6±0·03
DBA/2	LPS positive	74	26	7·2±0·1	12·4±0·1

PHA is phytohaemagglutinin and LPS is bacterial lipopolysaccharide.
From Dumont and Robert (1975).

Balb/c strain, 49% were slow moving and 51% fast moving, but from the DBA/2 strain, 74% were slow moving and 26% were fast moving. These results are compatible with their degree of responsiveness to PHA which stimulates T-cell responsiveness (Janossy and Greaves, 1972) and LPS which is known to be a B-cell mitogen (Andersson *et al.*, 1972; Dumont, 1974a, Shek *et al.*, 1976). Ault *et al.* (1976) have stated that the low mobility cells are true B-cells bearing IgM or IgD or both. These cells had all the characteristics of B-cells the most important being their ability to synthesise their own surface immunoglobulin (Rowe *et al.*, 1973). Ault *et al.* (1976) also observed another class of IgG-bearing cells which are distinct from B-cells. They found the T-lymphocytes to be a heterodisperse group and suggest the possibility of this group representing a spectrum of functional subclasses.

The possibility of using the differences in mobility to achieve a separation appears to have prompted most of these investigations. Free-flow electrophoresis appears to have achieved a partial separation of the lymphocyte subclasses (Stein, 1975, 1976; Häyry *et al.*, 1975). Griffith *et al.* (1975) have therefore developed the method of density gradient electrophoresis which seems to allow a much superior enrichment of the lower mobility B-cells. Ault *et al.* (1976) could recover only 30–50% of the cells, although up to 90% of the recovered cells were in the viable state. Most of the cell loss seems to have occurred in other procedures than electrophoresis. Neither was there any preferential loss of any subclass. Nonetheless, at this state, these experiments have had much greater success in characterising the subclasses than with their separation.

Zeiller *et al.* (1976) found also distinct electrokinetic differences in free-flow electrophoresis in direct PFC (plaque-forming cell) progenitor cells. These progenitor cells all possessed surface immunoglobulin and B-cell specific antigen. Zeiller *et al.* in fact have suggested that these electrokinetically different PFC progenitor cells may represent distinct subsets of B-cell population.

Mehrishi and Zeiller (1974) have examined the possibility of the differences in the mobilities of T- and B-lymphocytes reflecting differences in the expression of electrokinetic groups on their surfaces. Their results are summarised in Table 18. From this data it appears that

Table 18 *Ionogenic groups on the surface of T- and B-lymphocytes*

	Number of groups $\times 10^{-5}$ cell^{-1} (surface area 100 μm^2)	
Ionogenic group	T-lymphocytes	B-lymphocytes
α-COOH (NANase-susceptible)	24	12
Unidentified anionogenic groups	76	43
Amino groups	6	11
Thiols	Not detected	20
Phosphate groups (RNAase-sensitive)	Not detected	16

From Mehrishi and Zeiller (1974).

the striking differences in their EPMs may be attributable to the fact that the T-lymphocytes have twice the number of NANase-susceptible α-COOH groups and other unidentified anionogenic groups on their surface than do the B-lymphocytes. Weiss *et al.* (1976b) have reported that the difference in the sialic α-COOH content of T- and B-cells is much smaller than indicated by Mehrishi and Zeiller (1974). Weiss *et al.*

found 17 and 13% of NANase-susceptible groups on T- and B-cells. In addition, claim Mehrishi and Zeiller (1974), the density of cationogenic groups on the T-cells is half of that on the B-lymphocytes. Apart from these quantitative differences, Mehrishi and Zeiller (1974) have been unable to detect the presence of thiol groups on the surface of the T-lymphocytes, while a considerable number of them could be found on B-lymphocytes. They also failed to detect RNAase-susceptible phosphate groups on T-cells but did detect them on B-lymphocytes. On the contrary, Weiss *et al.* (1976b) have described that up to 28 and 63% of the anionic groups on T- and B-cells respectively are RNAase-susceptible. It is difficult to reconcile these two groups of observations. If B-cells have lower mobility and higher content of RNAase-susceptible anionic groups the number of cationic groups ought to be greater in B-cells than in T-cells. At any rate, the differences in sialic acid releasable by NANase has been confirmed by Wiig (1974). The enzyme is estimated to have released $5{\cdot}6\times10^{-8}$ μg of sialic acid from the T-cell and $5{\cdot}2\times10^{-8}$ μg from the B-cell. The non-ionic detergent Kyro EOB which disrupts plasma membranes also released equivalent quantities (5·3 and $4{\cdot}5\times10^{-8}$ μg cell^{-1}) of sialic acids (Wiig and Maehle, 1975). Thus most of the total sialic acids of the cell was present in the plasma membrane (60–70% in the T-cell and 70–90% in the B-cell). Nevertheless, when the Kyro EOB-disrupted plasma membrane of B-cells was subsequently treated with NANase, a further release of sialic acids was observed. This may indicate that some of the sialic acids may be located in the deeper regions of the plasma membrane where they are not accessible to the enzyme if treatment was carried out on whole cells.

The data of Wiig and Maehle (1975), it must be pointed out, give values of 100×10^{6} and 88×10^{6} sialic acid molecules for each T- and B-cell respectively. These are far in excess of the estimates given by Mehrishi, and Zeiller (1974). Fugita *et al.* (1975) have reported that a major glycoprotein of molecular weight 30 000 was associated with membrane fractions obtained from B-lymphocytes. This glycoprotein was not associated with the T-lymphocyte membrane. The significance of this discovery to the electrophoretic behaviour of T- and B-cells is still unclear.

A similar observation has been made by Andersson *et al.* (1976) who studied surface glycoproteins of T- and B-lymphocytes obtained from normal human subjects and blood lymphocytes from patients with chronic lymphocytic leukemia (CLL). These authors labelled the glycoproteins and separated them by polyacrylamide gel slab electrophoresis. They noticed that the glycoprotein pattern of T- and B-cells differed, and also that the pattern of normal cells was different from CLL cells.

For instance, a glycoprotein, GP4, of molecular weight 160 000 was found to be associated with T-cells but not with B-cells, nor with CLL cells, a majority of which are believed to be of the B-type. The component GP4 therefore seems to be specific for T-cells. On the other hand, a protein (NS) component of molecular weight 57 000 was strongly labelled in the CLL cells.

On the whole there is sufficient evidence that the processing of bone marrow cells by thymus and gut-associated central lymphoid tissue into T- and B-cells involves considerable differentiation of the plasma membrane. But the nuclei of these cells and those of thymocytes show similar EPM and apparently there are no sialic acids associated with the thymocyte nuclear membrane (Wiig and Maehle, 1975).

The mechanisms of adhesion and aggregation of platelets may involve the formation of intercellular "bridges" mediated by divalent cations like calcium, and/or by an effective reduction in the electrokinetic charge on the cell surface (see Pethica, 1961). It is known that platelet aggregation can be brought about in citrated or heparinised plasma by the addition of small quantities of adenosine diphosphate (ADP) (Gaarder *et al.*, 1961; Mitchell and Sharp, 1964) or by 5-hydroxytryptamine (5-HT) (Mitchell and Sharp, 1964; Swank *et al.*, 1963). Addition of ADP ($200\ \mu g\ ml^{-1}$) to platelets suspended in citrated plasma produces a decrease in their EPM with concomitant aggregation. For both these events to occur the presence of Ca^{2+} appears to be necessary since neither effect is noticeable when ethylenediaminetetraacetic acid (EDTA) is also present in the plasma (Seaman and Vassar, 1966). Adsorption of ADP on the platelet surface ought to increase their EPM, as it indeed does when very low concentrations ($0{\cdot}05\ \mu g\ ml^{-1}$) of ADP are used (Hampton and Mitchell, 1966a). The reduction of EPM at higher ADP concentrations may be associated with the release of 5-HT by the platelets, or by a counter adsorption of molecules when the net negative charge density has reached a critical maximum level.

It has been suggested that ADP may interact with the positively charged amino groups as a primary step in the process leading up to platelet aggregation (Merishi, 1970a,b). This suggestion appears to be based on two observations. Firstly, the number of ADP receptor sites ($2{\cdot}5 \times 10^5$ $cell^{-1}$) as estimated by Born (1965) agrees well with the number of amino groups ($2{\cdot}42 \times 10^5$) detected on the platelet surface (Mehrishi, 1970a). Secondly, *N*-ethylmaleimide, a reagent considered tto be specific for thiol groups but which can also react with amino groups, inhibits spontaneous and ADP-induced aggregation of platelets (Seaman, 1967).

Alone 5-HT, which is a basic compound, readily adsorbs on to platelets suspended in standard saline and causes a reduction in their

EPM. Similar reductions in EPM, i.e. reduction in the net negative charge density, occur also when platelets are treated with polycationic compounds such as poly-L-lysine, with a concomitant induction of aggregation. On the other hand, polyanionic substance like heparin and poly I:C increase EPM and keep platelet aggregation at a low level (Zbinden *et al.*, 1970). Mester *et al.* (1972) described the possible participation of carbohydrate groupings in ADP- and 5-HT-induced platelet aggregation. They showed that the velocity of aggregation was altered by changes in the sialic acid content of the platelets.

The surface components involved in the phenomenon of platelet aggregation have also been partially characterised by using plant lectins with different carbohydrate-binding specificities. Phytohaemagglutinins from *Phaseolus vulgaris* and *Lens culinaris* showed great affinity of binding to human platelets. Both lectins produced agglutination and functional changes mimicking the thrombin-induced aggregation and release reactions (Majerus and Brodie, 1972). Lectins with specificities towards α-D-mannopyranose or α-D-glucopyranose moieties of glycoproteins did not affect the aggregation process. Lectins with galactose specificity (*Ricinus communis*) reacted with platelets to a varying degree but wheat germ agglutinin (WGA) which binds *N*-acetylglucosamine interacted strongly. Both these lectins caused platelet aggregation and the release of both ADP and serotonin. At subthreshold levels WGA inhibited ADP-induced aggregation (Greenberg and Jamieson, 1974). The agglutination induced by WGA was only partially inhibited by the use of phosphoenol pyruvate kinase system to destroy ADP but agglutination induced by the galactose-specific lectin was strongly inhibited. Greenberg and Jamieson (1974) therefore believe that while WGA is acting as a true agglutinin, the *Ricinus* lectin mimicks physiological effects of aggregation presumably by releasing ADP. These authors isolated three glycopeptides from platelet membrane preparations by proteolytic digestion. Of these, glycopeptide I of molecular weight approximately 120 000 and containing both *N*- and *O*-glycosidic linkages, specifically inhibited platelet agglutination produced by WGA. The latter, it may be recalled, is also a strong inducer of ADP release by the platelets.

The mechanisms involved in these various effects are not, by any means, completely understood, but the investigations discussed here do underline the significant part that the electrokinetic groups on the platelet surface play in the phenomenon of platelet adhesion and aggregation.

Pathological conditions such as thrombasthenia and von Willebrand's disease appear to be associated with changes in surface properties and electrokinetic behaviour of platelets. Platelets from patients with thrombasthenia do not adhere to glass in a whole blood system or in platelet-rich plasma containing ADP. Neither do they adhere to each other in the presence of ADP, noradrenaline, thrombin or connective tissue extract (Hardisty *et al.*, 1964; Zucker, 1964). The EPM of platelets from patients with thrombasthenia and von Willebrand's diseases are normal (Zucker and Levine, 1963; Hampton and Hardisty, 1967). They do, however, differ from normal platelets in certain aspects of their behaviour *in vitro* and in their responsiveness to exogenous agents that induce aggregation. For instance, Hampton and Hardisty (1967) found that no changes could be induced in the EPM of platelets by contact with glass or after incubation with ADP. Such changes do take place in normal platelets and appear to depend upon the integrity of the thiol groups (Hampton and Mitchell, 1966b). If normal platelets were subjected to agents which modify or block the thiol groups, they appeared to behave like platelets from patients with thrombasthenia. On being further treated with L-cysteine to restore the thiol groups, they assumed normal pattern (Hampton and Hardisty, 1967). As mentioned earlier, Hampton and Mitchell, (1966a,b) did find that the ADP-induced aggregation was accompanied by increase in EPM of the cells when low concentrations of ADP are used. Hampton and Hardisty (1967) therefore suggest that the lack of ADP-induced aggregation in platelets from thrombasthenic patients and the absence of any increase in EPM on incubation with exogenous ADP may be due to the inability of ADP to bind to the platelet surface. They have further suggested that the loss of EPM of normal platelets on contact with glass may be attributed to the removal of ADP in the process. Whether the thiol groups are involved in the ADP-induced aggregation is a moot point. For it appears that *N*-ethylmaleimide, which Hampton and Mitchell (1966a,b) used in their investigations, may not be specific for thiols and there is a possibility that the reagent may also react with amino and imino groups (Mehrishi, 1972).

The processes of platelet adhesion and aggregation are surface-mediated phenomena and therefore the involvement of molecular components of the platelet membrane has been investigated. Mester *et al.* (1972) described the possible participation of carbohydrate groupings in ADP- and 5-HT-induced platelet aggregation. Human platelet membrane contains three major glycoproteins. As characterised by PAGE with SDS, these glycoproteins are of molecular weights 155 000, 135 000 and 103 000 (Nachman and Ferris, 1972; Nurden and Caen, 1974, 1975). Nurden and Caen (1974, 1975) examined the membrane glycoproteins from five patients with Glanzmann's thrombasthenia and noticed that the 135 000 MW glycoprotein component was consistently absent. The platelets from thrombasthenic patients do show the initial response to ADP in the form of alteration of their morphology (Caen and Michel, 1972). It appears, therefore, that this particular glycoprotein may be involved only in the later stages of the ADP-induced aggregation.

Nurden and Caen (1975) also examined the membrane glycoproteins of platelets from two patients with Bernard-Soulier syndrome which disorder is associated with an impaired platelet adhesion. In these platelets the 155 000 MW glycoprotein was conspicuous by its absence. They also showed that this glycoprotein accounted for the major part of trypsin-releasable sialic acid. This is consistent with the observations of Grottum and Solum (1969) that Bernard-Soulier platelets had reduced sialic acid content and also showed lower EPM. From this it appears that the 155 000 MW glycoprotein may be directly involved in the molecular interactions occurring between the platelet surface and surface components of the subendothelium.

SURFACE CHEMISTRY OF BACTERIAL CELLS

The study of surface chemistry of bacterial cells by electrophoresis dates back to the first two decades of this century. Much of the work in this field has been discussed in detail by Abramson (1934), Abramson *et al.* (1942) and James (1957), and will be briefly summarised here.

THE BACTERIAL SURFACE AND THE AGE OF CULTURES

The electrophoretic behaviour of bacteria appears to vary with the age of the cultures. Shibley (1924) found that the EPM of *Diplococcus pneumoniae* Type I and *Salmonella paratyphii* increases in the early phase of growth, reaching a maximum value after about 10 hours, which roughly corresponds with the logarithmic phase of growth, and then declines to a steady level. Similar variations are also seen in cultures of *E. coli* rough and smooth forms. But unlike the organisms investigated by Shibley, in *E. coli* a decrease in EPM is observed in the early growth phase. This decrease is maintained over a few hours before a stable normal value is reached (Moyer, 1939), which is then maintained over several days (Buggs and Green, 1935a,b; Pedlow and Lisse, 1936; Tittsler and Lisse, 1928). Although these changes are apparently related to changes in the surface of the cells, they are by no means a universal phenomenon. Not only have variations among strains been detected (Stearns and Roepke,

1941) but some species do not show any variations in their EPMs at all. There is a clear possibility therefore that these variations may reflect differences in the surface structure and composition of bacteria, provided, of course, that the EPMs of the various strains at different stages of growth were measured at a number of different ionic strengths of the suspending media. However, in *S. gallinorum* and *Aerobacter aerogenes* this was not found to be the case (James, 1957). Joffe and Mudd (1935) and Moyer (1936) observed that certain species of *Salmonella* showed no mobility in any phase of growth and even when tested at varying ionic strengths of the suspending media. It may be concluded from these early studies that the surface structure of bacteria is independent of its phase of growth. In mammalian cells, on the other hand, it has been shown that increased EPM is associated with cell division (Eisenbert *et al.*, 1962; Heard *et al.*, 1961; Mayhew and O'Grady, 1965; Ruhenstroth-Bauer *et al.*, 1962; Brent and Forrester, 1967).

BACTERIAL VARIATION

In the handling of vast numbers of rapidly reproducing bacteria, occasionally variants are encountered which possess properties different from the parent strain. The variants could be either phenotypic variants which are organisms that have undergone temporary adaptive change, or genotypic mutants which have undergone relatively more permanent genetic changes and have acquired a variety of new properties. The variation in pathogenic bacteria from the smooth to the rough form occurs fairly frequently and it is also an important form of bacterial variation. This variation is indicated by changes in the morphology of the colonies, and involves a loss of surface components and virulence.

Early investigations in this area have been aimed at examining possible correlations between electrophoretic behaviour of bacteria with variation, virulence, etc. The data summarised in Table 19 shows that in most instances the rough variants possessed considerably higher electrophoretic mobilities as compared with the smooth forms. Although the investigations rarely proceeded beyond the demonstration of increased EPMs, Dekker *et al.* (1942) have suggested that the transition from the smooth to the rough form in *S. typhii* involves two antigens and that the expression of these antigens may cause the overt behavioural changes.

In the smooth form both these antigens, designated IX and V_i are said to be present; but both are absent in the rough form. The transition may occur through an intermediate half smooth form where either of the two antigens is only present. Dekker *et al.* found that the rough variant cells show a pIE of 2–2·5. The variant half smooth and the smooth forms

Table 19 *Electrophoretic mobilities of smooth and rough variants of bacteria*

Organism	Electrophoretic conditions: Buffer	pH	*I* (M)	EPM (−TU)	Reference
Escherichia coli var. communis	Acetate	5·5	0·02	(S) 5·2	Moyer (1936b)
				(S) 6·9 (R) 36·0	Moyer (1936a)
Salmonella typhii	NaCl/HCl	4·5	0·1	(S) 0·6–3·1 (R) 19·0–26·0	Dekker *et al.* (1942)
	Acetate	5·0	0·02	(S) 0 (R) 26·0	Joffe and Mudd (1935)
Mycobacterium tuberculosis	NaCl(?)	7·2	0·015	(S) 5·0 (R) 6·0	Choucroun and Plotz (1934)
Mycobacterium leprae	Acetate	5·9		(S) 41·0 (R) 22·9	Reed and Gardiner (1932)
Bacillus calmette-Guerin	NaCl	7·2	0·015	(S) 4·0 (R) 4·5	Choucroun and Plotz (1934)

S: Smooth form; R: Rough form of variant.

which possess the V_i antigen appear to possess groups which ionise below pH 2, indicating that the V_i antigen might contain strongly acidic groups such as PO_4^{3-} or SO_4^{2-}.

BACTERIAL VARIANTS AND SURFACE STRUCTURE

Many bacteria possess distinct surface structures which have been described as flagellae or filaments (Houwink and Van Iterson, 1950; Bisset and Hale, 1950). These structures have been demonstrated with the use of the electron microscope. Brinton *et al.* (1954) have been able to show that the surface structure of these organisms correlates well with their electrophoretic behaviour. They have described a variant of *E. coli* B, which they have designated as the F variant. This variant occurs in two forms, the electrokinetically fast and slow forms. The two forms differ in their EPM by a factor of two. Under the EM the slower form is seen to have surface filaments which appear to cause the reduction in their EPM. When the filaments are stripped off the organisms by mechanical agitation, the EPM resembles that of the non-filamented form. The defilamented forms form, on incubation and division, the filamented electrokinetically slow variant.

VIRULENCE

A number of investigators have examined possible relationships between virulence and electrokinetic behaviour, but none has been revealed (Brown and Broom, 1932; Buggs and Green, 1935a; Jones, 1931; Thompson, 1931a; Frampton and Hildebrand, 1944). The only instance where a reasonable degree of correlation has been described is in *D. pneumoniae* (see Table 20). Thompson (1931b) found that the EPM of Types II, I and III of *Diplococcus* was 31–33, 41–45 and 57–64 TU respectively. These increasing EPM values are compatible with the chemical nature of the capsular material extracted from the organisms (Heidelberger, 1927). The capsular material obtained from Type II organisms was a nitrogen-free polysaccharide and weakly acidic. The material from Type I was strongly acidic and contained galactouronic acid. But in addition, it also contained an amino sugar derivative, and

Table 20 *Electrokinetic behaviour of* Diplococcus pneumomiae

Type	EPM (−TU)[a]	pIE[b]	Nature of capsular material[c]
I	41–45	3·5–3·9	Galactouronic acid; amino sugar derivative; 5% nitrogen, pI ~ 4; precipitated by 1:6 × 10^6 diluted antiserum against Type I but requires 1:400 diluted antiserum Type III.
II	31–33	3·9–4·1	Nitrogen-free polysaccharide made up chiefly of glucose, dextrorotatory weak acid. Precipitated by 1:5 × 10^6 diluted Type II antiserum but does not precipitate with even 1:400 diluted antisera against Types I and II materials.
III	57–64	4·3–4·6	Nitrogen-free polysaccharide made up by condensation of aldobionic acid such that reducing sugars disappear but only carboxyl groups remain free.

[a] Thompson (1931b).
[b] Thompson (1932).
[c] Heidelberger (1927).

therefore the electrokinetic property was a consequence of the combination of the strongly acidic and weakly basic groups of the amino sugars—giving these organisms an intermediate EPM value between Types II and III. The material from Type III was found to be a nitrogen-free polysaccharide and strongly acidic in nature. It was made up by the condensation of aldobionic acid. However, in the process of condensation the reducing sugars are shielded, and the electrokinetic behaviour of the Type III organisms is attributable exclusively to the acidic carboxyl groups. On the other hand, the pIE values obtained for the different types do not agree so well with the nature of the capsular material. Although the correlations are inconclusive, it is safe to assume that these differences in the electrokinetic properties of the three variants indicate differences in the make up of at least the outermost layers of the capsule. In fact, Thompson (1932) also noticed that the EPM of Type III decreased to that shown by Type I variant, when the former were extracted with 0·25 N hydrochloric acid. These cells were not affected by alcohol extraction. Type I and II variants were unaffected by acid treatment. Presumably treatment with acid is hydrolysing the outermost layer of capsular material which is conferring on the Type III variants their characteristic electrokinetic properties.

The differences in the composition of the surface material was also obvious from the degree of agglutination achieved using antisera raised against capsular material of the three types. Heidelberger (1927) found that antiserum against Type I material precipitated Type I organism at a dilution of $1:6\times10^6$, but Type III antiserum required a dilution as low as 1:400 to produce agglutination of Type I organisms. Similarly, antiserum against Type II organisms caused precipitation of Type II organisms at a dilution of $1:5\times10^6$. In comparison, even at 1:400 dilution antisera against Type I and Type III capsular material failed to cause agglutination of Type II organisms. Quite clearly the materials obtained from the surfaces of the three types of *Diplococcus* organisms not only give them their electrokinetic abilities but also distinctive antigenic properties.

These early experiments show that electrokinetic behaviour of bacteria is a valid and fairly reliable indicator of their surface constitution. Thus Lowick and James (1955, 1957) showed that cells of *aerogenes* trained to crystal violet have a predominantly lipid surface while in untrained cells it is mainly polysaccharide in nature. When these cells were grown in a synthetic medium containing crystal violet, a line of cells (called C-cells) which had a very low EPM as compared with the normal cells, was produced. The C-cells showed a pH-mobility curve characteristic of protein and lipid surface (with a pIE of 3·5) in contrast to the normal cells. In the normal cells the pH-mobility curve was independent of the pH above pH 5. In addition, the EPM of the C-cells was found to increase to that of the normal cells, on treatment with lipase at pH 7. Apparently growth in the presence of the dye had resulted in a change in the composition of the surface material. Similar changes were also seen by James and Barry (1954) when the organisms were grown in the presence of proflavine.

CHEMICAL MODIFICATION OF THE BACTERIAL SURFACE

Cell electrophoresis has been used in conjunction with method devised to modify particular surface groups by specific chemical treatments. This line of investigation involves the examination of the EPMs of the cells under a variety of experimental conditions of pH and ionic strength, or examining the mobilities of cells whose surfaces are selectively altered, in order to obtain information regarding specific ionogenic groups that reside on the surface. The pH-mobility curve at a given ionic strength indicates the nature of the electrokinetic groups. The cells are then subjected to specific treatments in order to alter the activity of the suspected ionogenic species. The pH-mobility relationship is then re-examined which will reflect the chemical modification that had been carried out. Such chemical modifications can also be detected by isoelectric equilibrium studies, with changes in the isoelectric points of the modified cells often indicating the kind and numbers of ionisable groups that have been modified. This aspect of characterisation of the bacterial surface is discussed in detail in Chapter 5 (pp. 185–188).

Some of the earliest work on chemical modification of bacterial cells and follow up of the substitution of ionogenic groups using cell electrophoresis was described by Cohen (1945). This investigation was aimed at discovering the antigenic groups responsible for the serological behaviour of *Bacillus proteus* OX-19 employed in the diagnosis of typhus by the Weil–Felix reaction. The reagent used by Cohen was benzene sulphonyl chloride which was known to react with amino groups at weakly alkaline pH (Miller and Stanley, 1942). It was found that treatment of the organisms at pH 8·4 over a period of 30 minutes to 3 hours caused a 2 to 2·5-fold increase in their anodic mobility. Cohen suggested that both amino and imidazole groups were affected. Douglas (1959) subsequently found that *p*-toluenyl sulphonyl chloride (PTSC) was much more efficient in blocking the charged amino groups of *B. subtilis*.

Gittens and James (1963) made a detailed study of the effects of PTSC on *E. coli* cells. A pH-mobility investigation of cells treated with PTSC for 1 hour at pH 11·0 indicated that the surface of these cells did not possess basic groups with pK > 5. The plateau mobility value indicated the possibility of secondary amino groups (pK ~ 4·5). The absence of primary amino groups was also confirmed by the use of fluorodinitrobenzene (FDNB), which in purely aqueous solutions produced no changes in the EPM of the treated cells. But even at very low pHs the organisms did not show zero mobility, which suggests the presence of phosphatidic groups. On extraction of the cells with alcohol the acidic component of the cell surface appears to have been removed leaving a protein-like surface possessing more basic groups than were detected on the surface before extraction with alcohol.

In *A. aerogenes*, neither PTSC nor FDNB produced any differences in EPM, although chromatographic examination of acid hydrolysates of the FDNB-treated cells revealed that FDNB had indeed combined with ε-amino groups of lysine. Since this binding was not reflected in changes in the EPM, it appears possible that these groups are located deeper in the surface, beneath the effective electrokinetic zone (EKZ). This demonstration of the absence of amino groups on the *A. aerogenes* surface thus confirms the observations of Plummer and James (1961) who came to the conclusion that the surface of these cells was purely carboxyl. The pK value of the carboxyl species is calculated using the Hartley–Roe equation is about 2·81 which compared well with the value of 3·19 for glucuronic acid determined electronically. The small difference may be attributed to the fact that glucuronic acid might occur with other monosaccharides. *Aerobacter aerogenes* cells modified with ethylene oxide (EO) and propylene oxide (PO), which are known to esterify carboxyl groups in proteins (Fraenkel–Conrat, 1944), had pH-mobility curves characteristic of a carboxyl surface. From the decreases in mobility it was estimated that about 50% of the groups were esterified. This observation led Gittens and James (1963) to suggest that only the outermost carboxyl groups are esterified by the reagent and that the remainder, presumably situated at deeper levels in the EKZ, are sterically protected from the reagent. This is also indicated by calculation of total charge due to carboxylic groups and those in the EKZ alone. According to Gittens and James, the surface charge density as obtained from the direct method is $2{\cdot}15 \times 10^5$ e.s.u. cm^{-2}, while that derived from mobility measurements is $2{\cdot}78 \times 10^3$ e.s.u. cm^{-2}. These values correspond to one charge per 22·3 and 1730 $Å^2$ respectively. Using isoelectric equilibrium method, Sherbet and Lakshmi (1973) showed that in *E. coli* one negative charge occurs every 350 $Å^2$. Since the area of a carboxyl group is approximately 20 $Å^2$, the only way by which such high charge densities can be explained is by invoking the occurrence of successive layers of charge as suggested by Haydon (1961). Of these layers, under low ionic strengths (of 0·05 M) employed by Gittens and James, the charges residing in the outermost ~15 Å of the surface will be able to exert their electrokinetic effect, while in the even lower ionic strength conditions of the isoelectric equilibrium studies of Sherbet and Lakshmi, ionisable groups occurring between 40 and 60 Å thickness of the surface are detected.

The suggestion of Haydon (1961) has in fact been substantiated. The bacterial surface has been shown, in ultrathin sections and in shadowed preparation, to have a multilayered arrangement (Weidel *et al.*, 1960; Martin and Frank, 1962; Murray *et al.*, 1965; De Petris, 1967; Hofschneider and Martin, 1968). Confirmation of such a three-dimensional arrangement has also come from freeze fracture studies (Bayer and Remsen, 1970; Van Gool and Nanniga, 1971). From the point of view of chemistry of these various surface layers, the situation may be summarised. The innermost layer, which is a trilaminar structure composed of lipids, proteins and carbohydrates, is analogous to the plasma membrane of the eukaryotic cell. External to the plasma membrane is a layer of peptidoglycan, which is made up of strands of polymers of disaccharides and a network of cross-lined peptides. Outside this peptidoglycan layer occur a variety of polysaccharides and teichoic acids in Gram positive organisms. In Gram negative bacteria occurs another layer composed of lipopolysaccharide, lipid and proteinaceous material termed the outer membrane. In *B. polymyxa* this outermost layer of the outermembrane appears to be about 120 Å thick and composed mainly of protein and has a regular array of 70 Å globules repeating at 100 Å intervals. This layer is easily separated using guanidine hydrochloride which indicates that it is linked through weak hydrogen bonds to the underlying lipo(?)polysaccharide layer. The globules can be released by treatment with SDS. This may indicate the involvement of hydrophobic bonds. The underlying polysaccharide layer, on the other hand, appears to be covalently bound to the peptidoglycan layer (Nermut and Murray, 1967).

ANTIGENICITY AND ELECTROPHORETIC BEHAVIOUR

It was mentioned in connexion with the question of bacterial variation how the expression of different antigens may result in overt changes in the electrophoretic behaviour of the cells. Not much work has been done on the actual characterisation of the antigens except that discussed by James (1965) which will be briefly summarised here. For, this work is an example of application of the method of cell electrophoresis used in conjunction with chemical and enzymic modification of the cell aimed at characterising a specific component of the surface.

Much of this work has been done on Group A *Streptococcus* strains. One of the antigenic components used for the purpose of classification are the M antigenic proteins, which may be associated with virulence. Most strains of *S. pyogenes* also have a T protein antigen. This is made use of in the typing of these organisms. Twenty-four hours after inoculation on to blood agar, the strains possessing both M and T antigens form small, semitransparent disc-shaped colonies which have a granular matt surface. Frequently strains which have a hyaluronic acid capsule form *glossy* colonies. These strains are known to be of low virulence and to contain little or no M antigen.

Hill *et al.* (1963a,b) examined the pH-mobility relationships of *S. pyogenes* after removal of the hyaluronic acid capsule and showed that surface had free carboxyl and amino groups and another ionisable group with pK 7·0 identified as the imidazole group of histidine. It appeared that the charge densities due to the carboxyl and amino groups were higher on the *matt* strain than on the *glossy* strain. But the ratio of carboxyl to amino is higher in the *glossy* strain than in the *matt* strain.

Treatment of the organisms with 1-fluoro 2,4 dinitrobenzene not only confirmed the carboxyl nature of the surface but also indicated the presence of two carboxyl species with pK values of 2 and 4. Treatment of the cells with a variety of decarboxylases showed that alanine and glutamic acid decarboxylases reduced the groups of pK 2 leaving the γ-carboxyl groups ($\sim$pK 4·0).

The M protein antigen can be easily stripped off the cell surface by treating the *matt* strains with trypsin for 16 hours at 37°C. When treated in this way, the pH-mobility curve coincided with that of similarly treated *glossy* strain. However, the trypsin-treated *matt* strain cells showed higher anodic mobilities. The protein material released by trypsin from both *matt* and *glossy* strains had the same amino acid composition. On the basis of these experiments it is postulated that both the strains have the M antigen, but that the protein in the *glossy* strains is considerably more cross-linked than in the *matt* strain thus giving rise to differences in M antigenicity and binding of hyaluronic acid on the comparatively less cross-linked *matt* strain.

ELECTROPHORETIC INVESTIGATION OF CELL–VIRUS INTERACTIONS

Cell electrophoresis has been found to be a useful tool in the investigation of the complex early cell–virus interactions. On mixing cells and virus under physiological conditions, a fairly rapid attachment of the virus to the plasma membrane occurs.

Hirst (1941, 1942a,b, 1948) and McClelland and Hare (1941) showed that the influenza virus adsorbed on to the erythrocyte membrane, as a first step in causing agglutination. They also showed that in a second step of the interaction, the virus was released from the membrane. This is now known to happen with many virus–host cell systems. When the virus has been adsorbed in the cold, elevation of the temperature causes elution of the virus from the surface (Joklik and Darnell, 1961; Holland, 1962; Mandel, 1967; Silverstein and Dales, 1968; Hochberg and Becker, 1968).

An electrophoretic examination of these initial events was made by Hannig (1948) who found that the EPM of human erythrocytes was reduced when they were mixed with PR8 influenza virus. These treated cells were not able to adsorb further viral particles, which indicated that the virus adsorbed on to specific receptors on the surface which were lost, at least temporarily, on account of the enzymes associated with the virus. Subsequently Klenk (1955) showed that the material released by the virus was neuraminic acid. The reduction in the mobility observed by Hannig (1948) and confirmed by Sachtleben and Straub (1959) and Straub and Sachtleben (1963) in the erythrocyte-fowl plague system is thus attributable to the loss of neuraminic acid moiety of the surface component following elution of the virus. The existence of enzymic activities in the superficial regions of the viral envelopes has been known for sometime. The involvement of neuraminidases of myxo- and paramyxoviruses in their attachment and elution from membrane has also been demonstrated (see Cohen, 1963). Pretreatment of cells with receptor-destroying enzyme (RDE) or interaction of the cells with viruses having neuraminidase activity eliminates or reduces subsequent adsorption of the virus (Zhdanov and Bukrinskaya, 1962).

The reduction of EPM caused by viral adsorption is independent of the concentration of the virus (Sachtleben and Straub, 1959; Straub and Sachtleben, 1963). This would suggest that the process of adsorption and elution of the virus occurs continuously as long as receptors are available. However, Zhdanov and Bukrinskaya (1962) found no such independence of the attachment of virus and the viral concentration employed. Using the Sendai virus and human and monkey tissue culture cells, they found that the degree of adsorption was proportional to the virus concentration and that the rate of adsorption entirely expressed by Brownian regularities of suspensions.

Also involved in the process could be the affinity of binding of the virus to the receptors, which would determine the speed of elution. Hirst

(1941) showed that virus inactivated by periodate is not eluted but continues to adsorb to the cell surface. This behaviour is quite clearly reflected in the fact that the reduction in EPM caused by the inactivated virus is strongly dependent upon virus concentration. In other words, one can dissociate the initial two events, namely the liberation of neuraminic acid, and elution of the virus. The inactivation of the virus does not affect the enzymic liberation of neuraminic acid but does affect the ability of the virus to elute. In fact, Zhdanov and Bukrinskaya (1962) found that the liberation of the virus may depend upon the susceptibility of the cells to infection. The virus is eluted by non-susceptible cells but achieves penetration into the susceptible ones.

The attachment of the virions to the surface may involve electrostatic interactions (Takemoto and Fabish, 1964; Benda, 1966). For it has been observed that polycations such as diethylaminoethyl cellulose (DEAE)-dextran enhances the binding, while the polyanionic substance heparin not only interferes with the binding but also causes a release of virions already bound (Miyamoto and Gilden, 1971; Vogt, 1967; Warren and Thorne, 1969; Hochberg and Becker, 1968). It would appear therefore that following binding the mucoprotein may be locally dissolved to provide a firmer attachment of the viral particle to the membrane prior to its penetration into the cell by viropexis or other mechanisms.

An involvement of electrostatic interactions between the viral particles and host cells, both of which bear a net negative surface charge, has not been borne out by an electrophoretic investigation described by Weiss and Horoszewicz (1971) (see Table 21). These authors used the Epstein–Barr (EB) virus and cell lines 64–10, 2367 and 8226 derived from human peripheral blood. Line 2367 was from a normal individual, while 64–10 and 8226 came from male patients with myelogenous leukemia and multiple myeloma respectively.

Weiss and Horoszewicz (1971) have argued that if electrostatic interactions were involved in the binding of viral particles, this would be determined by the average interaction energies between the viral and host cell surfaces. They noticed that line 8226 which did not adsorb the virus showed a maximum repulsive energy of 85×10^{-13} ergs, but the corresponding values for lines 64–10 and 2367 which did adsorb the virus, were 65×10^{-13} and 50×10^{-13} ergs. However, the random collisions of viral particles with cells is governed by Brownian theory (Allison and Valentine, 1960; Zhdanov and Bukrinskaya, 1962) and therefore a repulsive energy barrier above $1{\cdot}5kT$ would not allow adsorption of the virus by any of these cell lines.

Table 21 *Adsorption of EB virus and EPM of host cells*

Cell type/treatment	Virus adsorption	EPM (−TU)	Change in EPM (%)
RPM1/8226	−	18·9±0·9	
+RNAase	−	14·9±0·5	−21
+NANase	−	13·9±0·5	−27
+RNAase and NANase	−	9·4±0·3	−50
RPM1/64-10	+	15·3±0·5	
+RNAase	+	13·6±0·3	−11
+NANase	+	10·7±0·5	−30
+RNAase and NANase	+	10·2±0·4	−33
RPM1/2367	+	14·2±0·3	
+RNAase	+	12·2±0·4	−14
+NANase	+	10·5±0·4	−26
+RNAase and NANase	+	8·4±0·3	−41
RPM1/64-10	+	14·0±0·2	
+2,4,6-trinitrobenzene sulphonic acid	+	12·4±0·2	−4·6
RPM1/64-10	+	13·0±0·2	
+2,3-dimethylmaleic anhydride	+	13·8±0·2	+8

From Weiss and Horoszewicz (1971).

Treatment of these cells with ribonuclease or neuraminidase reduced the EPM of cell line 8226, but no adsorption of the virus was induced by these enzymic treatments. No adsorption was detected when both the virus and cells were subjected to treatment with the enzymes. On the other hand, such treatment did not prevent the virus from adsorbing to cell lines 64–10 and 2367. This may indicate that the components of the surface which act as receptors are not the substrates for the enzymes. Nonetheless, the differences seen in these cell lines may be due to non-uniform distribution of the high energy barrier areas interspersed with receptor sites where net negative charge density is low enough to allow close contact between the virus and cell surfaces and the formation of physical and chemical bonds. Further experiments by Weiss and Horoszewicz (1971) have attempted to exclude the involvement of electrostatic attractive forces suggested by Puck and Tolmach (1954), Allison and Valentine (1960) and Valentine and Allison (1959), in the interaction. They observed that neither 2,4,6-trinitrobenzenesulphonic acid nor 2,3-dimethylmaleic anhydride affected the virus adsorption by cell line 64–10, despite the fact that at least 2,3-dimethylamaleic anhydride had modified the surface amino groups. This reagent, in fact, produced an 8% increase in the anodic mobility of the cells. The modification of amino groups of the cells treated with the 2,4,6-trinitrobenzenesulphonic acid must, however, remain in doubt. For, in these cells, a decrease in EPM was observed even though the reagent appeared to have reacted with the amino groups. Thus the decrease in anodic mobility may have been a secondary effect of an unknown interaction. Nevertheless, it does seem reasonable to assume that only a small proportion of the amino groups may have been modified. In the absence of any quantitation of virus binding, there is not enough experimental evidence to rule out the involvement of surface amino groups in the process of virus adsorption.

The mechanisms by which the immunological system of the body neutralises viral infections is another aspect of the problem of host cell–virus interactions that is amenable to investigation using the cell electrophoresis method. The antibodies produced by the body in response to infection by a virus can neutralise viral particles already bound to cells and also neutralise the circulating virus. Coating with specific antibodies of cells with attached herpes simplex virus, for instance, results in freezing of the particles on the membrane (Morgan *et al.*, 1968). In other instances an accelerated elution or a failure to adsorb may occur as in the case of Sendai, polio, Newcastle disease virus, etc. (Howe and Morgan, 1969; Mandel, 1967; Rubin and Franklin, 1957).

In analogous situations, the electrophoretic behaviour of cells has provided some interesting information about the cell–virus interactions and the neutralising effect of the antibodies. Straub (1965) showed that erythrocytes showed reduction in EPM on treatment with a suspension of fowl plague virus. When the virus-coated cells were then treated with virus-specific antiserum further reductions in EPM were observed. In contrast, treatment of the cells with receptor-destroying enzyme caused a reduction in the EPM. But when these cells were subsequently treated with an antiserum against the enzyme the EPM actually showed an increase (Straub, 1965; Sachtleben *et al.*, 1973). This increase in EPM does not occur if the enzyme-treated cells are treated with antiserum raised against the virus. From these results Straub casts doubts on the previous finding that release of neuraminic acid is responsible for the reduction in EPM noticed in cells following treatment with virus.

This conclusion may not be completely valid. One may postulate two phases in the reduction of EPM by viral adsorption. In the process of attachment the virus will release neuraminic acid from the surface but at the same time the viral particles may introduce into the surface a smaller number of negatively charged groups than they have released. This will result in a decrease in the net negativity of the surface. The second phase of EPM reduction is now simply due to the elution of the virus from the cell surface by the neutralising antibodies. This postulate requires that treatment with receptor-destroying enzyme should produce a greater loss in the EPM than does either the active or inactive virus. The experiments described by Straub (1965) indeed show that under the specified experimental conditions, treatment with the enzyme lowers the EPM of chick erythrocytes to approximately 3·5 TU over a 60-minute period, while the lowest EPM observed on treatment with the active virus or the inactivated virus is about 5 TU.

In erythrocytes treated with the enzyme and antiserum against the enzyme, a completely different situation obtains. The loss of EPM on treatment with the enzyme may be due to a combination of release of neuraminic acid and adsorption of the enzyme molecules on the cell surface. It is known that the enzyme releases neuraminic acid. The suggestion that enzyme molecule may be adhering to the surface is actuated by the observation that a subsequent treatment with anti-enzyme antiserum actually causes an increase in the mobility. This possibility is suggested by Mehrishi (1972) who has argued that the enzyme masks negatively charged groups to produce a loss of EPM. He further suggests that treatment with anti-enzyme antibodies may have the effect of eluting the enzyme particles thus allowing the negatively charged groups, which had been masked by the enzyme, to exert once again their electrokinetic effect. Mehrishi, however, bases his argument on the premise that antibodies would not cause an increase in EPM if they were themselves to adsorb on to the enzyme molecule. The effect of antibodies on the EPM of cells is unfortunately still an open question. This aspect is considered in detail in a subsequent section of this chapter. It will suffice to say here that immunoglobulins have been found to be an heterogeneous population of molecules with pIs ranging from 5·95 to 7·5 (Freedman and Painter, 1971; Hoffman *et al.*, 1972). In the electrophoresis conditions adopted in the above work and much of other work involving the use of antibodies, the EPM were measured at pH 7.2. One would expect therefore that a substantial part of the population to be negatively charged. Thus increase in EPM in the experiments described by Straub (1965) and Sachtleben *et al.* (1973) on treatment with anti-enzyme serum, could be due to binding of the antibodies to the enzyme molecules adsorbed on the cell surface.

As mentioned before, a second means by which the body will defend itself against viral infection is by neutralising the circulating virus with specific antibodies. Although this problem has not received much attention, some preliminary studies by Straub (1965) indicate that the viral particles neutralised by specific antibodies fail to bind the cells. It is known that such antibody-inactivated particles are eliminated by the macrophages (Boland *et al.*, 1957; Hanson *et al.*, 1957).

The foregoing discussion, while giving a general picture of the initial events of virus–host cell interactions as deduced from information using cell electrophoresis, also indicates the scope of the investigation in this field that can be undertaken. The mechanism of action of anti-viral agents, the mechanisms involved in the mediation by viruses in cell fusion, are two problems in this field which appear to be amenable to investigation by cell electrophoresis. No doubt electron microscopy coupled with biochemical analyses has provided insight into virus–host cell interactions, but within its limitations cell electrophoresis has much to contribute to this important field of study.

EFFECTS OF ANTIBODIES ON ELECTROPHORETIC MOBILITY

Recent years have seen a synthesis of the dynamic, structural and functional aspects of cell membranes. In the wide spectrum of membrane phenomena, mechanisms relating to membrane functions in differentiation and morphogenesis, dedifferentiation, neoplastic transformation, growth, cell recognition, immunological specificity, etc. must involve the nature, composition and orientation of surface macromolecules. It was mentioned in connexion with discussion of the erythrocyte membrane and blood group antigens that glycoprotein molecules occurring on the cell membrane conferred electrokinetic properties and immunological specificities on the erythrocytes.

In this section we shall be concerned with effects of treatments with specific antisera on the electrokinetic behaviour of cells. The origin of this line of investigation dates back to the 1920s. Coulter (1921) observed that antisera reduced the EPM of erythrocytes. Freund (1925) showed that treatment of Tubercule bacilli with normal and immune sera produced a reduction in the ζ potential of these organisms. The immune serum was found to produce a comparatively larger loss in the ζ potential than did normal serum, and it also produced agglutination of the cells. Shibley (1924, 1926, 1929) showed that not only antiserum reduced EPM of *Pneumococcus* cells but that immunological specificity was also involved. Antiserum against *Pneumococcus* Type I caused reduction in the ζ potential of Type I cells only, whereas in Type II and III the reductions noticed were comparable to the effects produced by normal serum. After a lapse of three decades there has been a revival of interest in the effects of antibody binding to cell surfaces. Not only have the earlier effects of antibodies on EPM been confirmed but the investigations have taken a new direction towards possible characterisation of the receptor sites occurring on cell membranes for specific antisera.

EFFECT OF ANTIBODIES ON EPM OF ERYTHROCYTES

Most of the work carried out on erythrocyte mobilities on treatment with antisera has been described in minute detail by Sachtleben (1965). The discussion here will therefore be restricted to the salient features of this work.

The first general effect of treating cells with antisera is to reduce their EPM. This reduction occurs mainly in higher dilutions of the antisera. At low dilutions (1 : 4 of neat serum), however, an increase in mobility is produced. This is said to be due to components of normal serum itself. For, when erythrocyte EPMs were measured in various dilutions of compatible blood, their EPM actually showed increases at dilutions below 1 : 4 (i.e. at higher serum concentrations). But even after a correction was applied the EPMs were still showing an upward trend at lower dilutions. Cells from different donors belonging to the same blood group behaved similarly in these experiments.

The reductions in EPMs were observed when effects of serum dilutions were tested at other pH ranges than physiological. But for a given serum dilution, reduction of ionic strength produced the expected increase in EPMs. At serum dilutions greater than 1 : 20 000, the reduction of I from 0·16 to 0·016 M caused approximately 50% increase in EPM. At lower serum dilutions below 1 : 10, the increase in EPM was only about 15%. This observation is somewhat difficult to interpret. Sachtleben attributes this to the formation of a complete coat of antibody on the cells. Even if this were the case, by decimating the ionic concentration, the effective electrokinetic zone is extended from approximately 7·5 to 32 Å i.e. at the lowest ionic concentration the EPMs should have been effectively caused by negative charges of nearly the whole of the antibody molecule. As mentioned before at pH 7·2 of the electrophoresis buffer most of the antibody molecules will be negatively charged, and therefore one would have expected to see a much larger increase in EPMs as a result of lowering the ionic concentration from 0·16 to 0·016 M.

Sachtleben has also described experiments using sera of Rh blood group system and anti-M and anti-N sera. None of these sera affected the EPM of erythrocytes at high serum dilution. The serum from Rh system in fact caused an increase in EPM at lower dilutions than 1:8. The only conclusion one can draw is a differential effect on the mobility of the cells on treatment with different antisera. Although it is claimed that electrokinetic activity of sera is in the same range as their agglutinating activity, the experiments do not bear this out. In the experiments with A-group serum on B-group erythrocytes, a decrease in EPM is noticed from serum dilutions 1:250 and lower until 1:8 dilution. But in this range no agglutination occurs until the serum concentration is raised to 1:32, while half of the total loss of EPM has already occurred. In the Rh group serum no change in EPM is detected from 1:8 serum dilution to 1:125 during which the agglutination index has dropped from 4 to 0. A similar lack of correlation between loss of EPM and gain of agglutinability is noticed in the experiments with the M and N system. Therefore there seems to be little justification for suggesting, as Sachtleben has done, that the lack of such correlation may be a feature of anti-M and anti-N sera. However, Sachtleben quite legitimately points out that no great differences are known to exist between molecules of the Rh and ABO antibodies. This only indicates that a reappraisal should be made of the results obtained from electrokinetic experiments involving cells and their interaction with antibody molecules (see also pp. 205–211). What appears incongruous is that binding of antibody should show a reduction in EPM at all. Freedman and Painter (1971) and Hoffman *et al.* (1972) investigated the isoelectric characteristics of purified anti-*p*-azophenyl-trimethyl ammonium and anti-*p*-azobenzoate antibodies raised in rabbits. These preparations were found to be heterogeneous populations of molecules with pIs ranging from 5·7 to 7·4. Valmet (1968) examined the isoelectric pattern of lyophilised human γ-globulin preparation and noticed that their pIs ranged from 6 to 9. Howard and Virella (1969) obtained a partial separation of IgG based on the subclass of γ-chain. The isoelectric spectrum was as follows, IgG_4 was found below pH 6, IgG_3 and IgG_2 were isoelectric between pH 7 and 9, while IgG_1 covered a wide range of pH even above pH 9.

Now most of the electrophoretic measurements described by Sachtleben and his group appear to have been made in a buffer of pH 7·2. In these circumstances one would expect that the antibody molecules with pIs above pH 7·2 would bear net positive charges, while those with pIs below 7·2 would show net negative charges. Therefore treatment with antibodies will have made little difference to the net charge density on the surface of the treated cells, unless it is assumed that antibody molecules with $pI>7{\cdot}2$ have higher affinity to the surface receptors than those molecules with $pI<7{\cdot}2$. This is a possibility which is in fact indicated by experiments of Sachtleben in which variations in EPM were observed over a period of two hours after A-group serum was added to B-erythrocytes. In the initial 15 minutes of mixing there was an exponential decrease in EPM to a level which was more or less maintained over a period of two hours.

The pH-mobility curves for normal and B-erythrocytes treated with anti-B serum add to the difficulties of interpretation. Sachtleben found only a slight reduction of EPM of control as well as serum-treated erythrocytes between pH 7 and 5. But at pH 5 a sharp drop in EPM begins and at pH 4 greater than 50% loss of EPM had already occurred compared with approximately 10% loss of EPM of controls at this stage. This pH-mobility relationship seems to suggest that the serum contains immunoglobulins of the lower isoelectric spectrum.

To summarise, most of the antisera, with the exception of anti-D sera, have produced reductions in the EPMs of erythrocytes. Anti-D sera at high serum concentrations (i.e. low dilutions) actually caused an increase and were without any electrokinetic effect at lower concentrations of serum. Ironically, anti-D antibodies could conceivably decrease EPM. Templeton and Milne (1975) showed the presence of three main groups in the isoelectric spectra of anti-D IgG immunoglobulins. The first group had pI between pH 8·5 and 9·5 (main peak at pH 9·2) a second group between pH 7·7 and 8 (main peak at pH 7·9) and a third group between pH 3 and 6 (with main peak at pH 4·5). The second group (pH 7·7–8·0) showed higher binding constants than the other two. The three groups presumably correspond to IgG_1, $IgG_{2,3}$ and IgG_4 respectively. This third group could conceivably be anti-D IgA immunoglobulin which is said to occur in about 10% of all anti-D sera (Mollinson, 1972). The anti-D serum which caused increases in EPM of erythrocyte must have contained predominantly the low pI range IgG_4 antibodies or the effects were due to IgA-anti-D antibodies. This seems plausible, for, although all four IgG heavy chain subclasses are found in anti-D immunoglobulin (Morell *et al.*, 1973), the relative proportion of each subclass in the sera may vary considerably (Morell *et al.*, 1972).

EFFECTS OF ANTIBODIES ON EPM OF TUMOUR CELLS

Forrester *et al.* (1965) examined the effects of normal and immune sera on the EPM of EL4 ascites tumour cells. They were able to demonstrate the binding of antibody molecules to the cell surface. The effects of immunoglobulins on the EPMs of silica particles and EL4 cells were similar.

The EPM of both silica particles and the cells was found to be considerably reduced when exposed to high concentrations of γ-globulin.

When the dilution was above 1:32 and higher, the EPMs showed an upward trend until at 1:256 dilution the particles had attained their original electrophoretic mobility. They also examined the mobility characteristics of cells at ambient pH ranges from 4·0 to 10·0 (Table 22).

Table 22 *Effect of pH on the EPM of EL4 ascites cells exposed to antisera*

	EPM (−TU) at pH				
Incubation	4·0	5·0	7·0	9·0	10·0
EL4 cells only	10·7	11·1	11·2	11·3	12·4
+normal serum	11·0	12·0	13·0	11·4	11·6
+immune serum	3·2	4·5	6·2	9·9	10·4
+immunoglobulin	3·3	4·5	6·2	9·1	—

From Forrester *et al.* (1965).

While addition of immune serum did lower the EPM of the cells by 71% of the original EPM values, these reductions were gradually alleviated when the EPMs were measured at higher pH ranges so that at pH 9, the reduction in the EPM was 13·2%. Hartveit *et al.* (1968) also observed similar reductions in the EPM in BP8 ascites and Ehrlich ascites cells treated with respective antisera. While these changes in EPM in relation to the ambient pH may be genuine, comparisons of the EPM values of normal and serum-treated cells may be misleading. All these measurements were made at physiolgical ionic strength and no attempts appear to have been made to examine the effects of γ-globulin binding under different ionic conditions. Since at the physiological ionic strength only 10 Å of the cell surface is liable to be probed, in the control series one is looking at 10 Å zone of the actual cell membrane, but in the γ-globulin-treated cells one is probing only the adsorbed layer of antibodies (see Fig. 19). Therefore, while the evidence presented does indicate the adsorption of antibody molecule, it has no relevance to immunological phenomena involved in the interaction of cell surface and antibodies.

It was mentioned earlier that immunoglobulins focus in the pH range 5·7–7·4 in isoelectric focusing. However, when they are adsorbed on a polarised surface, they may manifest changes in their electrical characteristics. Thus Forrester *et al.* (1965) found that silica particles coated with immunoglobulins showed a pIE of around 4·6. It is to be expected therefore that increasing the ambient pH will result in an increase in the

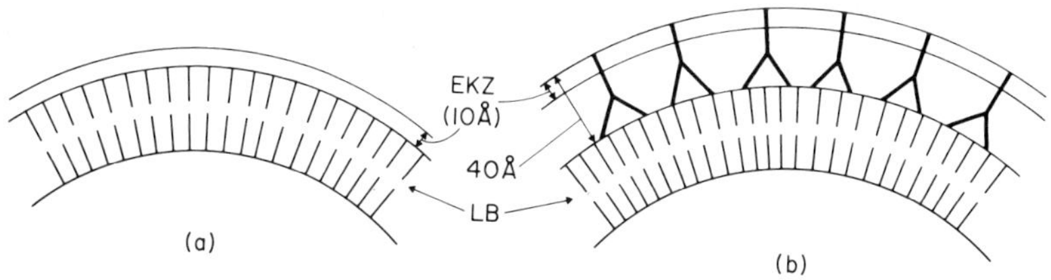

Fig. 19. Zones explored by cell electrophoresis in normal cells (a) and those with adsorbed antibody molecules (b) represented diagrammatically in order to show how in the latter case (b) the electrokinetic properties may be determined by the layer of adsorbed antibodies alone. LB represents the lipid bilayer; EKZ the electrokinetic zone. Surface components not shown.

net electronegativity of the immunoglobulin-coated cells. This series of experiments again proves that all these observations relate to the immunoglobulin coat.

ASSAY OF CELL SURFACE ANTIGENS

Four groups of antigens can be distinguished on the cell surface. These have been defined by Franks (1968) as follows: species antigens which are antigens detectable on cells of all members of a species; heterophil antigens such as Forssmann and Paul-Bunnel antigens, which characterise all members of a species but may be detected in some other species; alloantigens such as histocompatibility antigens detected in some but not all members of a species and cell-specific antigens, which are associated with certain types of cells, such as tissue or organ-specific antigens and tumour-specific antigens.

The molecular characterisation of surface antigens has received much attention lately because the expression of these antigens is governed by selective gene activity associated with the complex processes of cell differentiation.

Fetal- and Tumour-specific Antigens on Ascites Tumours

Woo and Cater (1972) have taken advantage of the changes in EPM produced by antibodies to assay the presence of possible fetal antigens on tumour cells. They raised antisera against fetal rat liver (anti-FL serum) and 14-day-old embryos (anti-E serum), and tested the effects of these sera on the EPMs of adult liver, fetal liver and hapatoma cells. They observed that anti-FL and anti-E sera reduced the EPM of adult liver cells by 14·3% and 19·3% respectively. The effects of these sera on fetal liver and heptoma cells were even more marked (see Table 23). The

Table 23 *Effects of anti-FL and anti-E sera*

Cell type	No serum	Anti-FL serum	Anti-E serum
		EPM (−TU)	
Adult rat liver	7·61±0·1	6·5 ±0·36	6·14±0·05
Fetal liver	9·42±0·15	4·38±0·06	4·74±0·09
Hepatoma cells	9·56±0·1	4·9 ±0·1	5·39±0·08

From Woo and Cater (1972).

reductions in the EPMs of fetal liver and hepatoma cells were comparable. These observations have been interpreted as indicating a comparable expression of fetal antigens on both types of cells, but a much reduced expression of these antigens on adult liver cells. These cross-reactivities were confirmed by testing the effects of the anti-FL serum after absorption, in separate serum samples, with adult liver (anti-FL-A), fetal liver (anti-FL-F) and hepatoma cells (anti-FL-H). Anti-FL-A serum now produced no significant effects on the EPM of adult liver cells but retained the ability to reduce the EPMs of fetal liver and hepatoma cells. Anti-FL-H serum, on the other hand, slightly affected the EPM of adult liver cells and about 9% reduction in EPM of fetal liver cells, but caused no change in the EPM of hepatoma cells. Anti-FL-F retained binding ability to adult liver and to hepatoma cells (see Table 24 for a summary of the EPM changes).

Table 24 *Reduction in EPMs of adult liver, fetal liver and hepatoma cells exposed to absorbed antisera*

Cell type	Reduction in EPM (%)		
	Anti-FL-A	Anti-FL-F	Anti-FL-H
Adult liver	1·8 (N.S.)	12·2	4·1
Fetal liver	51·0	2·4 (N.S.)	9·3
Hepatoma cells	45·0	20·2	2·5 (N.S.)

N.S. represents change statistically not significant; anti FL-A, anti-FL-F and anti-FL-H are antiserum against fetal liver cells, adsorbed with adult liver, fetal liver and hepatoma cells respectively.
From Woo and Cater (1972).

It appears from these experiments that hepatoma cells not only possess antigens which characterise their neoplastic state but also express to a lesser degree antigens detectable on fetal liver cells.

Antigens on the Lymphocyte Surface

Bert *et al.* (1970) found that the EPM of lymphocytes is altered by their exposure to immunosuppressive agents such as anti-lymphocyte sera (ALS) and prednisolone. Lichtman and Weed (1970) suggested the possibility that normal and leukemic human lymphocytes owe their surface electric charge mainly to the presence of sialic acid-containing glycoproteins and to sulphated mucopolysaccharides. In order to test the molecular components with which ALS binds to alter the EPM of cells, Bona *et al.* (1972) undertook a detailed investigation into the ALS-binding abilities of lymphocytes of which the surface components had been selectively removed by specific enzymic treatments (see Table 25 for summary of results). They found that NANase and sulphatase reduced the EPM of lymphocytes by about 20%. Trypsin increased the EPM from 13 to 17·5 TU, while maltase, β-glycosidase, β-galactosidase and lecithinase had no effect at all.

Exposure of cells to ALS before and after enzymic treatment revealed that up to 90% of NANase-treated cells had lost their ability to bind ALS. On the other hand, peroxidase-positive subpopulation (of ALS-binding cells) had increased between 50 and 100% after treatment with α-maltase, β-glycosidase and trypsin. These observations indicate that sialic acid-containing glycoproteins and sulphated mucopolysaccharides may be involved at the sites of binding of ALS antibodies. Although the authors have examined the cytotoxicity of ALS after the various enzymic treatments had been carried out and did find increases in cytotoxicity due to ALS in NANase, sulphatase and trypsin-treated cells, none of these changes are statistically significant. Pretreatment with maltase is said to have reduced the cytotoxic effect of ALS. This series of cytotoxicity test does not appear to tie up with the satisfactorily demonstrated inhibition of ALS binding by pretreatment with NANase and sulphatase. Nonetheless, alterations in agglutinability of treated lymphocytes at various ALS dilutions seem to be consistent with inhibition of ALS binding. Thus, both sialidase- and sulphatase- (and also trypsin) treated cells were comparatively less susceptible to agglutination. On the other hand, cells treated with α-maltase, β-glycosidase, β-galactosidase and lecithinase which enhanced the binding of ALS were agglutinable to the same extent as untreated cells. From these experiments, Bona *et al.* (1972) draw the reasonable conclusion that not only terminal sialic acids

Table 25 *Binding of ALS to lymphocytes in relation to electrophoretic mobility, cytotoxicity and agglutinability*

Parameter examined	Lymphocytes treated with							
	Control	NANase	Sulphatase	Trypsin	Maltase	β-Glycosidase	β-Galactosidase	Lecithinase
EPM (−TU)	13·0±0·5	10·5±0·2	10·0±0·2	17·5±0·5	13·0±0·2	13·0±0·2	13·0±0·5	13·0±0·5
Peroxidase-labelled ALS binding (%) positive cells	45	10	45	100	100	100	50	50
Cytotoxic index with[a] ALS after pretreatment		>C	≃C	>C	<C	>C	≃C	≃C
Agglutinability by[b] 1:4 diluted ALS after pretreatment	+++	++	++	—	+++	++	+++	+++

[a] >C, <C, ≃C represents percentage of dead cells was greater than, less than and approximately equal to control, respectively.
[b] Degree of agglutinability +++ high, ++ moderate, —none.
ALS is anti-lymphocyte serum.
From Bona *et al.* (1972).

are present in the antigenic components that bind ALS antibodies, but also that these antigenic sites may be masked by neutral glycoproteins. They suggest further that the susceptibility of the surface components at the binding sites, to treatments with trypsin, α-maltase and β-glycosidase resulting in increased ALS binding, indicates that the main polypeptide component of the masking molecule contains at least one terminal α-glycosidically linked glucose residue and one terminal β-glycosidically linked glucose or fucose residue belonging to a terminal β-glucoside or β-fucoside. Hydrolysis of any of the relevant components results in the exposure of the antigenic determinant.

THE ELECTROPHORETIC MOBILITY OF TUMOUR CELLS *in vivo*

While these experiments have tried to demonstrate changes of EPMs of cells treated with antisera *in vitro*, Hartveit *et al.* (1968) have found that changes occur in the mobility of implanted tumour cells with time. BP8 cells had undergone little change when examined for their EPM 2–4 days after inoculation into host animals. However, from the fourth to the eighth day after implantation the cells were found to show considerable losses in their EPM. The losses in EPM were greater, the larger the size of the original inoculum. Similar changes *in vivo* were also observed in the case of Ehrlich ascites cells. This is suggested to be due to the cells being coated by antibodies. This is a likely explanation of events. It is known, for example, that in breast tumour patients specific antibodies are detectable only in post-excision serum. Before it is excised the neoplasm probably mops up all the antibodies produced by the patient. Nevertheless, it is somewhat difficult to see why larger reductions in EPMs should occur when larger inocula are administered to the animal. One would have expected that if larger numbers of cells were present to mop up the antibodies, the changes would be correspondingly smaller.

ANTIGENIC STIMULATION AND ELECTROPHORETIC MOBILITY OF LYMPHOID CELLS

The entry of antigens into the body produces two immunological reactions, namely the production of humoral antibodies and the production of "sensitised" lymphocytes with cell-bound antibodies. The lymphocytes are involved in cell-mediated immune reactions such as delayed hypersensitivity and graft rejection. It could be anticipated therefore that antigenic stimulation may cause changes in the electrophoretic behaviour of lymphoid cells.

Hartveit *et al.* (1968) compared the electrophoretic behaviour of lymphocytes and thymocytes derived from C57Bl and C3H mice at various stages after inoculation with tumour cells. Initial experiments showed that lymphocytes obtained from C57Bl mice which had received four injections of BP8 ascites cells, could be divided into two subgroups based on their EPM described as "fast" and "slow" lymphocytes. A bimodal pattern emerged clearly during the ten weeks of immunisation. In normal mice the great majority of lymphocytes were slow lymphocytes. In the immunised animals not only was the appearance of fast lymphocytes detected but the relative proportion of the fast and slow cells also appeared to change. For instance, at two weeks after immunisation about 40% were fast lymphocytes compared with the few fast cells occurring in the normal mice. The increase of fast cells continued to attain over 50% of lymphocytes. Thymocytes also showed a similar bimodal pattern, but the number of fast thymocytes in normal animals was negligible. Between two to four weeks of immunisation about 20% of the thymocytes were describable as fast. Subsequently they formed up to 40% of thymocytes over the period of 4–10 weeks of immunisation.

In C3H mice again a bimodal pattern of fast and slow lymphocytes was noticed. When these animals were immunised with BP8 cells, the fast cells appeared to acquire higher EPMs until about seven weeks of immunisation but then disappeared altogether. In normal C3H mice no fast thymocytes were detected, but they appeared during 4–7 weeks of immunisation and constituted up to 28% of the thymocyte population.

In contrast with the lymphocytes and thymocytes, the macrophages obtained from the peritoneal cavity of normal as well immunised C57Bl or C3H mice, showed remarkably uniform electrophoretic mobilities.

It is now known that the functionally distinct T- and B-lymphocytes differ considerably in their EPMs (Hannig and Zeiller, 1969; Zeiller *et al.*, 1972; Norling *et al.*, 1972), with the T-cells possessing greater EPMs than B-cells (see pp. 71–75). These subpopulations occur normally and they proliferate and transform into lymphoblasts and plasma cells respectively on being stimulated by antigen. The fast and slow cells described by Hartveit do not appear to correspond to these normal subpopulations of lymphocytes. Their relevance to the immunological events taking place upon entry of the antigen, is still uncertain.

Somewhat different results have been obtained by Sundaram *et al.* (1967a,b) in Wistar rats stimulated with *V. cholerae* or with TAB antigens. They found that the lymphocytes from the axillary nodes had 8–14% of their EPM following antigenic stimulation. When these stimulated cells were incubated with the corresponding antigens further irreversible alterations in EPM were noticed. Sundaram *et al.* (1967) believe that this indicates an irreversible binding of the antigen by the sensitised cells. From this they conclude that the sensitised cells show the loss of EPM because they bear specific antibodies on their surface. In other words, there is an implied suggestion that the slow lymphocytes may be plasma cells.

These observations have subsequently been confirmed by Phondke and Sundaram (1971) who found that normal lymphoid cells from Chester Beatty strain and Wistar rats could be differentiated into fast and slow cells which the authors term as Class A and Class B cells. In animals immunised with TAB and *Salmonella* "H" antigen, a new class of lymphocytes with EPM lower than that of Class B lymphocytes appeared. There was also an alteration in the relative proportion of Class A and B cells. The overall number of Class B cells was unaffected but that of Class A cells was found to be significantly reduced, indicating the possibility that Class A cells are competent cells responding to antigenic stimulation by proliferation and differentiation into Class C cells with low mean EPM. Phondke and Sundaram (1971) also found a close relationship between the amount of antibody produced and the numbers of Class C cells. They also suggest that their Class B cells may be T-cells because these cells appeared to be identical with adult thymocytes.

THE EFFECTS OF COMPLEMENT ON ELECTROPHORETIC MOBILITY OF ANTIBODY-COATED CELLS

Forrester *et al.* (1965) showed that when EL4 cells are treated with an antiserum raised against them in the rabbit, the antibody molecules adsorb to the surface. This could be demonstrated by using immunofluorescence methods. There was also a concomitant reduction in the EPM of the EL4 cells. When the cells were incubated with antibodies in the presence of complement, the surface immunofluorescence disappeared, but the interior of the cells was stained. This process was accompanied by a swelling of the cells. With these events taking place, the cell also recuperated some of the loss of EPM it had sustained when treated with antiserum alone. Forrester *et al.* (1965) also investigated in these various events the ability of the cells to bind calcium ions. If calcium (0·01 M) was added to the electrophoresis buffer when cells treated with antiserum and complement were being examined, a marked reduction occurred. This indicated the binding of the Ca^{2+} ions. The addition of calcium had no effect when the cells were treated with complement alone or with normal serum plus complement.

These authors suggest that the effects of calcium may be due to interaction between Ca^{2+} and phosphate groups of membrane phospholipids. Since this appears to be a possible explanation, the experiments may be interpreted as suggesting that the action of complement involves the unmasking of phosphate groups of membrane phospholipids.

Nevertheless, it should be stated that this is a highly simplistic interpretation. The immune lysis of cells which occurs when antibody-coated cells are incubated in the presence of complement is a result of highly complex events beginning with an activation of Cl component of complement and progressing through activation and amplification in sequence of several other components of complement. Various enzyme activities are involved at the different stages. The terminal components may undergo transformation into an active phospholipase which causes membrane damage.

The membrane configurational changes occurring in association with the loss of viability of the cells are gradual and progressive. Sherbet *et al.* (1972) found that when a population of cells was put on an electrofocusing column, they resolved into two groups. One group of cells which possessed a higher pI value was up to 90% viable, while the second, a rather widespread, group with a lower mean pI value was made up of up to 60% non-viable cells. This group of non-viable cells was found to spread over 1·5 to 2 pH units. This suggests that cell death is accompanied by a progressive qualitative change in membrane characteristics at least one of which appears to be a progressive increase in the net surface charge. It is therefore difficult to justify the conclusion drawn by Forrester *et al.* that the increase Ca^{2+}-binding ability is directly involved with the interaction of complement with cell-bound antibodies.

THE CELL SURFACE IN MORPHOGENESIS

The complex phenomena of cellular organisation, differentiation and morphogenesis encountered in embryonic development; the morphogenetic mechanisms in regenerative processes and the phenomena of neoplastic development, dissemination and metastasis, are in part attributable to the surface properties of component cells of the different systems. This is an extensive field of study and has received considerable attention since the 1930s. It is clearly beyond the scope of this work to undertake a detailed discussion of the developments in this field, and besides, there are excellent accounts of the various aspects of this particular field (see for example, Curtis, 1966, 1967; Weiss, 1967; Collins, 1974; Easty, 1974; Carlson, 1974). The following discussion will therefore be restricted to cell electrophoretic studies relevant to the elucidation of the biological role and characterisation of the major components of the cell surface.

MORPHOGENETIC MOVEMENTS

It was recognised even in the 1930s that the cell surface was intricately involved in bringing about the cellular migrations which are so characteristic of the developing embryo. Such cellular migrations had been demonstrated in the preceding few years in amphibian (Vogt, 1925) and later in chick (Wetzel, 1929; Gräper, 1929; Waddington, 1932; Pasteels, 1935; Spratt, 1946) embryos by marking the cells with supravital dyes or with carbon particles. When Spemann and Mangold (1924) demonstrated that the differentiation of the brain tissue is triggered by "inductive" interactions with a patch of specialised mesodermal cells, it became clear that as a consequence of the migrations different cell types having reached their definitive positions are enabled to interact with the neighbouring cell types. As a result of such interactions either or both interacting components differentiate into more specialised cell types. Such cellular interactions constitute a temporal hierarchy of interacting systems, each interaction constituting a step in the process of differentiation (see Sherbet, 1970).

Scientists of the time did not require much persuasion to enter the field of study of the cell surface. In an attempt to explain the mechanisms which bring about the cell migration during the gastrulation phase of embryonic development, Holtfreter (1938, 1939, 1943a, 1944) examined the behaviour of embryonic tissue explanted and cultured *in vitro* in a number of combinations. The tissues exhibited affinities or lacked affinities, which could be predicted with a knowledge of the processes of development. Presumptive neural ectoderm adhered firmly to presumptive epidermal ectoderm. These are different regions of the continuous ectodermal layer of the embryo. When these closely adhering presumptive tissues were allowed to differentiate, the neural and epidermal tissues separated from each other and differentiated side by side. These events correspond remarkably well with those occurring during the formation of the neural tube in the embryo.

Holtfreter subsequently described other similar instances. He placed endoderm cells of the blastopore lip area (from early gastrula stage of amphibian embryos) on an explant of endoderm. The blastoporal endoderm cells invaginated into the endodermal explant exactly as they might have done had they not been excised from the embryo. The cells of gastrulae also showed considerable mobility and tendency to spread out when cultivated *in vitro*. It was clear from these early experiments that the morphogenetic cell migration was a property of the component cells—a property by which they could exhibit selective affinity and changing adhesiveness to their fellow cells in the course of embryonic development.

As a consequence of the cell migration the embryo acquires a degree of architectural organisation, with the component cell types occupying definitive positions in relation to one another. If one could illustrate this point using the arrangement of cell types in the amphibian embryo: the endoderm occupies the innermost regions of the embryo and the ectoderm forms the outermost layer; the mesoderm, when it does differentiate, appears in between the ectoderm and the endoderm.

Townes and Holtfreter (1955) observed migratory activity also in embryonic tissues explanted and cultivated *in vitro*. They combined embryonic tissues of known presumptive differentiative ability, either in the undissociated or in the dissociated state. In these co-cultivation or reaggregation experiments two kinds of cellular migration were noticed, namely an inward movement of cells as in the case of neural plate or mesodermal cells that had been combined with epidermal or endodermal cells and an outward migration of the epithelial or endodermal cells in the same tissue combination. The outcome of these cellular movements was a spatial organisation of the cells which closely corresponded with the arrangement of the respective tissue in the embryo itself. Townes and Holtfreter concluded from these experiments that the migration of the cells in embryonic systems, migration of one cell type in relation to the others, was an inherent property. Although the description of this ability as an inherent property might appear vague and noncommital now, these experiments remain remarkable because they clearly recognised the selectivity of interactions occurring in the embryonic system. Such selectivity could probably be explained in terms of differential adhesion between cells (see pp. 117–120).

SURFACE CHANGES IN EMBRYONIC DEVELOPMENT

This short introduction to the role played by the cell surface in morphogenesis underlines the importance of changes in the cellular adhesiveness and selective affinity between cells presumably expressed via differential adhesiveness among cells. Since these properties of the cell surface are determined by the chemical nature and conformation of the surface components, the cell surface must show differentiation in the course of embryonic development.

Garrod and Gingell (1970) observed a progressive change in the surface properties of the slime mould *Dictyostelium discoideum* in its transition from the feeding to the aggregating stage. They determined the surface charge densities using cell electrophoresis and observed a progressive decrease in EPM as the cells neared the aggregating stage.

Very little data on the electrophoretic behaviour of early embryonic cells is available in the literature, apart from some interesting studies described by MacMurdo and Zalik (1970) and MacMurdo-Harris and Zalik (1971). These authors dissociated embryos of *Xenopus laevis* at the blastula, early and mid-gastrula and neurula stages. When the dissociated cells were layered on a discontinuous density gradient and centrifuged, six bands were produced (Table 26). Microscopic examination of the morphology and microdissection of embryos into the animal pole, vegetal pole and marginal zone, followed by density gradient centrifugation under identical conditions, indicated that cells in Band A and B belonged to presumptive ectoderm, C and D to mesoderm and E and F to presumptive endoderm.

Table 26 *Ficoll density gradient distribution of* Xenopus *embryo cells*

Density	Animal pole	Marginal zone	Vegetative pole
Aqueous–1·07	+		
1·07–1·08 (A)	++		
1·08–1·09 (B)	++		
1·09–1·10 (C)	+	++	
1·10–1·11 (D)		++	+
1·11–1·12 (E)			++
1·12–1·13 (F)			++

From MacMurdo-Harris and Zalik (1971).

An electrophoretic investigation revealed that EPM decreases between the blastula and early gastrula stages, to attain an intermediate value at the neurula stage. Cells derived from embryos between early and mid-gastrula showed the highest EPMs. Cells from Bands C and D (presumptive mesodermal origin) derived from blastula and early gastrula embryos had much lower EPMs than the cells from Bands A, B, E and F from either embryonic stage (Table 27). It may be recalled that

Table 27 *EPMs of* Xenopus *embryo cells*

Stage of development	Band	EPM (−TU)
Blastula	A	15·9
	B	16·3
	C	15·9
	D	16·1
	E	16·1
	F	15·2
Early gastrula	A	15·2
	B	15·2
	C	15·4
	D	14·4
	E	14·8
	F	14·6
Mid gastrula	A	16·3
	B	16·5
	C	17·1
	D	16·5
	E	16·1
	F	15·4
Neurula	A	15·6
	B	15·9
	C	15·4
	D	15·4
	E	15·4
	F	15·4

From MacMurdo-Harris and Zalik (1971).

during this phase of development the marginal zone cells begin to invaginate at the lip of the blastopore into the interior of the embryo. This morphogenetic activity is initiated by a group of cells which possess a characteristic flask shaped form. The necks of the cells are long and contractile (Holtfreter, 1943a, 1944) and contain cytoplasmic material exhibiting birefringence indicating a fibrous structure (Waddington, 1940). Electron microscopy has revealed the presence of microtubules in the vesicular neck region (Perry and Waddington, 1966) and EM studies of blastopore region of *Hyla* gastrulae revealed the presence of cells of three distinctive shapes (Baker, 1965). From these investigations it appears reasonable to conclude that the invagination activity involves considerable and constant alteration in the shape of the cells of the blastoporal region. According to Weiss (1965) cells are easily deformed when their surface charge is reduced by the removal of sialic acids. The observations of MacMurdo and Zalik (1970) and MacMurdo-Harris and Zalik (1971) that marginal zone cells of the presumptive mesoderm show reduced EPM, is consistent with the events occurring during the process of involution of cells of the marginal zone at the blastoporal lip.

At the neurula stage most cells will have formed stable contacts and adhesions and it may be expected that this will be reflected in the generally low EPM values of all cells derived from neurula stage embryos. On the other hand, it is somewhat more complicated a question why the cells of Band C and D show the highest EPM values in the mid-gastrula stage.

The EPM measurements of presumptive epidermal cells made by MacMurdo-Harris and Zalik (1971) are in partial agreement with the observations of Ave *et al.* (1968) and Dasgupta and Kung-Ho (1971). Ave *et al.* observed that presumptive epidermal cells of *Triturus pyrrhogaster* showed considerable heterogeneity using column cell electrophoresis. They reported that cells of presumptive epidermis of mid-gastrulae formed three distinct bands, designated as N, M1 and M2, in the cathode-anode direction. When presumptive epidermis was chemically neuralised and then subjected to electrophoresis only the Band N remained. After mesodermisation with bone marrow extract, only Bands M1 and M2 remained. The authors therefore concluded that the original heterogeneity of the presumptive ectoderm reflects the developmental potentialities of the components of the ectodermal cells. The cells recovered after neuralisation and mesodermisation formed 34 and 66% the original cells of the presumptive ectoderm. This compared very well with the composition of the spectrum of cell banding of the presumptive ectodermal cells, where 35% of the cells were in Band N and 65% in Bands M1 and M2. The heterogeneity of presumptive epidermal cells was confirmed by Dasgupta and Kung-Ho (1971). They found that the cells of presumptive epidermis from early gastrula embryos of *R. pipiens* formed two distinct bands, designated as N and M2. Under similar electrophoretic conditions, presumptive epidermal cells from mid-gastrula embryos also contained a third intermediate band, Band M1 which were, as judged from their position in the column, more negatively charged than the cells in Band N. At the late gastrula stage only Band N remained. These changes also appeared to occur in presumptive epidermal cells *in vitro* on being incubated for the same duration as taken by early gastrula embryos to reach the mid- and late gastrula stages in normal development. The formation of Band M1 cells during development from the early to the mid-gastrula provides reasonable grounds for supposing that the process of development is accompanied by changes in the surface properties of the epidermal cells. Disappearance of both Band M1 and M2 by the time development of the embryo has reached the mid-gastrula stage may either indicate a selective lysis of the cells or a true disappearance of the cells from the presumptive epidermal zone due to the processes of invagination taking place at the time.

In the development of the chick embryo from the unincubated blastoderm to the head process stages, changes in electrophoretic behaviour on similar lines, have been described (Zalik *et al.*, 1972). The EPMs were rather widespread but average EPMs decreased from Stage 1 (unincubated blastoderm) through to Stage 5, which was also accompanied by a decreased ability to bind colloidal iron (Sanders and Zalik, 1972). In Stage 1 embryos, the majority of cells had high EPMs and possessed pIE of 4·2. In Stages 3 (intermediate primitive streak) and 5 (definitive primitive streak), the average EPMs were reduced because of the appearance of a subpopulation of cells with much lower pIE value at 3·8. These observations are consistent with the findings of Dasgupta and Kung-Ho (1971) who reported the appearance of Band M1, except for the fact that the new band of cells of Dasgupta and Kung-Ho were more negatively charged while the new subpopulation of cells discovered by Zalik *et al.* (1972) possessed much lower net charge than the original population. Zalik and colleagues have made the further interesting observation that only a small proportion, if at all, of the electrical charges on cells from Stage 1 and Stage 4 chick embryos were attributable to the terminal carboxylic groups of sialic acids. For, the EPMs of the cells was reduced only by about 6–13% by treatment with NANase, as compared with 36% reduction of EPM of avian erythrocytes and 70% of reduction of EPM of human erythrocytes. It may be recalled that in erythrocytes, the surface negative charge is almost entirely due to the sialic acid moieties (see pp. 57–60). It ought to be pointed out, however, that although the reduction in EPMs by NANase treatment of human erythrocytes is comparable to that reported by Seaman and Cook (1965), the loss of EPM (36%) found by Zalik *et al.* (1972) falls very much short of the reduction (by 72%) in EPM of avian erythrocytes reported by Eylar *et al.* (1962). The comparative lack of NANase effect on the EPMs of chick embryo cells may not on its own constitute sufficient evidence to warrant the conclusion that the surface negativity is not due to sialic acid carboxylic groups. Nonetheless, another piece of indirect evidence is the relatively high pIE of 3·8 and 4·2 of the embryonic cells, which points to the possibility that the carboxylic groups on these cells have a much higher pK value than the sialic acid carboxylic groups. In contrast, cells obtained from chick embryonic primordia possess low pIE (see pp. 111–112).

Zalik *et al.* (1972) also attempted to separate cells of the chicken epiblast from the cells of the hypoblast on density gradient and examine their EPMs in relation to the developmental stage, as they had previously done with the amphibian embryonic system. However, unlike the amphibian system, cells of the different layers of the chicken blastoderm showed no remarkable differences in the electrophoretic behaviour apart from showing a generally reduced EPM.

If lectin-binding characteristics are any indication, even the small differences in EPM of cells derived from different stages of the early chick embryo may be highly significant. Zalik and Cook (1976) for example showed that unincubated chick blastoderm as well as primitive streak stage blastoderm cells are agglutinated by wheat germ agglutinin, *Ricinus* lectin and Con A, but not by fucose binding protein. But 10-to-12-day-old chick embryo liver cells showed completely different agglutination behaviour. Moscona (1971) and Kleinschuster and Moscona (1972) had earlier observed that with progressive differentiation, alterations were also noticeable in the pattern of lectin binding.

Whether the pattern of lectin binding at these early stages is in any way related to the morphogenetic movements occurring at the time is a matter of considerable interest. Undoubtedly the process of invagination of the epiblast cells at the primitive streak leading to the formation of the mesoderm ought to involve alterations in the surface components because in this process, not only do changes occur in the differentiative status of cells but also in their form and motility. It may be of interest in this context that Robertson *et al.* (1975) found that the Con A receptors of the migratory micromeres of sea urchin 32–64 cell stage embryos show a greater mobility than those of the meso- and macromeres.

ELECTROPHORETIC CHARACTERISATION OF CELLS OF EMBRYONIC PRIMORDIA

Collins (1966a) examined the pH-mobility relationships of cells derived from chick embryo heart ventricle, liver and neural retina. The curves were roughly similar with steep sections at either end corresponding to pH below 4 and above 9. These sections may indicate the dissociation of carboxylic and amino groups respectively. Between pH 4 and 9 the increase in charge density was gradual and probably represents the dissociation of imidazole groups of histidine ($pK \sim 5{\cdot}97$) and thiol groups with $pK \sim 8{\cdot}3$. Liver cells showed much greater charge densities than heart cells at pH above 4, despite the similarity in the general shape of the curve. This may be interpreted as indicating that these cell surfaces have similar ionogenic groups and that they differ only in their relative quantities. Further the curves also indicate that the higher net negative charge density on liver cells is due to the presence of comparatively larger number of anionogenic groups than to a relative deficiency of cationogenic ones.

A comparison of the pH-charge density relationship reveals even more striking differences. Neural retinal cells bear not only much higher numbers of negative charges but their pIE is less than 2. The pIE of heart cells on the other hand is much higher at pH 3. This suggests the possibility that the neural retinal cells possess acidic groups of low pK such as phosphate or sulphate groups. The presence of Ca^{2+} in the electrophoresis buffer produces a dramatic change in the shape of the pH-surface charge density curve of neural retinal cells and a progressive reduction of charge density of the cells with increasing Ca^{2+} concentration (Collins, 1966b). This is presumably due to binding of Ca^{2+} to phosphate groups of membrane phospholipids. Such ionic specificity of Ca^{2+} binding is not a certainty, for Ca^{2+} appear to bind firmly to chondroitin sulphate (Schubert, 1964) and a small proportion of ions may also bind carboxyl groups (Carr, 1953). Nevertheless, Forrester *et al.* (1965) found that the EPMs of particular preparations of the synthetic phospholipids di-palmitoyl lecithin and di-palmitoyl phosphatidyl ethanolamine were significantly reduced by Ca^{2+} but the EPMs of human and mouse erythrocytes which have a predominantly sialic acid carboxyl surface were not affected.

The EPMs of all the primordial systems examined by Collins (1966a) showed reductions, but to variable degrees. Thus epidermal and trypsin-dissociated neural retina cells showed reductions of 14%. In the case of liver primordial cells the loss of EPM was 10%. The cells of heart ventricle showed the highest loss of EPM of 22%. The difference in Ca^{2+} binding once again seem to point to difference in the relative composition of acidic groups on the cell surface.

The loss of calcium binding ability by older embryonic primordia has also been observed by Armstrong (1966), Armstrong and Jones (1968) and Harris and Zalik (1974). Harris and Zalik (1974) in fact studied the effect of increasing Ca^{2+} concentrations on various developmental stages of the chick embryo. Cells from early developmental stages before and during gastrulation show reductions in EPM with increasing ionic concentration. Further, some of the surface components which bind Ca^{2+} are susceptible to trypsin treatment, but a majority of them are not affected by the enzyme. Harris and Zalik (1974) draw attention to similar effects, namely the lack of charge reversal in the presence of Ca^{2+} observed in a variety of adult cells (Weiss and Mayhew, 1967; Ward and Ambrose, 1969; Bangham and Pethica, 1960; Wilkinson *et al.*, 1962) and suggest that the calcium response may be a characteristic property of embryonic cells. The binding pattern of other divalent cations such as Mg^{2+} and Mn^{2+} also appears to be a distinguishing feature of embryonic cells. Harris and Zalik (1977) observed a reversal of polarity of cells derived from chick embryo blastoderms with Mg^{2+} and Mn^{2+} ions. But the cells did not respond to the presence of strontium or barium ions. Harris and Zalik also claim that while embryonic tissues show great affinity for magnesium and manganese ions, neither differentiating nor adult tissues show this specificity when identical ionic concentrations are employed. Without doubt these experiments demonstrate that changes occur in the surface properties of cells in the course of development and differentiation.

Bernard *et al.* (1969), however, caution against a rigid interpretation of results of investigations where embryonic primordia are disaggregated using enzymes such as trypsin, or even chelating agents. They examined the electrophoretic behaviour of chick neural retina cells as a function of age and time after dissociation with trypsin. The EPM of cells immediately following dissociation was between −10·8 and −11·5 TU irrespective of the age of the embryos from which the primordia were obtained. When the EPMs were measured 24 hours after dissociation increases in the EPMs of between 70 and 96% were observed. But Maslow (1970) found that in trypsin-dissociated liver cells the EPM decreased over the 24-hour period and remained constant thereafter. Adam and Adam (1975) reported decrease in the EPM of 3T3 mouse fibroblasts on treatment with 0·007% trypsin. Bernard *et al.* (1969) attribute the low mobility of freshly dissociated cells to adsorption of trypsin molecules to the cell surface. Although these authors consider it unlikely that the reductions in EPM were due to proteolytic activity, there does not appear to be any direct evidence for this. Poste (1971) also believes that in tissue dissociation by enzymes, the latter adsorb to the surface of the dissociated cells and remain enzymically active for as long as 24 hours after dissociation. On the other hand, Sherbet and Lakshmi (1974) observed that human fetal brain cells showed lower pIs when they had been harvested from culture flasks with the aid of trypsin than cells which were separated by mechanical means. Since the pI of trypsin is 10·4 at 0·1 M ionic strength (Bier and Nord, 1951), it is difficult to see how its adsorption on the fetal brain cells could have reduced their surface pI. Recently Fike and Van Oss (1976) found that in RPMI 1846 (hamster melanoma cells) trypsin treatment increased the surface potential. However in two other cell lines, namely HEP-2 (transformed human epithelial) cells and BGM (transformed Buffalo green monkey kidney) cells the ζ potential was found to have been reduced as a result of trypsin treatment. Obviously the effects produced by the enzyme are very variable and no generalisations can be made as regards the mechanisms by which the enzyme brings about surface changes. No attempts have yet been made to investigate the effects using covalently bound enzyme.

Trypsin and other proteolytic enzymes are known to be capable of changing the chemical composition of cell surfaces, such as the removal of sialomucopeptides and acidic mucopolysaccharides (Cook *et al.*, 1960; Mäkela *et al.*, 1960; Shea and Ginsburg, 1968; Kemp, 1969; Codington *et al.*, 1970; Price, 1970; Forstner, 1971; Yamamota *et al.*, 1973; Hynes, 1974) and alter the distribution of lectin-binding sites at the cell surface (Burger, 1969; Inbar and Sachs, 1969a; Sela *et al.*, 1970; Moscona, 1971; Kleinschuster and Moscona, 1972; Nicolson, 1972a; Nicolson and Blaustein, 1972). In other words, dissociation procedures may radically alter the physicochemical features of the cell surface.

Although embryonic primordia and cells derived from early developmental stages show considerable change in their EPM, it may be true to say that these cells generally possess much higher electrophoretic mobilities than do cells from tissues of adult animals (Simon-Reuss *et al.*, 1964; MacMurdo and Zalik, 1970; MacMurdo-Harris and Zalik, 1971). These differences are probably related to the degree of organisation achieved by the tissues. Lefford (1972) demonstrated that differences existed in the migratory ability of cells derived from 9- and 19-day-old chick embryos and has suggested that the differences are inherent and associated with differences in their degree of organisation. On the other hand, lower EPMs may be a characteristic of the fully differentiated cell surface (but see below). MacMurdo and Zalik (1970) observed, for instance, that even adult amphibian erythrocytes possessed lower EPMs than cells from embryos at early developmental stages.

Whether or not higher surface charge may be characteristic of embryonic cells, there is some evidence which suggests that the surface characteristics reflect the differentiation status of a tissue. Cellular metaplasia occurring in the regeneration of lens in the newt has afforded an excellent experimental system for investigations into alterations and variations in cell surface components associated with these changes. The experimental system has also the additional advantage that much information is available in the literature about the biochemical events taking place in the metaplastic change. When the lens is excised, the pigmented iris cells undergo dedifferentiation which is characterised by an increased RNA synthesis and replication of DNA and cell division. In the phase of redifferentiation DNA synthesis stops and the cells begin to synthesise lens-specific proteins (see Yamada, 1967, 1972).

It is known that surface components are lost during cellular dedifferentiation (Dumont and Yamada, 1972; Zalik and Scott, 1972). The EPM of normal differentiated iris cells is decreased by treatment with NANase and RNAase (Zalik and Scott, 1972, 1973). A similar effect is produced also by chrondroitinase and hyaluronidase (Zalik and Scott, 1973). Such effects are also produced by these enzymes in iris cells until four days after lentectomy. After five days of lentectomy, however, these enzymes (with the exception of NANase) are ineffective. The lack of effect continues for a further 5–10 days, when the susceptibility returns. These effects coincide with the known time scale of dedifferentiation and redifferentiation processes. The observations thus indicate that dedifferentiation of iris cells is accompanied by the disappearance from the surface of components which are susceptible to hyaluronidase, chondroitinase and RNAase. Decrease in EPM by NANase continues to occur until 8–10 days after lentectomy but not in regenerates 10–15 days after lectectomy. This indicates that the NANase-susceptible component disappears from the surface after hyaluronidase, chondroitinase and RNAase-sensitive components. Although a detailed characterisation of these components has not been attempted, these investigations clearly support the view that the cell surface accurately reflects the differentiated state of the cell.

It is also probable that alteration in the surface charge is simply related to the need to form specific cell contacts and stable adhesions, as in the process of implantation of the fertilised egg. Nilsson *et al.* (1973) found that surface charge of the blastocyst shows a decrease prior to implantation which makes a close physical contact between the trophoblast cells and cells of the uterine epithelium easier. Jenkinson and Searle (1977) have examined the changes in the surface of the blastocyst prior to implantation, using colloidal iron binding. They have shown that sialic acid containing glycoproteins make a major contribution to the surface charge of the blastocyst surface. These components are apparently lost before implantation. A loss of sialic acid component may also lower the rigidity of the membrane thus enabling the trophoblast cells to invade the maternal tissues. Similarly the surface charge density is involved in the process of platelet adhesion and aggregation, but unlike the events of blastocyst implantation, the biochemistry of platelet adhesion and aggregation is much better understood (see pp. 75–78).

CELL SORTING

The processes of organisation of cells into tissues and tissues into organs may be considered as most complex and therefore least amenable to experimental analysis. It was known for some time that disaggregated vertebrate tissue cells on being brought into random contact rearranged themselves to construct a tissue reaggregate which simulated the organisation of the original tissue (Holtfreter, 1944b; Moscona and Moscona, 1952; Townes and Holtfreter, 1955; Weiss and Taylor, 1960). Cell sorting has therefore been used as a model to analyse the processes of morphogenesis, and such analyses may be the closest approximation yet of the *in vivo* processes of morphogenesis.

Three main hypotheses have been put forward to explain the mechanism of *in vitro* cell sorting, namely directed migration; the timing hypothesis and the differential adhesion hypothesis (see Collins, 1974; Curtis, 1967 for reviews).

Directed Migration

This hypothesis suggests how cells may sort out in reaggregation rather than offer a mechanism for the underlying processes. It simply implies that cells move towards a specific target or zone i.e. to the interior or exterior of a cell aggregate. An outstanding example of directed migration *in vivo* is the migration of neural crest cells. The migration of these cells appears to be determined by the neural tube (Weston, 1963, 1970; Weston and Butler, 1966). Weiss (1958) postulated that the precise path along which the crest cells migrate might be determined by some form of contact guidance (see below). On the other hand, the migratory ability may be due to adhesive properties of the crest cells among themselves and of the cells which form the environment in which crest cells migrate, with changes in the cellular environment resulting in the impairment of the migratory behaviour (Weston and Butler, 1966). This may also be true of the mobility and distribution of pigment cells (Rawles, 1948) and of the distribution of melanophores (Watterson, 1942).

Alterations in the adhesiveness of migratory mesenchyme as a direct result of mutation may cause severe abnormalities, which has indeed been found to be in the case of chicken carrying the homozygous talpid^3 mutant gene (Ede and Kelly, 1964a,b).

The postulate of "contact guidance" proposed by Weiss (1947, 1961) contained the notion that the migration of embryonic cells was guided by cues received by the migrating cells from the substratum over which they were moving. It was shown, for instance, that the orientation of cells cultured on plasma clot was related to the orientation of the fibrin molecules in the semi-solid plasma clot. Cells in culture even showed orientation which was related to the grooves scratched on glass.

De Haan (1961, 1963a,b,c, 1965) has attempted to apply the contact guidance postulate to explain the migration of the pericardial mesoderm cells from the presumptive areas to their final position in the process of formation of the heart. The presumptive heart-forming areas occur at the Hensen's node level on either side of the primitive streak chick embryo. In the early stages of migration the heart-forming cells form a crescent with cells on either side of the embryo becoming confluent in the anterior regions. De Haan (1963) has shown that these migratory cells actually use the endodermal layer as a substratum. When the embryos were treated with sodium citrate the endoderm separates leaving the ectodermal and mesodermal layers almost intact. When such embryos were cultured the heart-forming cells were unable to migrate. Instead they formed clusters and differentiated into contractile heart tissue in a crescentric patch around the brain. This arrest of heart cell migration is directly attributed to the absence of the endoderm. De Haan has shown also that the heart-forming cells move relative to the movements occurring in the endoderm itself. This was achieved by labelling the heart cells. Thus it appears that the initial migration of these cells makes use of the endoderm as a substratum, but up to this stage the direction of migration is not orientated. Subsequently however these cells move to a median position where they actually form the heart. It is not known how this directed migration occurs. De Haan believes that contact guidance may be involved here. He has suggested that the final orientation of these migrating cells is dictated by a special arrangement noticed in the endodermal cells.

In vitro, directed migration in cellular reaggregates has been suggested to be due to chemotaxis. It is suggested in general that the metabolic activity of the cells creates a pH gradient in the environment of the aggregate and that the cells detect and respond to such gradients, with local immobilisation (contraction?) of the cell membrane in sections exposed to both extremes of pH, while the unaffected regions presumably expandable move the cell by traction (Weiss and Scott, 1963). It is known that the leading edge of a migrating amoeba has a higher surface charge than the tail and that the direction of migration can be changed by injecting a polycationic compound such as polylysine in front of the leading edge (Ambrose, 1967). Thus the chemotactic mechanism appears plausible although it may not answer all the questions relating to *in vitro* sorting of cells.

The Timing Hypothesis

The timing hypothesis proposed by Curtis (1960, 1961, 1962) states that a given cell type changes its behavioural properties and becomes "trappable" in fixed adhesion with other cells. The trappable properties are achieved by different cell types of the reaggregating mass at different times. This temporal change in the cellular properties is attributed to the organisation and behaviour of protein molecules of the cell surface. The protein layer behaves as a non-Newtonian liquid when the area of each molecule falls within a certain range; that is, their viscosity changes as a function of the shear forces. A reduction of the shear forces causes increase in cellular adhesiveness. Trapping is supposed, and most likely, to begin at the periphery, with further trapping occurring in relation to the already trapped cells and a consequent segregation of the cells internally.

The Differential Adhesion Hypothesis

This hypothesis for cell sorting proposed by Steinberg and cogently presented in a number of publications (Steinberg, 1962a,b,c, 1964, 1970; Martz *et al.*, 1974) attributes the sorting out behaviour of reaggregating cells to differences in the strengths of adhesion between different cells i.e. the magnitudes of energies of adhesion. Cells from different tissues are said to adhere with different strengths. Steinberg proposed that the cells of a mixed population tend to rearrange themselves spontaneously to achieve an equilibrium configuration to maximise the adhesive energy evolved, i.e. maximise adhesive contact area and to exchange weaker adhesions for stronger ones. In the equilibrium configuration less cohesive cells tend to envelope those which are more cohesive that segregate internally.

The Specific Adhesion Hypothesis

Moscona (1961a,b, 1963a,b, 1968) and Lilien (1968) believe that cell sorting in a reaggregating mixed population of cells is due to chemical specificity in the adhesion process. Moscona has proposed the working hypothesis that the recognitive and adhesive interactions are mediated by specific cell-ligands located at the cell surfaces.

Specific cellular adhesions were first demonstrated by Wilson (1907) with dissociated red and yellow marine sponges. When the cell suspensions from the two species were mixed and allowed to aggregate the cells sorted out into species specific aggregates. This phenomenon of homologous cell aggregation was reported to be due to the release of aggregation-promoting factor (APF) (Humphreys, 1963; Moscona, 1962, 1963b; Lilien and Moscona, 1967; Pessac-Pejsachowicz and Alliot-Mayet, 1968; Kuhns *et al.*, 1973). Surface material has been isolated from embryonic neural retinal cells (Lilien, 1969; McClay and Moscona, 1974) and mouse cerebellum (Garber and Moscona, 1972) which has properties of causing specific aggregation of homologous cells.

The specific adhesion phenomenon not only requires the mediation of specific aggregation-promoting factors but also the postulation of specific receptors at the surface of the cells. The aggregation-promoting macromolecules recognise these binding sites and then form molecular bridges between contiguous cells. Pessac and Defendi (1972) have indeed provided considerable evidence for the existence of these receptors. Cells which spontaneously aggregate release an aggregation-promoting factor, and probably also possess receptors because their aggregation can be induced by supernatants from culture of homologous or heterologous cells. But aggregation is not promoted if the cells are pretreated with trypsin. A murine cell line P388 obviously released no APF for supernatants derived from them had no ability to promote aggregation. On the other hand, these cells do possess receptors as they could be induced to aggregate using supernatants from other systems known to contain the aggregation-promoting factor.

Very little is known about the chemical nature of the receptor other than that they are either inactivated or destroyed by trypsin treatment (Pessac and Defendi, 1972). Weinbaum and Burger (1973) have reported the isolation of a heat-labile, pronase-sensitive, non-dialysable component from *Microciona prolifera.* This component is necessary for reaggregation in addition to the presence of the APF, and it presumably represents material derived from the APF receptors. The receptor component was isolated from the medium after hypotonic shock of chemically dissociated cells. Earlier Burger (1968) had shown that hypotonic shock caused a release of some surface components. Weinbaum and Burger (1973) found that these cells show no aggregation, presumably because they lack both APF and their receptors. When the receptor components were added to the suspending medium, subsequently removed and then followed by the addition of APF, reaggregation occurred. If however APF was added first and then removed, and this was followed by the addition of the receptor component, no reaggregation took place. These experiments indicate that the alleged receptor components do indeed bind to the cell surface and subsequently bind the APF macromolecules to result in cellular aggregation. These authors also demonstrated that the receptor component could inactivate APF when incubated with it. Essentially similar conclusions were drawn from experiments with acellular systems of sepharose beads coated with APF or the receptor component.

The APF appears to be a high molecular weight glycoprotein which requires the presence of Ca^{2+} for its activity (Gasic and Galanti, 1966; Margoliash *et al.*, 1965). The APF obtained from 10-day-old chicken neural retina had a molecular of weight of 50 000 as determined by SDS-PAGE electrophoresis (McClay and Moscona, 1974). The biological activity of APF derived from *M. prolifera* appears to reside in the glucuronic acid moieties. Removal of these residues leads to the inactivation of the APF (Turner and Burger, 1973). Weinbaum and Burger (1973) have suggested that the termini containing these residues may be recognised by the receptor sites on the cell surface and each APF macromolecule may have more than one such recognition-binding site.

It is not intended to discuss the merits of these various views on the phenomenon of cell sorting except to state that the various hypotheses are probably not mutually exclusive. The basic strain which runs through these hypotheses dealing with cell sorting or the actual *in vivo* processes of morphogenesis is the part played by adhesive interactions whether they be mediated by specific chemical components of the surface, or by quantitative differences engendered by differences in the work of adhesion performed, or by a molecular reorganisation mechanism involving the surface components.

MECHANISMS OF CELLULAR ADHESION

The morphology of cell contacts and their temporal change are obviously implicated in the adhesion-dependent processes of histogenesis (Mercer, 1965) and cell aggregation (Rossomando *et al.*, 1974; Alwan and Lawn, 1974; see also Sherbet and Lakshmi, 1967, pp. 160–162). Three different mechanisms have been suggested, namely bivalent cation bridge concept; intercellular cement and the lyophobic colloid concept.

Bivalent Cation Bridges and Cell Adhesion

Steinberg (1958) and Pethica (1961) postulated an attractive theory to account for cell adhesion. They envisaged bivalent cations as forming bridges with negatively charged groups of interacting contiguous cells to hold the cells together. The strength of adhesion depended upon the density of the cationic bridges. It is a well known and well documented fact that removal of bivalent cations from the external environment greatly aids, and in many tissues causes, their dissociation and that for their reassociation the presence of bivalent cations is essential (Herbst, 1900; Allison and Lancaster, 1964; Feldman, 1955; Galtsoff, 1925;

Moscona, 1962; Steinberg, 1962d). But it has been argued that cells should approach closer than 10 Å for calcium bridges to form. An alternative suggestion is that binding of Ca^{2+} to the carboxyl groups reduces the net negative charge densities and thus reduces the electrostatic repulsive forces occurring between the cells thus facilitating adhesion. Curtis (1963) found that cell adhesion is maintained even when the ionisation of carboxyl groups is suppressed. It may be well to remember, however, that probably only a small proportion of Ca^{2+} ions will bind carboxyl groups (Carr, 1953). Besides, it is known that sialic acid carboxyl groups have a lower affinity to Ca^{2+} than do carboxyl groups of amino acid residues (Weed *et al.*, 1969; Forstner and Manery, 1971). Therefore Ca^{2+} ions may principally bind to phosphate groups (Forrester *et al.*, 1965). On the other hand, Armstrong (1966) has argued that, if the role of calcium ions is principally to lower the negative surface charge as a prelude to cellular adhesion, different cations should cause cellular aggregation at identical rates, when the concentrations of the cations are so adjusted as to bring the net negative charge of the cells down to the same level. Apparently this does not happen. Armstrong (1966) showed that magnesium was the most effective cation in promoting aggregation, calcium less so, and strontium and barium exerted little adhesion-promoting effect. Armstrong therefore believes that calcium ions participate in other "adhesion-related reactions" than merely promote adhesion by reducing the electrostatic repulsion between cells.

Intercellular Cement

A second view is that cell adhesion is carried out by extracellular material which combines with the surface of one cell and then forms molecular bridges with components of the neighbouring cell surface to cause adhesion. This is a very old theory presumably inspired by observations of tissues such as cartilage where chondrocytes are embedded in cartilaginous matrix. There is much electron microscopic and biochemical evidence which suggests the existence of surface coat material which may participate in intercellular adhesions (Martinez-Palomo, 1970; Rambourg, 1970; Winzler, 1970). Besides cells produce microexudates, which are produced as a result of active metabolic processes and are not a result of mere deposition of the material following leakage through the cell membrane (Poste *et al.*, 1973; Weiss, 1974). A temporal correlation between the synthesis of microexudates and alterations in adhesive properties has also been reported (Maslow and Weiss, 1972; Poste, 1971; Takeichi, 1971). When cells grown in tissue culture are detached from the culture vessel, they leave behind material that is obviously derived from the cell surface (Poste and Moss, 1972; Weiss, 1961a; Weiss and Lachman, 1964; Boland *et al.*, 1970; Malucci *et al.*, 1972; Rambourg, 1972; Takeichi, 1971; Weiss and Neiders, 1971; Yaoi and Kanaseki, 1972).

Weiss (1974) found that cells that had not been exposed to cycloheximide were easily detached with the aid of the enzymes, RNAase, NANase and hyaluronidase. Using the electrophoretic method Weiss showed that these enzymes acted on the surface of the cells. If the cells were pretreated with cyclohexamide, detachment of the cell is facilitated by hyaluronidase and trypsin but not by RNAase or NANase. This appears to suggest that the cell surface polysaccharides play an important role in the processes of adhesion. Mallucci (1971) showed that a clear correlation existed between the formation of microexudates by cells which had been stripped of their original surface coats by enzymic treatment, and the appearance of receptor sites for plant lectins which have specificity towards polysaccharides. Concanavalin A (Con A), which specifically binds polysaccharides that contain α-D-gluco and mannopyranosyl residues, is said to affect adhesion of melanoma cells B16-C2W and human lymphoma and mastocytoma cells to the substratum (Sato and Takasawa-Nishizawa, 1974; Mori *et al.*, 1973) and the aggregation of rat hepatoma cells (Kuroda, 1973). Very little is known how lectins affect these processes. Sato and Takasawa-Nishizawa (1974) noticed that cultured melanoma cells which were incubated with Con A were resistant to detachment using trypsin and pronase. In the presence of Con A, human lymphoma and mastocytoma cells adhered rapidly to culture vessels and the spreading cells were resistant to trypsinisation or EDTA. The adhesion of mastocytoma cells was not affected by actinomycin D or cycloheximide, but colchicine did slightly affect their adhesion and caused alterations in their morphology. Sato and Takasawa-Nishizawa suggest that the bonds formed by Con A as a cross bridge between the cell surface and the substratum are resistant to trypsinisation. Sato and Takasawa-Nishizawa (1974) also refer to some of their unpublished work in which they observed a decrease of net negative charge in the lectin-treated cells. While this is bound to make for stronger adhesion between cell and the substratum, it is doubtful if Con A does indeed reduce the net negative charge on the surface. Experiments on the binding of Con A to normal and virus-transformed 3T3 mouse fibroblasts as assessed by isoelectric equilibrium method (see pp. 213–216) have indicated the contrary. In rat ascites hepatocarcinoma cells, Con A showed a biphasic effect. The EPM of these cells increased by treatment with Con A up to a concentration of 20 μg ml^{-1}.

Above this concentration the EPM decreased (Yamada and Yamada, 1973). On the other hand, Kuroda (1973) found an inhibition of aggregation as a result of Con A treatment of hepatoma cells. Steinberg and Gepner (1973) have reported that Con A has no effect on the adhesion-dependent processes of cellular aggregation, sorting and histotypic rearrangement of embryonic cells.

The Lyophobic Colloid Theory of Cellular Adhesion

Curtis (1960, 1962, 1964, 1966) has made a notable contribution to our understanding of the mechanisms of cellular adhesion by applying to cell systems the behaviour and laws relating to the interaction of lyphobic colloids. The most important forces operating in the interaction of lyophobic colloids are the electrostatic forces of repulsion and the London dispersion forces of attraction (Verwey and Overbeek, 1948; Derjaguin and Landau, 1941). The stability of a colloid is maintained when the repulsive forces are of sufficient magnitude. If the repulsive forces are absent or the attractive forces predominate, stability of the colloid is lost and flocculation results. This theory of the stability of lyophobic colloids is known as the DLVO theory. The repulsive forces are due to charged groups of like sign. These forces decrease exponentially with increasing distance between the interacting particles. The Van der Waals–London dispersion forces also decrease with increasing distances, but this happens considerably less dramatically. For small values of H, these forces are inversely proportional to H where H is the distance between the two particles, but at higher values of H, the attractive forces are inversely proportional to $1/(H+2r)^6$ where r is the radius of curvature of the particles.

Repulsive energy of interaction. For similar spherical particles which have small (less than 25 mV) and similar surface potential, and whose radius of curvature (r) is large compared with thickness of the double electrical layer ($1/K$), and with dielectric constants dissimilar to that of water forming the bulk phase, the repulsive energy of interaction (V_R) is given by the equation (Overbeek, 1952b):

$$V_R = \frac{Dr\psi^2}{2} \log_e [1 + \exp(-KH)] \tag{29}$$

where D is the dielectric constant, r the radius of curvature of particle, ψ the surface potential, K the Debye–Hückel function and H the distance between particles.

Attractive energy of interaction. For small values of H, the attractive energy of interaction (V_A) is given by the equation (Overbeek, 1952b):

$$V_A = \frac{-Ar}{12H} \tag{30}$$

where A is the Hamaker constant.

When two particles made of the same substance are suspended in a fluid of different ionisation potential, A is given by the equation (Hamaker, 1937):

$$A = \frac{3\pi^2}{2}\left[\frac{\mu_1^2}{2I_1} + \frac{\mu_0^2}{2I_0} - \frac{2\mu_0\mu_1}{I_0 + I_1}\right] \tag{31}$$

where I_0 and I_1 are ionisation potentials of the fluid between the particles and the particles themselves and $\mu_0 = \alpha_0 q_0 I_0$ and $\mu_1 = \alpha_1 q_1 I_1$ where α_0 and α_1 are polarisabilities of particles and the medium, q_0 and q_1 being numbers of atoms cm^{-2} of the fluid and the particles. Weiss (1972) has drawn attention to the difficulties associated with the assignment of numerical values to the Hamaker constant of biological systems.

The total interaction energy is given by a summation of the attractive and repulsive energies. On account of the characteristic decay of the repulsion and attraction energy with increasing distance between the interacting particles, the curve for total interaction energies at different levels of separation between the particles, assumes definite characteristics, which have formed the basis of the application of the DVLO theory to cellular adhesions.

When two particles or cells are separated by distances over 200 Å, the attractive forces of the Van der Waals–London dispersion interactions are far greater than the electrostatic repulsive forces. When the particles approach within 100–200 Å of each other, a stability of the particles at the secondary attractive minimum obtains which has characteristically a low interaction energy level. The adhesive bonds formed at the secondary minimum are weak and reversible. Gap adhesions with spacing between 100 and 200 Å exist in biological material, although gap junctions are also formed with spacing of 20–90 Å (Goodenough and Revel, 1970; Revel, 1972; Johnson and Sheridan, 1971).

At closer approximation i.e. between ~16 and 100 Å a high repulsive energy barrier obtains between the particles. If this barrier is overcome and the particles approach within ~10 Å of each other, a primary attractive minimum exists which is characterised by a high interaction energy level. This energy level is compatible with the formation of chemical adhesive bonds between the particles (see Table 28). These

Table 28 *Stability and adhesive interaction of particles in relation to interaction energies*

Distance between particles		Interaction energy level	Nature and type of adhesive bond
>200 Å $V_A > V_R$			
100–200 Å	Secondary attractive minimum	Low	Weak reversible adhesion, slow. Adhesions of type known as gap junctions
~ 100–16 Å $V_A < V_R$	Repulsive barrier		
<10 Å $\simeq 1/K$ $V_A > V_R$	Primary attractive minimum	High	Chemical bonds, rapid, irreversible. Adhesions of type zona occludentes (tight or close junctions)

adhesive bonds are formed rapidly and are irreversible. The tight junctions or zonulae occludentes occurring in early embryos are presumably adhesive bonds formed at the primary attractive minimum. The zonulae occludentes are thus formed by the fusion of closely approaching cells by meshwork of branching and anastomosing filamentous structures.

The colloidal approach to the question of cell adhesion made by Curtis has enabled a much better understanding of adhesion phenomena and has much predictive value. However, Trinkaus (1969) has pointed out that, although gap junctions 100–200 Å were predicted by Curtis' lyophobic colloid theory, the smaller gap junctions of the dimensions 20–90 Å are inconceivable on account of the potential repulsive energy barrier existing between cells separated by ~16–100 Å. The dimensions of the gap or the closest approximation will depend upon the level of the repulsive energy barrier which will be determined by the actual numerical values of the attractive and repulsive forces operating at a given separation. Nonetheless, it may be well to remember that the colloidal approach can provide a theoretical basis relevant to biological systems and only a rough approximation is permissible from the DLVO equations. In biological systems the conditions obtaining are completely different from those under which the stability of pure colloidal systems is achieved. Thus both Van der Waals–London dispersion forces and the electrostatic repulsive forces are affected by a variety of factors which are characteristic of biological systems. As Weiss (1972) has pointed out, the parameter of dielectric constant may lead to a considerable underestimation of the repulsive forces, as does also the calculation of the repulsive force using ζ potential rather than the potential at the interphase. In addition, the radius of curvature of the approaching particles determines the level of the potential repulsion barrier. On the other hand, increasing ionic valency and/or strength will reduce the repulsive force (see eqn 29). Considering the attractive forces, Vold (1961) has shown that the greater the resemblance between the ionisation potentials of the suspending medium and the particles, the smaller will be the values of the Hamaker constant A and therefore the smaller will be the magnitude of the attractive forces operating between the particles.

Surface Potential and Cellular Adhesion and Adhesion-dependent Phenomena

The discussion of the last three sections distinctly underlines the involvement of surface potential in cellular adhesion and adhesion-dependent phenomena. Cell surfaces bear a net negative charge and electrophoretic measurements of surface potentials of cells at between 0·1 and 0·2 M ionic strength and physiological pH, have given values of −8 to −23 mV (see Curtis, 1967, pp. 60 and 61). These negative potentials which are, in the main, responsible for the repulsive forces operating between cells, are predominantly attributable to the sialic acid moieties of surface polysaccharides. In the highly non-adhesive cell type such as erythrocytes the negative electric charges are mainly contributed by carboxylic groups of sialic acids and α-carboxylic groups of proteins, and probably with minor contributions from β- and γ-carboxylic groups (see pp. 57–60). In general, approximately two-thirds of the total sialic acids may be located at the cell surface (Kraemer, 1966) as indicated by biochemical estimation, using Warren's (1959) method, of sialic acid release by NANase, and by the reduction by 60–90% in the EPMs of erythrocytes by the enzyme (see Table 15). The carboxylic groups of sialic acid residues may be expected to exert considerable repulsive forces.

The adhesiveness of a number of varieties of echinoderm eggs has been found to be inversely proportional to their ζ potential (Dan, 1936, 1947). Wilkins *et al.* (1962) found that the rate of flocculation of sheep erythrocytes by cations was also inversely proportional to their ζ potential. Increasing the net negativity of the surface of quail liver cells by adsorbing the polyanionic compound, dextran sulphate, causes an inhibition of aggregation (Kuroda, 1974). On the other hand, if the net surface negative charge is reduced by treating cells with both high and low molecular weight poly-*l*-lysine, which is a polycationic compound, aggregation of HeLa cells is enhanced (Deman and Bruyneel, 1974). Both long- and short-chain polymers produced a linear promotion of aggregation. The linearity in this relationship was obtained between 0 and 5 $\mu g\ ml^{-1}$ of short-chain (MW 4000–23 000) and 0–3 $\mu g\ ml^{-1}$ of long-chain (MW 70 000–100 000) poly-*l*-lysines. In cells which had been stripped of sialic acid using NANase, a linearity of aggregation was obtained irrespective of the length of the polymer used. These experiments suggest that the promotion of aggregation is due to the lowering by the polymer of the surface negative charges contributed by sialic acids. Binding of poly-*l*-lysine reduces the ζ potential of cells (Katchalsky *et al.*, 1959; Nevo *et al.*, 1955). Polycationic substances like nuclear histones have also been found to increase the adhesion of HeLa cells to glass and plastic substrata. Although the mode of action of histones is not clearly understood, the effects are mainly at the cell membrane surface–interphase (Bases *et al.*, 1973).

Cationic anaesthetics, on the contrary, are known to inhibit substrate adhesion, spreading, cellular motility, virus-induced cell fusion, surface capping of macromolecules and increase the susceptibility of cells to agglutination by Con A (Rabinovitch and DeStefano, 1974, 1975; Gail and Boone, 1972; Poste and Reeve, 1972; Ryan *et al.*, 1974a; Poste *et al.*, 1975c). These anaesthetics are also known to produce a reversible contraction and rounding-up of cells (Rabonovitch and DeStefano, 1976). Gail and Boone (1972) observed reduction in cellular motility and proliferation with procaine hydrochloride, which, they suggest may be due to alterations in membrane permeability to ions resulting in changes of membrane potential. The method of determining the degree of cellular motility adopted by Gail and Boone appears to be somewhat arbitrary. Besides, they have made no measurements of the membrane potentials of their treated cells. The difference in the effects of histones and the cationic anaesthetics may be found in their mechanism of action. Histones may reduce the surface potential by non-specific neutralisation of the negatively charged groups on the surface. But the mode of action of cationic local anaesthetics may be more complicated. Poste and Reeve (1972) found that these agents displace Ca^{2+} from the surface. Poste *et al.* (1975) have further proposed that such displacement of Ca^{2+} may alter the functions of proteins on the inner surface of the membrane. They envisage Ca^{2+} as linking these proteins to the receptors residing at the surface. Nicolson *et al.* (1976) found that local anaesthetics not only cause cells to contract but also at the same time induce significant reductions in the membrane-associated cytoskeletal structures, the microtubules and microfilaments, and their attachment to the plasma membrane. In other words, local anaesthetics can cause surface alterations which presumably are transduced into changes in the cytoskeletal structure of the cell.

The involvement of sialic acids in adhesion is indicated by the observations of Weiss (1961b, 1963) that NANase aids detachment of cell from glass substratum. The ease with which cells could be detached varied in different cell lines. And finally, according to Deman and Bruyneel (1974) cells harvested at the stationary phase of growth show the least mutual adhesiveness and display the strongest repulsive effects which are generated by surface sialic acids. The strength of these long range repulsive forces, as measured by an agglutination assay, can be altered by raising or lowering the tonicity of the suspension medium (Deman *et al.*, 1976).

A most intriguing example of the possible involvement of surface charge in adhesion-dependent phenomena is the mobility of cells and tissue fragments implanted into chick embryo blastoderms. When homografts or heterografts are implanted between the ectodermal and endodermal layers of the chick primitive streak embryo, they show a varying degree of movement in consonance with the morphogenetic movements of cells occurring in the ectodermal layer. This phenomenon has been termed as morphogenetic displacement by Sherbet and Lakshmi (1974b). The involvement of surface charge in this phenomenon was suggested by the observation that the morphogenetic displacement of Morris minimum-deviation hepatomas was closely correlated with their growth rates (Sherbet *et al.*, 1970), and increased growth rates are known to be associated with increased negative charge density at the surface. Sherbet and Lakshmi (1968) and Sherbet *et al.* (1070) showed that when the surface of the cells of Hensen's node grafts was altered by treatment with lysozyme, histones and poly-*l*-lysine, a reduction in morphogenetic displacement occurred on implantation into chick blastoderms. Since these substances are cationic in nature, it may be expected that these treatments may have resulted in a decrease in the negative charge density at the surface of the implanted cells. On the other hand, higher morphogenetic displacement appears to accompany higher negative charge density as found in the case of HeLa cells, γ-globulin-treated HeLa cells and Yoshida ascites sarcoma cells (Sherbet *et al.*, 1970; Lakshmi and Sherbet, 1974).

Much of the evidence presented above argues in favour of the idea that the surface potential owing to the sialic acid generates repulsive forces that determine the equilibrium between approaching particles and between the particle and the substratum, to enable adhesive bonds to form. Nonetheless, it should be pointed out that there is also an equally large body of contradictory evidence.

Kemp (1968) showed that NANase inhibited the aggregation of chick embryo fibroblasts. A complete inhibition of aggregation took up to two hours of rotation in a gyratory shaker. The time lag, it is suggested, may represent the time taken for a complete removal of sialic acids by the enzyme. The aggregates formed in the presence of the enzyme were very much smaller than those obtained in the absence of the enzyme. Even human erythrocytes, whose electronegativity has been shown to be entirely due to the presence of sialic acids, do not show increased adhesiveness following treatment with NANase (Deman *et al.*, 1974).

In an interesting study of neural retinal cells, Steinberg *et al.* (1973) showed that the net negativity of the cell surface was unrelated to their ability to form aggregates. In this investigation they found that exposure of chick embryo neural retinal cells to trypsin designed to obtain dissociation, rendered them unable to reaggregate for at least 30–40 minutes. At this stage the cells showed the first signs of aggregation which subsequently resulted in the formation of compact aggregates on further incubation. Using cell electrophoresis, Steinberg *et al.* (1973) showed that the recovery of adhesiveness and the ability to form aggregates was unrelated to the EPMs. Only in two out of five experiments statistically significant increases in the EPM in association with recovery of adhesiveness, were observed. EDTA-dissociated and trypsin-dissociated cells initiated adhesions at identical rates despite large differences in their EPMs.

If surface potentials were the prime mover of cell adhesion and aggregation, at a lower surface potential Steinberg's trypsinised cells should be capable of greater adhesion than EDTA-dissociated cells which have higher surface potentials. But the latter do not show the time lag in aggregation which has been demonstrated for trypsinised cells. It is, nonetheless, worthwhile pointing out that Bernard *et al.* (1969) found that, while the EPMs of freshly trypsin-dissociated chick neural retinal cells were low, with the passage of approximately 24 hours the EPMs increased by between 70 and 96%. Besides, Steinberg *et al.* (1973) have omitted to say how accurate their EPM measurements were, nor provided any evidence that the reduction in the electronegativity was not due to adsorption of trypsin molecules to the surface, as indeed shown by Bernard *et al.* (1969) and Poste (1971). Trypsinisation alters several physicochemical features of the cell surface (see pp. 113–114). One could not be cautious enough in interpreting these data in relation to the DLVO theory of cellular adhesion.

A lack of correlation between surface negative charge density and ability to aggregate has also been argued by Kemp and Jones (1970). They found that pretreatment of chick embryonic muscle cells by *p*-benzoquinone and tannic acid inhibited cellular aggregation to the same degree, but these substances caused increase in EPM of 20 and 84% respectively. In both the compounds the increase in net charge density was produced by the binding of the compounds to the positively charged groups of the cell surface. Kemp and Jones (1970) suggest that the inhibitory effects are due to the blocking of movements of surface components involved in adhesion since quinones are known to increase rigidity of protein films (Pankhurst, 1968). In a similar vein, ADP and ATP also inhibit cellular aggregation and produce a concomitant increase in EPM (Jones and Kemp, 1970). A lack of correlation between surface potential and aggregation, one may add, is suggested solely on the basis that equivalent quantities of *p*-benzoquinone and tannic acid produced the same level of inhibition despite raising the surface potential to different levels. This is an inadequate premise on which to investigate the existence or lack of correlation. It would be interesting to see if cellular aggregation was inhibited to different degrees when the increase in net negative charge caused by the two compounds were made comparable by adjusting the concentrations employed for treating the cells.

It appears possible that raising the surface potential of the cells enhances the magnitude of the repulsive forces and as a result the already low adhesive interaction energy levels at the secondary attractive minimum are even further reduced thus making the cells liable to dispersion by the action of Brownian motion. For adhesion at the primary attractive minimum the cells have to penetrate the strong repulsive barrier which will be further raised by the higher surface potential of the cells. In other words, if the surface potential is increased by experimental means above an optimum threshold, reaggregation would fail irrespective of the actual magnitudes of increase in the surface potential.

Kemp and Jones (1970) point out that in accordance with the DLVO theory, lymphocytes which have a high surface charge density (4090 e.s.u. cm^{-2}) (Bangham *et al.*, 1958) should be non-adhesive (Bangham and Pethica, 1960), but they do adhere to target cells irrespective of their surface charge. But surface charge density determination from electrokinetic data provides an average value and assumes a uniform distribution of the charges on the surface. When cells make contacts with other cells or the substratum, this is not done by approximation of large areas of the surface but by means of protrusions which have low radii of curvature. On electrostatic considerations Bangham and Pethica (1960) believe that approximation and contact of two cells is facilitated by the mediation of such protrusions of low radius of curvature. It is conceivable that cells make such protrusion-mediated contact in areas of low electrostatic charge density, where the repulsion barrier would be less than if calculated using average ζ potential values and contact between cells energetically favourable (Weiss, 1968). Besides, a reduction in the radius of curvature reduces the magnitude of both attractive and repulsive forces. But if the effective distance between the particles is over 200 Å (which is the zone where the repulsive forces V_R preponderate over V_A, see Table 28), reduction of radius of curvature will cause an absolute reduction in V_R.

In the cells of Kemp and Jones (1970) immobilisation of surface protein may indeed have occurred with consequent immobilisation of the cell membrane. This may be reflected in the loss of ability to reaggregate. But the experiments do not necessarily suggest a repudiation of the applicability of the DLVO theory to cell contact phenomena.

Among other objections to the concept is the role played by Ca^{2+} in promoting cellular adhesions. This has been discussed earlier (pp. 120–121). One view is that Ca^{2+} ions reduce the surface potential as a prelude to adhesions (Curtis, 1963). The kinetics of aggregation of erythrocytes in the presence of trivalent lanthanum ions also supports Curtis' view. Lerche *et al.* (1976) found that increasing the concentration of La^{3+} ions raised the ability of erythrocytes to aggregate. The effect also correlated with La^{3+} bound to the surface. But Armstrong (1966) provided convincing evidence that calcium ions may participate in adhesion-related reactions other than lowering of surface potentials of interacting cells.

ELECTROPHORESIS OF SPERM CELLS

The study of sperm cells over the past fifty years, whether making use of their electrokinetic properties or sedimentation characteristics, has been occasioned by the desire to control sex ratio in mammals, and for the purposes of selective breeding of economically and hereditarily advantageous breeds of animals. Separation of X and Y sperms has been attempted by taking advantage of the possible differences in their specific gravities. Some success has been reported in altering sex ratios using this technique (Bhattacharya, 1958, 1962; Krzanowski, 1970; Lindahl, 1958; Schilling, 1971; Stambaugh and Buckley, 1971a,b).

Electrophoresis of sperm cells was attempted four decades ago with a view to separating X and Y sperm cells. Schröder (1934, 1942) claimed that rabbit sperm drawn off from the region of the cathode produced female progeny in 80% of conceptions. Subsequently Gordon (1957, 1958) also made similar claims. Nevertheless, one ought to draw attention to the fact that a number of investigators have been unable to detect any differences in the surface charge characteristics between X and Y sperm cells (Hafs and Boyd, 1971, 1974; Manchowska and Schegaloff, 1935; Sevinç, 1968). Bangham's (1961) experiments indicate that there may not be clearly defined phenotypic differences between the sperm types. Bey (1965) found that sperm cells from the bull were homogeneous as regards their surface electric charge.

Bey (1965) has examined the EPMs of sperm cells from a number of mammals (Table 29). There appear to be considerable differences in the values for cells from the different sources, unlike, for instance, the EPMs for erythrocytes of various animals (see Tables 11 and 13). These differences noted by Bey are so great as to make one doubt the accuracy of these determinations. It should be remembered that sperm cells have considerable ability for free swimming, and their observed mobilities can be varied by cooling (Bangham, 1961).

Table 29 *Electrophoretic mobilities of sperm cells*

Donor species	EPM (−TU)
Bull	13·02 ± 0·3
Rabbit	11·2 ± 0·22
Cock	15·45 ± 0·2
Boar	4·95 ± 0·34
Man	13·62 ± 0·46

From Bey (1965).

The electrical negativity of sperm cells appears to be in part attributable to the occurrence of sialic acids. Bey (1965) observed an approximately 30% loss in the EPM on treatment of bull sperm cells with NANase and the loss of EPM paralleled the free neuraminic acid released from the cells. If this value is assumed to be accurate, sialic acid-containing components may be considered to contribute much less significantly to the surface negative charge than in the case of erythrocytes (see Table 15) where NANase produces EPM losses of between 60 and 90%. Treatment with trypsin also results in reductions of EPM. Incubation of bull sperm with their own seminal plasma causes reductions in EPM which correspond with similar alterations produced by treatment with trypsin. Bey (1965) has also shown that the effects of seminal plasma on EPM can be counteracted using trypsin inhibitors such as soya bean inhibitor and trasylol. Besides, the EPMs of sperm cells from the epididymis and first and second ejaculations are found to differ a good deal and this has been attributed to the varying degree of effects of the seminal plasma. This is consistent with the observations of Moore and Hibbit (1975) who found that normal spermatozoa have an higher pI than those derived from boars without seminal vesicles. The pI of the latter spermatozoa could be increased by incubating them with seminal plasma which indicates that seminal plasma alters the pI of spermatozoa on ejaculation.

Surface changes, presumably pertaining to the glycoprotein components, have been reported in the process of sperm maturation and capacitation. A number of investigations involving the binding of plant lectins to the sperm cell surface have shown that carbohydrate-rich components are located on the sperm cell surface (Nicolson *et al.*, 1972; Nicolson and Yanagimachi, 1972; Edelman and Millette, 1971; Gordon *et al.*, 1974). Bedford (1963) observed that the EPM of sperm cells changed during their passage through the epididymis. Cooper and Bedford (1971) showed that most of the surface charge of the cells is acquired during epididymal maturation of the cells. The acquisition of charge also correlated with differences in the auto-agglutination behaviour of sperm cells.

Unfortunately there is little new activity in this field, especially in the characterisation of the sperm cell surface. This is presumably because there is little progress to be made. Where information is required for the purposes of controlling sex ratio and related problems, the tendency has been to develop immunological methods (Bennet and Boyse, 1973), in which area there have been several interesting developments.

Sperm cells also undergo changes in the female reproductive tract. Vaidya *et al.* (1971) found that the EPM of sperm cells decreased after they had been incubated with oestrous uterus for about six hours. Gordon *et al.* (1974) reported that the process of capacitation of sperm cells also resulted in a reduced localisation of Con A. Vaidya *et al.* (1971) have pointed out not only that NANase is known to reduce EPM of sperm cells (Bey, 1965) but also that this enzyme is usually present in the rabbit uterus and may be responsible for the loss of EPM by sperm cells (Hartree and Srivastava, 1965).

ELECTROPHORESIS OF TUMOUR CELLS

More than three decades ago Coman (1944) attributed the invasive ability of tumours to reduced cell to cell adhesion between tumour cells as opposed to that between normal cells. It was found to be easier to separate tumour cells than normal cells (Coman, 1944; McCutcheon *et al.*, 1948; Tjernberg and Zajicek, 1965). Much work has since been carried out to discover the cause, whether structural, biochemical or molecular, of this reduced adhesiveness of tumour cells. The cell surface components are obvious mediators of cell–cell interaction and adhesion. Cell electrophoresis has played a considerable part in the characterisation of these surface components and in understanding the involvement of surface charge in tumour cell behaviour *in vivo* and *in vitro*, and in the examination of cell surface changes in relation to progressive neoplastic development. In this final section of this chapter we shall discuss some of these problems in relation to the neoplastic state.

SURFACE CHARGE OF NORMAL AND TUMOUR CELLS

The involvement of surface potential of cells in cellular adhesion and adhesion-dependent phenomena has been discussed in detail in this chapter (pp. 120–131). Lower surface potentials may be conducive to cellular adhesion. In fact Coman (1944, 1953) suggested that the decreased adhesiveness of tumour cells may be due to a decreased Ca^{2+} content of these cells, although this has subsequently been disputed (Hickie and Kalant, 1967). Ca^{2+} may form bridges between negatively charged groups of adjacent particles or by lowering the surface potentials of approaching particles. Ambrose (1965, 1966) has put forward the thesis that tumour cell surfaces have intrinsically higher charge densities than normal cells. Ambrose *et al.* (1956) and Lowick *et al.* (1961) noticed that stilbesterol-induced hamster kidney tumour cells showed greater EPM than corresponding normal cells. A higher EPM was found also in butter yellow(DAB)-induced hepatoma cells as compared with normal liver cells (Fuhrman, 1965).

A detailed study of the electrokinetic properties of normal blood cells and their deviation in the pathological state has been described by Ruhenstroth-Bauer (1965). Rueff *et al.* (1963) established the electrokinetic pattern of normal blood cells, and compared it with the pattern from patients with leukemias. In 30 cases of myeloid leukemia the normal granulocyte group which has an EPM of $-8{\cdot}74 \pm 0{\cdot}35$ TU was found to be missing but a new group with EPM of $-12{\cdot}09 \pm 0{\cdot}35$ TU appeared. In lymphatic leukemia, instead of the normal lymphocyte group, a faster moving fraction appeared. In cases of remission a normal electrophoretic pattern returned. In a number of other pathological states also additional cellular bands have been reported (Rueff, 1963; Ruhenstroth-Bauer, 1965) e.g. lymphosarcomas, erythematoses, reactive leukocytoses, etc. Despite the extensive nature of these studies, the main observation that a population of cells with higher EPM appears in association with the neoplastic state in leukemic patients, has not been confirmed (Mehrishi and Thomson, 1968; Thomson and Mehrishi, 1969; Lichtman and Weed, 1970; Patinkin *et al.*, 1970a,b; Schubert *et al.*, 1972). Cook and Jacobson (1968) indeed claimed that leukemic cells from mice had lower EPM than the corresponding normal cells. The peripheral blood lymphoblasts from patients suffering from acute lymphatic leukemia have been described to possess 7–25% lower EPMs than cells from normal subjects. The cells from normal subjects show a bimodal distribution while leukemic cells show a single symmetrical distribution. Although normal B-lymphocytes are known to have lower mobilities than T-lymphocytes, the lowest EPM of leukemic cells was far below that of the B-cell population (B. A. Smith *et al.*, 1976). This is consistent with the observation of Andersson *et al.* (1976) that the cells of patients with chronic lymphocytic leukemia possess a surface component not detectable on either T- or B-cell surface.

Using isoelectric equilibrium method, Sherbet and Lakshmi (unpublished work) found that the net negative surface charge density on normal liver cells of the Chester Beatty strain of rat was nearly twice that on the hepatoma cells induced in this strain by the administration of DAB—the respective charge densities were found to be $3{\cdot}98 \times 10^{12}$ charges cm^{-2} and $2{\cdot}0 \times 10^{12}$ charges cm^{-2} respectively.

SURFACE CHARGE OF NORMAL AND VIRUS-TRANSFORMED CELLS

The transformation of normal cells *in vitro* by oncogenic viruses is accompanied by a variety of changes. Prominent among these are altered cultural requirements for optimum growth, changes in growth pattern, colonial morphology, changes in the karyotype, appearance of specific nuclear antigens and also a diversity of changes associated with the cell surface. The surface-associated changes include the appearance of virus specific and virus coded antigens, carcinoembryonic antigens and transformation-associated surface antigens. Transformed cells have been reported as possessing a thicker surface coat compared with untransformed cells (Burger, 1969, 1973; Poste, 1970, 1973; Martinez-Palomo, 1970; Martinez-Palomo and Braislovsky, 1968; Vorbrodt and Koprowski, 1969). Differences have also been reported in the activities of enzymes such as glycosyl transferases (Den *et al.*, 1971; Grimes, 1970; Meezan *et al.*, 1969). Transformation is also often associated with changes in the surface components such as the disappearance of normal components and/or the appearance of new protein and glycoprotein components (see pp. 139–140) and in the form of increases or decreases in the sugar components of the surface (Chiarugi and Urbano, 1972; Grimes, 1970). Considerable changes in the transport of sugars (see Hatanaka, 1974; Kletzien and Perdue, 1974a,b; Bradley and Culp, 1974), amino acids (Foster and Pardee, 1969; Isselbacher, 1972) and phosphate (Cunningham and Pardee, 1969) have been reported to occur following transformation.

Alterations in surface components have also been demonstrated using plant lectins as a molecular probe. Thus differences have been shown to exist in the susceptibility to agglutination by lectins, between normal and transformed cells. There also appears to be a difference in the mobility of lectin-binding sites on the cell surface (see pp. 211–212). Such a wide variety of surface changes undoubtedly influence the electrokinetic behaviour of transformed cells, and within the technical constraints and theoretical limitations, cell electrophoresis has provided much information about these surface changes.

Forrester (1965) examined the electrophoretic behaviour of a line of BHK-21 (neonatal hamster kidney) cells transformed *in vitro* by infecting cell monolayers with polyoma virus (Macpherson and Stoker, 1962). As a result of incubation with the virus two types (I and II) of transformed clones appeared. These were distinguishable by their colonial morphology. While Type I clones appeared to be stable, Type II changed into Type I during further propagation. Electrophoretically the transformed cells were distinguishable into two groups: one group with EPMs comparable with the untransformed cells and the second group showed 25% higher EPMs. However, there is no evidence to suggest that these low and high mobility groups correspond with the transformed Types I and II. Latner and Turner (1974) working with BHK-21 cells and their polyoma virus-transformed (Py6) cells, however, found no difference in their EPMs although the transformed cells showed a greater spread of mobilities. Nor does any difference appear to exist in the EPMs of untransformed 3T3 murine fibroblasts and those transformed by simian virus 40 (SV-40) (Adam and Adam, 1975). Sherbet and Lakshmi (1976) (see also p. 199) found that the charge densities on SV-40- and polyoma virus-transformed cells were marginally, if not significantly, lower than untransformed cells.

Earlier Weiss *et al.* (1975) examined the electrophoretic mobilities of normal and SV-40-transformed 3T3 cells and two revertant transformed lines. They observed that the EPMs of transformed cells were about 25% lower than the untransformed cells. Interestingly enough the revertant transformed lines showed increases in EPM and one of the two revertant lines had EPM very close to the untransformed cells. These differences were confirmed by the colloidal iron-binding pattern shown by these cells. In SV-40-transformed cells the surface between the microvilli bound less colloidal iron than the untransformed cells. If read in conjunction with the pattern of electrophoretic behaviour, this would suggest that the total anionic component of the virus-transformed cell is lower than in the untransformed cells. In a recent study, Subjeck *et al.* (1977) have investigated the intermicrovillar binding of colloidal iron by chick embryo fibroblasts transformed by the temperature sensitive *ts* mutant of Rous sarcoma virus (RSV). Subjeck *et al.* detected a higher binding by *ts* mutant-transformed cells at the permissive temperature of 36°C. But at the non-permissive temperature of 41°C the transformed cells bound as much colloidal iron as did untransformed cells. These observations constitute the most convincing evidence available to-date that viral transformation is associated with changes in the negative charge. Nonetheless, it is difficult to see why the 3T3—SV-40-transformed 3T3 system and the chick embryo fibroblast—*ts* RSV-transformed cell system should show such radical differences in the binding pattern, since in both these systems the transformed cells are known to have increased fucose-containing glycoproteins (Warren *et al.*, 1972a,b).

From this rather limited experimental evidence, one may conclude that there is not sufficient justification to believe that higher surface charge densities are somehow intrinsically associated with transformation by oncogenic viruses, or that higher charge densities are characteristic of neoplastic cells.

SIALIC ACIDS AND MALIGNANCY

The proponents of the view that higher surface charge densities are associated with malignant transformation have tended to suggest that such increases are due to the higher sialic acid content of tumour cells. Forrester *et al.* (1964) and Forrester (1965) found that, although polyoma-transformed BHK-21 cells had higher EPM than untransformed cells, treatment of the transformed cells with NANase reduced their EPMs to the level of the untransformed cells.

The investigations of Weiss *et al.* (1975) on normal 3T3 and SV-40-transformed 3T3 cells completely contradict these observations of Forrester and colleagues. Weiss *et al.* not only found that normal cells had higher EPMs than untrasformed cells but also that treatment with NANase and RNAase caused greater reductions in the EPMs of normal cells than of the transformed type. In addition, transformed cell lines which had revertant phenotypic properties showed greater (but comparable to normal cells) loss of EPM on treatment with the enzymes than did the transformed cells.

The observations of Latner and Turner (1974) indicate that the normal and polyoma-transformed BHK-21 cells have the same amount of sialic acid associated with the surface. It ought to be mentioned here, however, that the use of NANase in conjunction with measurements of EPM values must be regarded with circumspection. It is known that sialic acids confer structural rigidity to membrane glycoproteins (Gottschalk, 1960a) and probably thereby contribute to the rigidity of the cell membrane (Weiss, 1963; Ray and Chatterjee, 1975). Removal of sialic acids may result in conformational changes in the membrane making a comparison between cell types following NANase treatment somewhat suspect. At any rate, malignancy is not always accompanied by increased sialic acid content (Patinkin *et al.*, 1970a,b; Vassar, 1963a,b). On the contrary, lymphocytes from patients with chronic lymphocytic leukemia have been found to have lower levels of neuraminic acids than corresponding normal cells (McClelland and Bridges, 1973). Weiss and Haushka (1970) examined the EPMs of TA3 cells in relation to their malignancy in a number of mouse strains and observed no correlation between EPM and malignancy, nor between malignancy and surface sialic acids. In human astrocytomas, the more malignant the tumour (as judged by its histological appearance) greater is the net surface charge density (Sherbet and Lakshmi, 1974). Nonetheless, meningiomas, which possess surface charge roughly comparable to the highly malignant astrocytomas, are non-malignant tumours. In the same vein, normal fetal brain cells were found to possess surface charge densities similar to the malignant astrocytomas. From this discussion of electrokinetic evidence one arrives at an inescapable conclusion that there is no correlation between surface charge densities or sialic acid content with malignancy. One often wonders if the basic premise, which leads one to examine such a relationship itself is wrong. For malignancy, according to Foulds (1967), is a mere clinical concept, not a biological entity. The surface charge is probably related to or a reflection of other biological attributes or functional state of cells (see pp. 222–225).

It might be relevant to discuss in this context the recent biochemical evidence about surface glycoproteins of normal and transformed cells. Wu *et al.* (1969) examined membrane fractions of 3T3 and SV-40-transformed 3T3 cells and found that fractions derived from the latter had lower amounts of *N*-acetylgalactosamine and sialic acid but increased amounts of *N*-acetylglucosamine than untransformed 3T3 cells. Normal cells also appeared to have higher amounts of complex glycolipids (Hakomori and Murakami, 1968; Mora *et al.*, 1969). According to Grimes (1970) transformed (SV-3T3) cells had only 60% as much sialic acid as did untransformed cells. The levels of activity of sialytransferase in transformed cells was only 31% of the normal level.

A number of investigators have also observed the loss of large protein components on transformation of the cells by means of viruses (Hynes, 1973; Gahmberg and Hakomori, 1976; Gahmberg *et al.*, 1974; Critchley, 1974; Hynes and Wyke, 1975). Vaheri and Ruoslahti (1974) and Keski-Oja *et al.* (1976) have demonstrated a surface antigen(s) on the fibroblast cell surface. The expression of the antigen appeared to be reduced on transformation. This antigen has been described to be made up of two or possibly three components (Kuusela *et al.*, 1975; Vaheri *et al.*, 1976) and that one of these components (MW 210 000) appears to be a glycoprotein.

Recently it has been reported that the fucose-containing glycoproteins derived from virus-transformed cells and mitotic cells have a higher sialic acid content (Buck *et al.*, 1970, 1971; Warren *et al.*, 1972a,b; Van Beek *et al.*, 1973). Van Beek *et al.* (1975) have concluded from a study of human leukemic and lymphosarcoma cells that their glycoprotein composition differs considerably from that of normal cells. They have attributed the difference to an increased content NANase-sensitive sialic acid in all cases studied with the exception of two cases of chronic myelocytic leukemia. Glick *et al.* (1974) have examined glycopeptides of hamster embryo cells infected with polyoma virus, and have observed a clone of delayed transformants with respect to a specific glycopeptide which is increasingly seen over a period of 85 days. The presence of this glycopeptide seems to coincide with the appearance of the transformed phenotype and the ability to produce tumours *in vivo*.

Investigations on the levels of sialyltransferase activity in relation to the transformed state or the mitotic cycle have helped to support the findings discussed above. Warren *et al.* (1972a) found increased levels in transformed BHK-21 cells. Bosmann (1968, 1972) also found increased sialyltransferase activity in 3T3 cells transformed by polyoma, Rous sarcoma virus and murine sarcoma virus. An increase in sialytransferase activity is also noticed in BHK cells in mitosis (Warren *et al.*, 1972a).

Even though higher net surface charge and higher sialic acid content may not be a universal characteristic of neoplastic cells, there might yet occur increases in both acidic and basic components, or in acidic components other than sialic acids. Surface changes of this nature probably occur in polyoma virus-transformed Py6 cells which have EPM similar to the untransformed BHK-21 cells. However, a chemical modification of amino groups by formaldehyde, shows that the transformed cells have more anionic groups than untransformed cells. For the EPM of the latter rose by 4% following formaldehyde treatment while that of Py6 cells rose by 22% (Latner and Turner, 1974). Such an increase in anionic groups in the transformed cells is presumably counterbalanced by an increase of equal magnitude in cationic groups. Unfortunately, Latner and Turner have not attempted a modification of the anionic groups. Positively charged groups are undoubtedly present on normal cell surfaces (see pp. 181–184) and in significant quantities also on the surface of murine CL3 leukemia and ascitic cells (Cook *et al.*, 1962; Mehrishi, 1970a), benzpyrene-induced sarcoma BP8 cells (Mehrishi, 1972) and a variety of laboratory tumour cell lines such as HeLa, Yoshida ascites sarcoma, Ehrlich ascites cells (Sherbet *et al.*, 1972) and SV-40-transformed 3T3 fibroblasts (see pp. 188–190). In none of these studies, however, was it possible to say if their densities were higher than corresponding normal cells, for there were no normal cells to compare with. Cook and Jacobson (1968) have shown that a new basic component appears on the surface of acute lymphoblastic leukemia which was absent in normal lymph node cells. In the 3T3 and SV-3T3 system, isoelectric equilibrium analyses have shown that the same kind and number of negatively charged groups occur but there are differences so far as the basic groups are concerned. The basic groups on SV-3T3 cells are much weaker and in addition there are about 30% more of these groups as compared with untransformed cells (Sherbet and Lakshmi, unpublished work, see pp. 188–190).

The presence of sulphated polysaccharides, glycosaminoglycans, may also contribute to the surface charge of cells. Kojima and Maekawa (1970) suggested that the negative surface charge in certain rat ascites heptamoas may be due to other residues than sialic acids. Kojima *et al.* (1970) and Suzuki *et al.* (1970) have demonstrated that the occurrence of sulphated mucopolysaccharide as a surface component. In the rat ascites hepatoma AH-130, cellular EPM is reduced by chondroitinase by approximately 20%, but NANase has no effect on the EPM. Assays have revealed a concomitant release of chondroitin A from the cell surface (Kojima and Yamagata, 1971). However, in the same paper Kojima and Yamagata have given corresponding data on two other ascites hepatoma lines. In these, both NANase and chondroitinase produced approximately a 36% reduction in EPM.

EFFECTS OF ULTRASOUND AND IONISING RADIATION ON TUMOUR CELL SURFACE

Among other areas of research in tumour biology where cell electrophoresis could contribute is in the monitoring and assessment of interaction of ultrasound and ionising radiation with the tumour cell surface.

Ultrasound has been claimed to be capable of selectively destroying tumour cells (Horvath, 1944; Woeber, 1949). Ultrasound has also been combined with ionising radiation in the treatment of tumours (Woeber, 1959) and it has been claimed that this has enabled a reduction of 40% in the total irradiation.

Sato and Kojima (1971) and Sato *et al.* (1972) noticed that X-irradiation caused a loss of EPM of three cell lines, with a concomitant loss in the ability of the cells to form colonies. But this was not accompanied by a release of sialic acids. Gersten and Bosmann (1975) however failed to detect any changes in the charge characteristics of L5178 mouse leukemic cells and two cell lines derived from a mouse melanoma. Neither did the partition behaviour of the cells show any alterations following irradiation. Nonetheless, the results reported by Gersten and Bosmann cannot be interpreted as repudiating the argument that changes in EPM without the release of sialic acids may indicate a rearrangement of the surface components (Sato *et al.*, 1972; Sato and Kojima, 1974). The difficulty in resolving this question using cell electrophoresis is clearly due to the dimensional limitations of the electrokinetic zone. The isoelectric equilibrium method may be able to provide the evidence which may support or repudiate the suggestions of Sato and colleagues.

Rapochioli *et al.* (1971) found that the mean EPM of Ehrlich ascites cells is reduced by 16% when subjected to low intensity ultrasound. Loss of EPM was also seen in cells exposed to X-irradiation (600–1000 rad). An additive loss of EPM occurred when ultrasound and X-irradiation were combined, and this produced up to 30% loss in EPM with a combination of ultrasound and 1000 rad of X-irradiation. Gram *et al.* (1975) found that ultraviolet irradiation of Chinese hamster cells also causes losses in EPM. At lower u.v. doses (below 250 erg mm^{-2}) a lag period of 4 hours occurred before losses in EPM could be detected.

When the irradiated cells were incubated for 24 hours after irradiation, a striking loss in EPM was noticed. Cells exposed to u.v. doses of 500 erg mm^{-2} show a maximum loss of EPM around 10 hours following irradiation. The loss of EPM is abolished if the irradiated cells are treated with actinomycin D and cycloheximide. The electrophoretic mobilities are restored to normal levels about 65 hours after irradiation.

While these experiments indicate alterations in the expression of surface components as a result of the treatment, there does not appear to be much worthwhile information in these papers about the mechanisms by which the treatments produce the surface changes.

Some indications as regards the mechanisms are given by the work of Sato *et al.* (1976) who found that the loss of EPM caused by irradiation could be abolished by pretreatment with Con A at low concentrations. But this abrogation effect is temperature dependent. Since binding of Con A at low concentrations is not likely to alter membrane fluidity and since there is no abrogation at low temperatures, it seems likely that the loss of EPM on irradiation without concomitant release of sialic acids may be due to conformational changes induced by irradiation.

It would be useful to examine the nature of any materials that may have been released by the treatments. A number of other questions could be asked. Have the different polymers different susceptibility to degradation or excision? Do ultrasound and irradiation act on same or different components? And finally does the loss of expression of any given component(s) correlate with the cultural behaviour of cells?

DIAGNOSTIC TEST FOR CANCER: ASSAY OF LYMPHOCYTE SENSITISATION BY MACROPHAGE ELECTROPHORETIC MOBILITY (MEM) TEST

One of the most exciting claims made in recent years is that peripheral lymphocytes from patients with malignant neoplasms are sensitised to a basic protein derived from human brain (the encephalitogenic factor, EF) (Field and Caspary, 1970, 1972). Caspary and Field (1971) have argued that this observation suggests the possibility that malignant transformation results in the appearance of the EF determinant or a component antigenically similar to the EF. Carnegie *et al.* (1972, 1973a) have identified a water-soluble basic protein of MW ~16 000 in the membranes of tumour cells. Although this protein resembles EF in possessing a single tryptophan residue required for its antigenicity, the cancer basic protein is said to differ from EF (Caspary and Field, 1971; Caspary, 1972; Field *et al.*, 1971). The protein is described to have proteolytic properties and its sensitising ability is said to reside with a small fragment of the protein (Dickinson and Caspary, 1973).

The MEM test has shown that peripheral lymphocytes are sensitised to a variety of antigens such as a protein derivative of tubercle bacillus (PPD) (Carnegie *et al.*, 1973b), saline extracts of muscle and nerve from myasthenia gravis (Field *et al.*, 1973b) and basic protein derived from diseased human sciatic nerve (Field *et al.*, 1973a). In addition patients with systemic lupus erythematosis and ulcerative colitis show sensitisation to a number of unrelated antigens as well as to EF and tumour basic protein (Field, 1973). That myelin basic protein, tumour basic protein and PPD share their activity by virtue of possessing similar antigenic determinants has been suggested by a number of workers (Field *et al.*, 1971; McDermott *et al.*, 1974; Coates and Carnegie, 1975). On the other hand, Johns *et al.* (1973) found that histones, which are basic in nature, could also reduce macrophage mobility in MEM tests. These experiments have provided grounds for suspecting that the reduction in mobility may be a function of the electrical charge associated with the antigens. But Shaw *et al.* (1976) tested the effectiveness of EF, PPD and tumour basic protein with other proteins of comparable basicity and showed that the electric charges occurring on the antigens did not determine their activity in the MEM test.

An incubation of sensitised lymphocyte with EF or basic protein obtained from tumours, appears to release a factor (the macrophage slowing factor, MSF) that has the ability to reduce the electrophoretic mobility of macrophages isolated from peritoneal exudates of Hartley albino guinea-pigs.

The test is performed by suspending peripheral lymphocytes (from normal subjects or cancer patients) and guinea-pig macrophages in medium containing the antigen, namely the EF protein or the tumour basic protein (this will be referred to as the antigen). The suspension is introduced to the cell electrophoresis chamber and the EPM of macrophages is determined. Control macrophage EPM values are measured in a similar suspension in the absence of the antigen. In the presence of lymphocytes (sensitised) from cancer patients, the macrophages show decreases in EPM of 15–20% in tests with EF as antigen, compared with less than 5% reduction in EPM in the presence of lymphocytes from normal subjects. When tumour basic protein is used as antigen in the place of EF, an even greater reduction of macrophage EPMs is noticed when tests are performed with sensitised lymphocytes. Pritchard *et al.* (1972, 1973a) have not only verified the claims of Caspary and Field (1971) for the MEM test as an *in vitro* technique for screening for malignancy, but they have also provided indirect evidence that sensitised lymphocytes do release MSF into the medium. Thus, with EF maintained at a concentration of 100 μg ml^{-1}, varying the numbers of sensitised lymphocytes caused clear alterations in the loss of EPM by the macrophages, the larger the number of sensitised lymphocytes used, the greater was the loss in macrophage EPM. Pritchard *et al.* (1973b) have given a modified protocol for the MEM test.

On the other hand, in the case of patients with degenerative diseases of the central nervous system, the MEM test was more effective with EF as antigen rather than with tumour basic protein.

Caspary and Field (1971) claim that not only does the MEM test distinguish between normal subjects and those with malignant disease but that it can also be applied to early as well as advanced stages of neoplasia. There does not appear to be an overlap between loss of EPM in normal subjects and patients with neoplasms (Pritchard *et al.*, 1972; Field *et al.*, 1973; Preece and Light, 1974), and this implies that early stages of sensitisation may not be detected in the MEM tests (Pritchard *et al.*, 1973). In addition, a number of conditions such as advanced tuberculosis, asthma, sarcoidosis and systemic lupus erythematosis are known to interfere with the MEM diagnosis of cancer (Field *et al.*, 1973a). Field *et al.* (1974) suggested that the MEM test may be useful in the diagnosis of multiple sclerosis. It is now known that the MEM test is positive in a spectrum of diseases (Pritchard *et al.*, 1976).

The general applicability of the MEM test at the clinical level is in considerable doubt since Lewkonia *et al.* (1974) and Crozier *et al.* (1976) reported that no clear separation between subjects with malignant and non-malignant disease could be achieved using the test. Glaves *et al.* (1977) investigated 44 patients with adenocarcinomas of the colon and the rectum and 33 healthy subjects. In 54 out of 60 MEM tests involving 33 (ex 44) patients the MEM test was positive. The test was positive also in 12 out of 33 normal subjects. Glaves *et al.* (1977) obtained several false positive and false negative results despite the fact that the lymphocytes were isolated under standard conditions and the same EF was used. In addition, the measurement of mobilities was also standardised using a reference particle.

One important factor that may contribute to the inadequacy of the MEM test is the considerable variability in the EPMs of the macrophages. Therefore the selection of healthy guinea-pigs to serve as donors of macrophages is of considerable significance. Glaves *et al.* (1977) also stress the importance of using a standard reference particle while making mobility measurements. And finally, some of the variations in the test could be attributed to variations in encephalitogenic factor itself.

ASSAY OF CYSTIC FIBROSIS SERUM CILIARY INHIBITORY FACTOR

Serum and secretions from patients with cystic fibrosis are known to contain a factor(s) of cationic nature, which inhibit a variety of ciliary tissue. Khan *et al.* (1973) and Litt *et al.* (1974) have sought to quantify the levels of this factor(s) in patients by measuring the reductions in the ζ potential of negatively charged colloid particles coated with sera containing the cystic fibrosis factor. They have claimed that the method is able to distinguish sera of cystic fibrosis patients from sera of normal subjects. They have also claimed that levels of factor measured in this way correlates with the clinical status of the patients. However, serum proteins themselves will non-specifically and substantially coat the colloid particles, and therefore it is highly doubtful if the presence of an additional factor such as the cystic fibrosis factor will produce significant differences in the ζ potential.

NOTES ADDED IN PROOF

(See Pages 71–75) Cells from Peyer's patches are also distinguishable into two subpopulations of 66% with low electrophoretic mobility and 34% with a considerably higher EPM. If the B-cell population is depleted by the use of cyclophosphamide (5% sIg+), fewer low mobility cells are encountered in the Peyer's patch cells. On the other hand, in B-enriched populations (nylon adherent) of Peyer's patch cells (80% sIg+), 92% of the cells were found to have low EPM. In contrast, in nylon non-adherent and cyclophosphamide resistant Peyer's patch cells, the great majority formed the higher mobility type. And finally, in athymic nude mice 70% of the cells derived from Peyer's patches were sIg+ and mainly of the low mobility type (Dumont and Robert, 1977). These investigations of Dumont and Robert not only confirm the electrophoretic distinction between T- and B-cells but also support the view that Peyer's patches constitute a part of the lymphoid system equivalent to the bursa Fabricius and are involved in the processing of B-cells.

(See Pages 134–135) An area which might yield interesting valuable information is the "null-cell" acute lymphoblastic leukemia (ALL). Most childhood ALL cells have neither T- nor B-cell markers i.e. they neither form cytoadherence rosettes with sheep erythrocytes nor have they surface immunoglobulins. Hence, they are described as null cells. Although a number of haemic lines originating from patients with ALL are available, there is some doubt about the malignant origin of the cells. Most lymphoblastoid lines established from null-cell ALL express EB virus nuclear antigen (EBNA) and surface antigen positive i.e. of B-cell origin. But leukemic cells obtained directly from patients are EBNA negative. This may suggest that the former are derived from EBV-infected B-cells of the patients (Karpas *et al.*, 1977b). Karpas *et al.* (1977a) have described a null-cell leukemic line which is EBNA negative, which shows 100% positive staining with periodic acid schiff (PAS) reagent, as compared with 10% PAS-positive reaction from T-cell ALL (Karpas *et al.*, 1977b). Cell electrophoresis may provide a useful tool for the investigations into the origin of malignant leukemic cells.

5. Isoelectric Equilibrium Studies of Cell Surfaces

INTRODUCTION

NOTION OF ISOELECTRIC POINT OF CELL SURFACES

The basic concept of the structure of biological membranes proposed by Danielli and Davson (1935), and refined and restated by Singer and Nicolson (1972) includes two essential components, namely the lipid bilayer which confers certain physicochemical properties of biological membranes and proteins. Danielli and Davson (1935) suggested that the lipid bilayer is coated on both sides with protein, for the surface tension of membranes is lower than that is expected if the membrane surface were made of lipid alone. The proteins, it is believed, are immersed to different degree in the lipid and some may extend across the membrane. The proteins may move or rotate in the plane of the lipid bilayer (see pp. 3–5). Protein molecules are multivalent ions containing a number of positively and negatively charged groups belonging to the side chains of amino acid residues and to carbohydrate moieties of glycoproteins. These various groups ionise as a function of the ambient pH and consequently at a certain value of pH the molecule will contain an equal number of positive and negative charges i.e. bear no net electric charge. This pH is the isoelectric point (pI) of the protein and is a characteristic feature of the protein. The cell surface can, by analogy, be considered as a multivalent ion, with its own characteristic isoelectric point. To quote Abramson (1934) "The isoelectric point is a representative property of living and nonliving surface equilibria".

EARLY EXPERIMENTS

The importance of the surface pI as a determinate physical characteristic of cells was recognised early, but its potential value as a source of information of the physicochemical make up of the surface was not appreciated. Thus the early experiments have been limited to a demonstration of differences in the pIs of different cell types and to attempts at correlating the isoelectric characteristics of bacteria to their virulence.

As early as in 1914, Kozawa showed that the pIs of erythrocytes of different mammalian species were different and that they were characteristic of those species. Subsequently Coulter (1920), Eggerth (1924) and Netter (1925) determined the pIs of human, rabbit and horse erythrocytes and obtained values between 3·5 and 4·7. Joffe *et al.* (1933) found that rough and smooth strains of *Bacterium flexneri* and *S. typhii* (Joffe and Mudd, 1935; Dekker *et al.*, 1942) had different isoelectric points, while Thompson (1932) found differences in the pIs of three different strains of *Diplococcus* (see also pp. 81–83). *Haemophilus influenzae* and *Brucella aborta* were isoelectric at the low pH ranges of 2·8 and 2·7 respectively (Mudd, 1933).

Most of these investigations involved the addition of acid and lowering the ambient pH until the cells showed no electrophoretic mobility. In reviewing much of this early work Abramson (1934) has pointed out that these pI values may be erroneous, since the acid may have hydrolysed some of the surface components. Besides, Abramson (1934) stipulates that the isoelectric state should be an "electrical state in which the total sum of its positive and negative charges at the surface *over a time average* is equal to zero". Most of the above data depends on EPMs measured soon after the addition of acid, no measurements being made over a length of time. Most important, these are more accurately described as isoelectrophoretic points (pIE) rather than as true isoelectric points. There seems to be a considerable discrepancy between the pI and the pIE (see pp. 234–235).

ISOELECTRIC FOCUSING

The only method available until recently for the determination of isoelectric points of ampholytes, such as proteins and cells has been free electrophoresis. This involves the measurements of EPMs in a buffer whose pH is adjusted to cover a wide range. EPM measurements are made both before and after reversal of sign and hence the direction of EPM changes. From such an experiment, a pH-mobility relationship is plotted, from which the point of zero mobility is taken as the pIE. Isoelectric focusing (IEF), on the other hand, involves the imposition of a potential difference on an electrolyte system constituting an increasing pH from the anode to the cathode, the system being stabilised by a density gradient. An ampholyte loaded on to such a pH gradient will lose protons and acquire a negative charge if the ambient pH is higher than its pI, or conversely accept protons and become positively charged if its pI is higher than the ambient pH. Depending upon the nature of the charge on the surface, the ampholyte will show an anodic or cathodic mobility until it equilibrates at a pH where it bears no net charge and hence comes to rest. This method of isoelectric focusing has found important application in the analytical and preparative separation of proteins. Sherbet and Lakshmi (1969, 1972) and Sherbet *et al.* (1973) first adapted and applied the method to the study of cell surfaces and found that it has considerable potential use in the physicochemical characterisation of cell surfaces.

THE THEORY OF pH GRADIENTS

ARTIFICIAL pH GRADIENTS

Kolin (1954, 1955a,b) showed that a pH gradient can be formed using two buffer solutions of different pH which can be stabilised against convection by means of a density gradient. Such an artificial pH gradient could be used for the separation of proteins, by connecting the gradient to an electrical field. Such a system is not in equilibrium and hence liable to change in the course of experimentation. It has been called an artificial pH gradient system (Svensson, 1960, 1961; Kolin, 1958) which helps to distinguish it from natural pH gradients which are produced by the electric current itself (see below).

NATURAL pH GRADIENTS

A pH gradient can be created in an electrolysis cell, such as for example an electrolysis cell containing sodium sulphate in low concentrations, provided that there are no other ions than those of water that can undergo oxidation and reduction at the electrodes. With electrolysis in such a set-up hydrogen and oxygen evolve at the cathode and anode respectively as also sodium hydroxide collects at the cathode and sulphuric acid at the anode resulting in a partial or incomplete separation of the acid and base and the system therefore provides an unstable and unbuffered pH gradient. A gradient of this type generated by means of passage of electric current alone is termed a natural pH gradient. Ikeda and Suzuki (1912) described a method for separating glutamic acid from hydrolysates of plant proteins using the natural pH gradient created in a sodium sulphate electrolytic cell, a rectangular apparatus divided by membranes into three chambers. They achieved a coarse separation of the amino acids in groups. Glutamic acid was found to be localised near the anode, while the remaining amino acids were found arranged in accordance with their individual pIs. However, the system suffered from a number of disadvantages like electro-osmotic effects, convection, etc. In addition, amino acids have a low conductivity and have poor buffering ability around their pI values.

If a mixture of low molecular weight ampholytes with adequate buffering ability and conductivity are dissolved in an electrolyte system and a potential difference is applied to the system, the ampholytes will show free electrophoresis, and depending upon the nature and magnitude of the net electric charge they will move towards the anode or the cathode and occupy definitive positions in relation to one another which is dictated by their respective isoelectric points. Due to their buffering capacity the ampholytes thus distributed give an ambient pH which corresponds with their pI. If this system is stabilised against convection currents and hydrodynamic back flow, a natural pH gradient is obtained with pH decreasing from the cathode to the anode.

Natural pH gradients of any restricted pH range can be created provided that suitable ampholytes are available. In the sodium sulphate electrolyte cell the gradient ranges from pH 2 to 12. A gradient of pH 3–4 can be obtained using a mixture of picolinic acid (pI 3·16), glutamic acid (pI 3·23), nicotinic acid (pI 3·44), anthranilic acid (pI 3·53) and *m*-aminobenzoic acid (pI 3·93) (Svensson, 1961). Svensson has also pointed out the limitations on the use of certain ampholytes due to their inadequate conductance at the pI which results in excessive heating of the system. He has also pointed out that the non-availability of ampholyte to generate gradients of a variety of pH range. These problems have been tackled in a systematic manner and a mixture of low molecular weight ampholytes have been synthesised by Vesterberg and Svensson (1966) which appear to meet the basic requirements of generating the gradients.

An interesting recent development is the use of Hepes buffer for the generation of pH gradients (Rao, 1977). Satisfactory pH gradients have been generated by a linear mixing of heavy and light density gradient solutions, each containing 10 mM of Hepes buffer, their pH adjusted to the range required. From the description given by Rao (1977) the generation of the gradient takes the same course as with ampholines, and a suitable pH gradient may be formed in a LKB 8101 Column in 4–5 hours.

Hepes buffer, which is *N*-2-hydroxyethylpiperazine-*N*′-2-ethanesulphonic acid, was first described by Good *et al.* (1966). The molecule behaves as a zwitterion. However, the mechanism by which the pH gradient is generated is uncertain. The initial mixing of the two density gradient solutions itself will provide a fairly linear pH gradient. Artificial pH gradients can be obtained by such a linear mixing of buffers but they will not be in equilibrium. In the Hepes system, however, Rao has described a gradual reduction in the conductivity, which, by analogy with the ampholine system, may indicate a progressive formation of a pH gradient.

Rao (1977) has used the Hepes system for the determination of the pI of haemoglobin. The work on this system is at an early stage and much needs to be done on the mechanics of the generation of the pH gradient. One can see a number of advantages with the system, the more important ones being: Hepes buffer would be more economical than ampholine and the Hepes buffer system is non-toxic up to about 30 mM. It ought to be borne in mind in this connexion that it is very sensitive to changes in temperature ($\Delta pK_a/°C = 0{\cdot}014$, Good *et al.*, 1966). In addition, its pK_a (the pH at the midpoint in the buffering zone) at 20°C is 7·55, and therefore probably it is not suited for the study of cell surfaces. However, the success which Rao (1977) has achieved indicates that exploration of other organic buffers in this way may be worthwhile.

ESSENTIAL PROPERTIES OF CARRIER AMPHOLYTES

The minimum essential properties of ampholytes suitable for the generation of pH gradients have been discussed in a number of papers (Svensson 1961; Haglund, 1970, 1975; Wadström *et al.*, 1973) and are summarised in Table 30.

Table 30 *Minimum essential properties of carrier ampholytes*

Good conductivity at pI
Good buffering ability
Not form complexes with sample under investigation
Separable from sample
Non-toxic
Good solubility
Not react with supporting density media

CONDUCTIVITY OF AMPHOLYTES

Let us consider a diprotic ampholyte a which has a zwitterionic and unprotolysed form represented as a_0, the anionic and cationic forms a^- and a^+ respectively. The equilibrium of the system may be represented as follows:

$$a_0 \rightleftharpoons a^- + H^+ \tag{32}$$

$$a_0 + H^+ \rightleftharpoons a^+. \tag{33}$$

Ignoring the activity coefficient one may write the following equations:

$$\frac{[a^-][H^+]}{[a_0]} = K_1, \qquad \text{i.e. } [a^-] = [a^0] \times 10^{(\log K_1 - \log [H^+])} \tag{34}$$

and

$$\frac{[a_o][H^+]}{[a^+]} = K_2, \qquad \text{i.e. } [a^+] = [a^0] \times 10^{(\log [H^+] - \log K_2)} \tag{35}$$

where $[a^-]$, $[a^+]$, $[a_0]$ and $[H^+]$ represent molar concentrations of a^-, a^+, a_0 and H^+ respectively. Substituting $-\log K_1$ and $-\log K_2$ with pK_1 and pK_2 and substituting pH for $-\log [H^+]$, eqns (34) and (35) may be rewritten as:

$$[a^-] = a_0 \times 10^{(pH - pK_1)} \tag{36}$$

$$[a^+] = a_0 \times 10^{(pK_2 - pH)}. \tag{37}$$

The degree of protolysis (α) can be given as:

$$\alpha = \frac{[a^+] + [a^-]}{[a]} \tag{38}$$

where

$$[a] = [a_0] + [a^-] + [a^+] \tag{39}$$

Dividing eqn (39) by $[a^-]$, substituting values for $[a^-]$ and $[a^+]$ from eqns (36) and (37) and inserting pI for $(pK_1 + pK_2)/2$:

$$\frac{[a]}{[a^-]} = 1 + 10^{(pK_1 - pH)} + 10^{2(pI - pH)}. \tag{40}$$

Now dividing eqn (39) by $[a^+]$, substituting values for $[a^-]$ and $[a^+]$ once again from eqns (36) and (37) and inserting pI in place of $(pK_1 + pK_2)/2$:

$$\frac{[a]}{[a^+]} = 1 + 10^{(pH - pK_2)} + 10^{2(pH - pI)}. \tag{41}$$

Having eliminated the factor $[a_0]$, and since in the isoionic state pH = pI:

$$\alpha = \frac{2}{2+10^{(pI-pK_2)}}. \tag{42}$$

In other words, smaller the difference between pK_2 and pI the greater is the conductivity of the ampholyte, since the degree of protolysis is directly related to conductivity, and in polyprotic ampholytes high conductivity at isoionic and isoelectric point is dependent upon how close to the pI are the pair of pK values on either side of the pI of the ampholyte (Haglund, 1970).

BUFFERING ABILITY

The buffering ability of ampholytes:

$$\frac{dQ}{d(pH)} = \frac{2 \log_e 10}{2+10^{(pI-pK_2)}} \tag{43}$$

where Q is the electric charge of the ampholyte. This shows that, just as in the case of conductivity, the smaller the difference between pI and pK_2 the greater is the buffering ability of the ampholyte at its isoelectric pH (see Fig. 20) (Svensson, 1962).

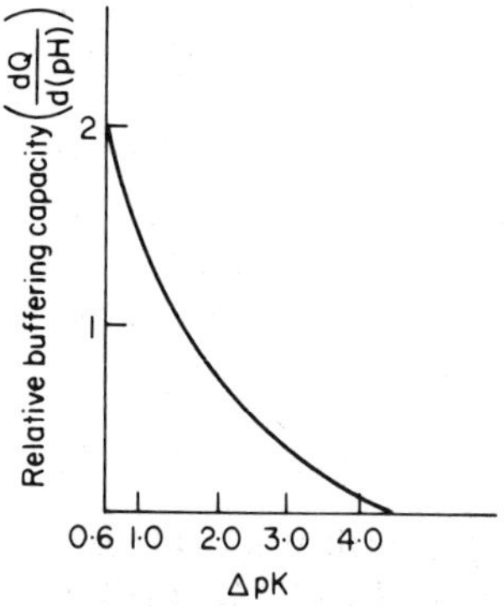

Fig. 20. Relationship between buffering ability of ampholytes and the difference between their pI and pK_2 (from Haglund, 1970).

FORMATION OF COMPLEX WITH SAMPLE

It is essential that the ampholyte should not form complexes with the sample under investigation nor with the supporting density gradient material. In either case, isoelectric focusing will fail due to failure of the sample to equilibrate at its pI or due to inadequate pH gradient. It appears that ampholines (LKB, Bromma, Sweden) do not form complexes with protein sample; even if they did, the process is in a reversible equlibrium. Vesterberg (1969a) showed this using ^{14}C-labelled ampholines. Vesterberg in fact achieved a complete separation of ampholines and the basic enzyme egg white lysozyme. According to Salton (1957) lysozyme and other basic proteins bind acidic substances especially firmly. This provides strong argument against a firm complex formation between ampholine molecules and proteins. Dean and Messner (1975) performed experiments with purified proteins such as albumin, ferritin and β-glucuronidase and mixtures of serum proteins and ^{14}C-labelled ampholytes. In none of these experiments binding between ampholytes and proteins was detected.

It is also known that ampholines do not denature proteins (other than those which are unstable at their pI). On the contrary, Wadström and Hisatsune (1970) and Wadström (1970) found ampholines actually stabilised proteins such as *Staphylococcus aureus* hexosaminidase. Although there is no direct evidence that ampholines do not bind to cell surfaces, it appears reasonable to assume this in view of the evidence that they do not form complexes with protein.

SEPARATION OF AMPHOLYTES FROM SAMPLE

The separation of ampholines from proteins following isoelectric focusing has turned out to be a simple procedure. Molecular sieving on Sephadex for the separation of high and low molecular weight substances (see Determan, 1967; Fischer, 1969) has been employed for the separation of ampholines from human serum albumin, γ-globulin and egg white lysozyme (Vesterberg, 1969). Nilsson *et al.* (1970) used repeated precipitation of proteins by ammonium sulpate.

TOXICITY OF AMPHOLINES

Very little work appears to have been done on the biological effects of ampholines. This is especially important when dealing with cells in isoelectric equilibrium studies. Ampholines produced no toxic effects when administered intravenously to rabbits. New born mice which were injected intracerebrally with ampholines showed no neurological symptoms. Neither did ampholines produce any cytotoxic effects on tissue culture cells (Wadström *et al.*, 1970). Sherbet *et al.* (1972) recovered HeLa cells after isoelectric focusing for 2–3 hours, and inoculated them on to plastic petri dishes. A plating efficiency of 55–75% was observed. This indicates that ampholines produced minimal toxic effects on HeLa cells. Longton *et al.* (1974) successfully grew 30-day fetal rabbit cells which had been subjected to isoelectric focusing. After an initial period of dormancy, the cells grew to form monolayers. However, the isoelectric pattern described by Longton *et al.* casts considerable doubt on their observations on the post-focusing culture of cells. They observed three cellular peaks in their focusing experiments, of which two were in the alkaline range. All cell types so far examined are known to have negative surface potentials and should therefore be isoelectric in the acid range of pH.

CHEMICAL AND PHYSICAL PROPERTIES OF AMPHOLINES®

It was mentioned earlier that Vesterberg and Svensson (1966) have synthesised a mixture of low molecular weight ampholytes that meet these requirements. Earlier Svensson (1962) showed that haemoglobin can be hydrolysed to yield a mixture of peptides which could be used as carrier ampholytes to create a pH gradient of well defined range. Such a gradient was, however, found unsuitable for the separation of proteins, for the proteins in haemoglobin hydrolysate resembled the proteins being separated. Further, the conductivity of the pH gradient varied from region to region.

The ampholytes synthesised by Vesterberg and Svensson (1966) are a mixture of aliphatic polyamino-polycarboxylic acids represented by the following general formula:

$$\begin{array}{ccccccc} -CH_2- & N & -(CH_2)_n- & N & -(CH_2)_n-NR_2 \\ & | & & | & \\ & (CH_2)_n & & R & \\ & | & & & \\ & NR_2 & & & \end{array}$$

where $n = 2$ or 3 and $R = H$ or $-CH_2-CH_2-COOH$. These ampholytes are synthesised by coupling acrylic acid to a mixture of polyethylene–polyamines under conditions where no amide bonds are formed. Acrylic acid is coupled to the nitrogen atom of the amine. The numbers of carboxylic groups introduced can be controlled by altering the ratio of acrylic acid to amine (Vesterberg, 1969b). The resultant mixture contains a variety of compounds with different isoelectric points ranging between pH 3 and 10, and molecular weights ranging between 300 and 1000 (but see Haglund, 1975). This mixture is then fractionated using a multichambered apparatus, with the chambers separated by membranes of polyvinylchloride paper, and which has platinum electrodes attached to the ends. When an electric current is passed through the mixture for about 24 hours the mixture of ampholytes separates out into the different chambers, with the most acidic ampholytes being found in the chambers near the anode while the more basic ones are found closer to the cathode. Using this technique ampholine mixtures covering smaller ranges such as pH 3–6, 4–7, 7–10, etc. could be prepared. Such a fractionation can also be achieved using Valmet's (1969) zone convection electrofocusing.

The carrier ampholytes are similar to amino acids in peptide linkage, but differ from peptides in the absence of actual peptide bonds. They have been found to react with ninhydrin as do proteins. They also are positive in the biuret and Folin tests (Davies, 1969). They show low absorbance in the ultraviolet range especially at 280 nm. This property is very useful especially when electrofocusing is used for the characterisation and separation of proteins.

The conductance and buffering ability of ampholine carrier ampholytes are minimum when their pI is pH 6–7 and rise with increase or decrease of pI from pH 7. Viscosity changes along the pH gradient also affect conductance, the conductance being lower the higher the viscosity. On the other hand, the supporting density gradient such as sucrose also affects the buffering capacity, for at pH ranges above 9 sucrose has appreciable buffering properties (Davies, 1969). Variations in conductance along the pH gradient are an important factor since this implies variations in the heat dissipation along the pH gradient resulting in thermal convection. Besides, conductance determines the distribution of field strength which in turn determines the degree of focusing. Svensson (1961) in fact showed that the degree of focusing is proportional to the square root of the field strength.

The carrier ampholytes are now supplied under the commercial name of ampholine® by LKB-Produktor AB, Sweden, as a 40% sterile solution covering a general wide range and several ranges encompassing two pH units.

CHOICE OF AMPHOLINE RANGE

In the new ampholine programme, LKB-Produktor AB offer sterile 40% w/v solution of carrier ampholytes in the following pH ranges: 3·5–10, 2·5–4, 3·5–5, 4–6, 5–7, 5–8, 6–8, 7–9, 9–9·5 and 9–11. Although these various ranges are extremely useful in the characterisation and purification of proteins, not all the ranges are useful in the study of cell surfaces. Since cells are negatively charged at the physiological pH range, ampholine below pH 7 are generally used. The choice of range depends upon the isoelectric point of the cell, the nature of the experiments i.e. whether pI of the surface is being determined in its native state or whether this is done following enzymic excision or chemical modification of the surface groups. The most suitable approach is to determine the pI of the cell using a pH gradient of 3·5–10, and to choose a restricted pH gradient range depending upon the pI value obtained in the preliminary experiment, in order to make a more accurate determination of pI. As an example one may cite the work of Sherbet and Lakshmi (1973) on *E. coli* cells, which has been discussed in later sections (pp. 185–188). These cells were found to have a pI of 5·6. This was determined using a 3·5–10 pH gradient. When the carboxylic groups present on the surface of these cells were modified resulting in a purely amino surface the pI was found to increase to 8·5. After preliminary experiments in a 3·5–10 gradient, accurate determinations were made using pH 7–10 gradient. Similarly, modification of the amino groups caused a reduction of the pI to ~3·85. When such a low pI was indicated in initial experiments with the 3·5–10 range, further experiments were performed with ampholines of pH range 3–6.

Exercising a choice of narrow pH range in this way has also the additional advantage that the cells are not loaded at ambient pH ranges far different from their isoelectric points, which may adversely affect their viability, and the time required for equilibration of the cells can be kept to a minimum, contributing to the viability of the cells.

CONCENTRATION OF AMPHOLINES

The ampholines provide the ionic environment for isoelectric focusing. In the study of proteins using this method it is necessary to keep the ionic concentration at the optimal level, since the ionic concentration contributes to the solubility and stability of the proteins. Isoelectric point, incidentally, is also the pH in which proteins show the least solubility and therefore it is not surprising that Valmet (1968) found that the precipitation of γ-globulins in isoelectric equilibrium could be controlled by altering the concentration of ampholine employed in the experiments. Valmet found that the proteins precipitated at a concentration of 0·25 and 0·5%. At 1% ampholine, protein precipitation was still observed, but at 2% precipitation was found to be minimal. A working concentration of 1% ampholine is generally considered suitable for most isoelectric focusing. Ionic concentrations are an important consideration in the isoelectric studies with cells, for the ionic atmosphere influences the surface potential of the particle (see pp. 45–47). Although the concept of ionic concentration and double electrical layer assumes a different significance as far as isoelectric equilibrium studies are concerned, it will be seen that the ionic concentration enters into the equation which has been evolved for the calculation of electric charges, numbers of ionisable groups, etc. It is therefore imperative that while describing isoelectric data of cells, proteins, etc. and any other quantitative data derived thereof, the ionic conditions are defined along with other conditions of experiment such as the temperature at which focusing is performed, etc.

ISOELECTRIC FOCUSING EQUIPMENT

PHYSICAL DESCRIPTION OF LKB FOCUSING COLUMN

The LKB isoelectric focusing column (Fig. 21) consists mainly of two components, namely one, an outer column (1) in the wall of which a water jacket is built in (2). Both its upper and lower ends have ground joints. The bottom ground glass joint takes a plug assembly made of teflon and stainless steel (3). A capillary tubing (4) is connected to a channel in the teflon plug. The column is drained through this capillary tubing and the bottom plug assembly. Two, an inner column which has a central tube and the wall has a water jacket built in (5). The inner column fits into the interior of the outer column. A part of the upper half of the inner column is ground to fit the upper ground joint of the outer column. The water jackets of the outer and inner columns are interconnected by rubber tubing (6). The central tube of the inner column can be shut or opened by a teflon valve and gasket (7) which is connected to a teflon rod which runs the whole length of the inner column and is attached to a spring loaded cylindrical teflon block (8). When this teflon block is depressed the valve (7) opens. The valve can be left in open position by depressing the spring loaded teflon block and locking it in the depressed position. A platinum wire is wound round the teflon rod and this is attached at the top of the inner column to an electrode terminal (9).

The electrofocusing chamber which is formed when the inner column is fitted into the outer column is essentially the annular compartment delimited by the inner wall of the outer column and the outer wall of the inner column. At the top of this focusing chamber a loop of platinum wire is located which in fact goes around the circumference of the outer

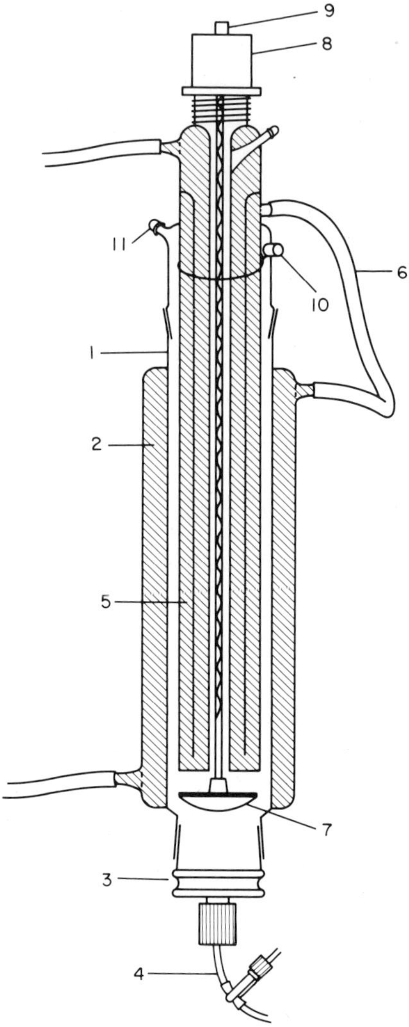

Fig. 21. Isoelectrofocusing column LKB 8101. (1) Outer column, (2) water jacket of the outer column, (3) bottom plug assembly with connected capillary tubing (4) by means of which the column is drained, (5) water jacket of the inner column, (6) connecting tube between water jackets of outer and inner columns, (7) teflon valve and gasket to shut off the central electrode chamber, (8) spring loaded teflon block connected to the central electrode. The teflon block can be locked in the depressed position to keep the central electrode chamber open at (7) and provide continuity with the focusing chamber. The focusing chamber is connected to the outside by a nipple (11) through which the column is layered with density gradient, (9) electrode terminal connected to the central electrode, (10) second electrode terminal connected to circular platinum electrode wound around the outer wall of the inner column.

wall of the inner column, and is connected to the second electrode terminal (10). The focusing chamber opens to the outside via a nipple (11) which is used to layer density gradient, ampholines and sample. The LKB instruction manual which accompanies the focusing column gives a more complete physical description. The schematic diagram of the column given on p. 158 is based on this manual which also contains another diagram that gives the intricate structure of this piece of equipment.

MODIFICATION OF LKB COLUMN FOR POST-pH EQUILIBRIUM LOADING OF SAMPLE

It is evident from the above description that the sample under investigation has to be loaded at the beginning of the experiment i.e. in the course of the layering of the supporting density gradient. The formation of a stable pH gradient requires 24–72 hours depending upon the pH range of the ampholine chosen. While this prolonged experimentation has no obvious detrimental effects on proteins, loading of cells at the beginning and allowing them to remain in the column for such extraordinarily long durations, was found to be a most unsuitable procedure. A side arm attachment (Fig. 22) was therefore devised for the LKB equipment, which connects the focusing chamber to the exterior, about the middle of the column. This side arm is sealed at the end by a special rubber septum, which consists of a layer of silicone rubber sandwiched between two more layers of ordinary rubber. The samples can be introduced through this specialised loading port. The procedure of loading is

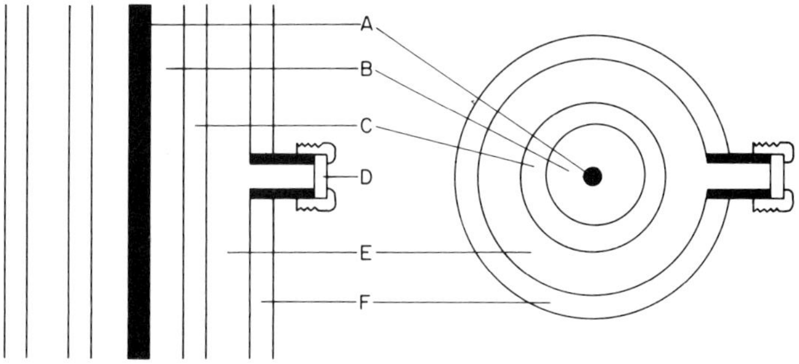

Fig. 22. Post-pH loading device in modified isoelectric focusing column, vertical view and sectional view passing through the post-loading port. (A) Inner electrode, (B) inner electrode chamber, (C) inner water jacket, (D) side arm sealed by septum for post-pH loading, (E) electrofocusing chamber and (F) outer water jacket (modifications by Sherbet and Lakshmi, unpublished work).

described fully elsewhere in this chapter. The importance of this modification is obvious, since we have here a means by which the sample could be introduced after the pH gradient is formed and this enables one to minimise the time required for equilibration of the sample.

MICROANALYTICAL ISOELECTRIC FOCUSING

The LKB column described above is made in two sizes. The smaller column (LKB-8101) has a capacity of 110 ml and the larger column 400 ml. These columns are eminently suitable for preparative isoelectric focusing. For the characterisation of small samples of proteins and for rapid investigation of a large number of samples, however, the LKB column is not suitable. These limitations have inspired attempts at scaling down the focusing equipment to suit microanalytical procedures. The use of miniature-scale focusing equipment has also the additional advantage that the running costs are considerably lower.

Koch and Backx (1969) have described a 12-ml capacity U-shaped focusing assembly. One of the arms 34 cm long and 5-6 mm internal diameter serves as the focusing chamber which is connected by a silicone rubber tubing to the other arm which serves as an electrode chamber. The second electrode is inserted at the top of the focusing arm. Koch and Backx (1969) analysed haemoglobin from *Salmo salar* in 1–10 μg quantities. Osterman (1970) and Godson (1970) have described even smaller focusing columns of capacity of approximately 10 ml. Osterman's column is built using generally the same principle as the LKB column. On the other hand, Godson (1970) used a J-shaped column for rapid focusing of protein samples. Besides, Godson's columns were made to fit into standard polyacrylamide gel electrophoresis equipment so that a large number of columns could be run at one time.

While the volume of the focusing chamber in these designs is between 7 and 14 ml, Jonsson *et al.* (1973) demonstrated that a commercial standard spectrophotometric flow-through cell (Hellma type 167-QI) of capacity 1·7 ml can be used as microanalytical IEF column. This cell is equipped with a mantle for cirulating the coolant and generally meets the requirements of an IEF column. Not only could samples as small as 100 μg be analysed, but the experiments could be completed in two hours. The cell could be set up in an optical scanning device and the absorptions at a given wavelength recorded at intervals so that the separation and the progress of the experiment could be monitored. In the original arrangement platinum electrodes were used arranged as described in IEF-M3 column (see below). The volume between the electrodes i.e. the effective capacity of the focusing chamber is 1·5 ml.

In order to reduce the focusing time from two hours to 90 minutes by increasing the electrical load, Fredriksson (1975) has used non-gassing electrodes made of palladium (Neihof and Schuldiner, 1960) 0·1 mm thick to cover the bottom of the focusing chamber to serve as cathode. Using this modified microanalytical column and heavy water to form supporting gradient, Fredriksson has been able to cut the time of experimentation to a mere 75 minutes. Degassing of the palladium electrode is achieved by leaving the cell empty overnight or by rapidly running the palladium sheet as anode.

Most of our initial investigations on cells were made using the LKB-8101 column in its original form and in its modified form (IEF-PPHE column) for post-pH loading. While these columns were suitable in the investigations with established cell lines such as HeLa, Ehrlich's ascites and others where one can culture large quantities of cells, a need was felt not only to scale down the experiments as in the study of the antigenic expression of normal and transformed cells and in the study of surface properties of cells of human astrocytomas of various grades of malignancy, but also to incorporate the post-pH equilibrium (PPHE) loading of the cells in order to improve viability and to keep to the minimum alterations in surface conformation which may result from isoelectric focusing procedures.

A perspex column with a 15-ml electrofocusing chamber fitted with platinum electrodes and with PPHE loading port was constructed (Fig. 23) and this (designated as IEF-M3) is routinely used in our laboratory and has been for the past five years. The apparatus consists of two vertical chambers opening at the top and communicating at the bottom by a narrow channel which can be obliterated by a needle valve. One compartment is of small volume and acts as the chamber for the first electrode while the second chamber, which may be of varying capacity, serves as the focusing chamber into which the second electrode is inserted at the upper end, and which is drained through a tap at the lower end. The specimen-loading port opens into the focusing chamber and is sealed with a septum (see Fig. 22) which can be pierced by a fine needle to allow the withdrawal and introduction of material.

DENSITY GRADIENT

For the stabilisation of the pH gradients against convection currents actuated by heat dissipation in the generation of pH gradient, and also for the stabilisation of protein and cellular zones, a density gradient of non-ionic solute is used. In general the solute must meet the following

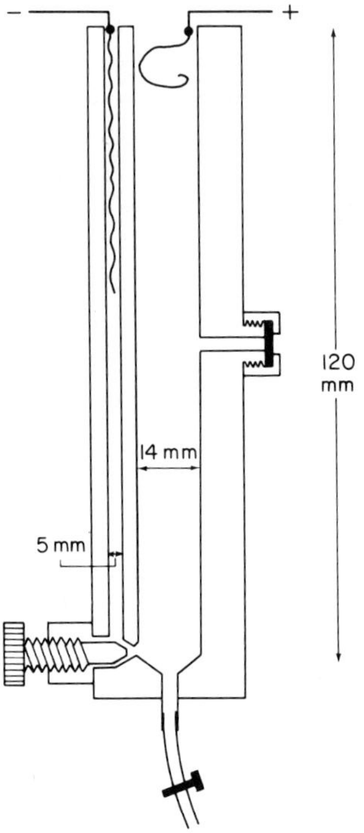

Fig. 23. IEF-M3 focusing column for cells with post-pH loading device (Sherbet and Lakshmi unpublished design).

requirements enumerated by Haglund (1970): good solubility and low viscosity in water; density should be high enough to provide a density difference of $0{\cdot}12\,g\,cm^{-3}$; solute should inert i.e. not interact with sample proteins or cells and solute should be non-ionic.

In isoelectric focusing of cells some additional criteria need to be satisfied. For example, the solute should not exert untoward osmotic effects on the cells, nor affect their viability, and ideally the solute should not be antigenic. It is also imperative that the solute should not affect the aggregation of the cells. Perhaps it is impossible to find a solute for density gradient, which meets all these conditions. A partial compromise with some of the criteria at least seems inevitable.

In our laboratory the suitability of glycerol, polyethylene glycol 6000, sucrose and ficoll which is a synthetic polymer made by the co-polymerisation of sucrose and epichlorohydrin, has been investigated. Glycerol, while quite inexpensive, needs to be used as a 60–10% concentration gradient in order to obtain a suitable density range. A 60–10% concentration range would give a density range of 1·15–1·02. At this concentration range, especially around the 30–60% region, glycerol is too viscous to be suitable for cell focusing.

Polyethylene glycol (PEG) of molecular weight 6000 was found to be toxic to cells. It also caused cellular aggegation and disruption. The cellular disruption appears to be due to the high osmolarity of the supporting gradient. Loos and Roos (1974) found a non-linear relationship between concentration and osmolarity of PEG. A solution of PEG with a density of $1{\cdot}09\ g\ ml^{-1}$ far exceeded the osmolarity of 290 mOsm.

We have used sucrose gradients of concentration range 50–5% routinely, and have found it second only to Ficoll (Pharmacia) as to its suitability in the focusing of cells. Cell viability was usually reasonably high and varied between 70 and 80% (Sherbet *et al.*, 1972). Sucrose is highly soluble in water and non-ionic except at high pH ranges. Not only has it proved harmless but even positively protective as far as proteins are concerned (Fredriksson, 1975). There is also the additional advantage that sucrose is available in grades of high purity. However, Fredriksson (1975) has also pointed out that the viscosity of sucrose solution increases greatly at higher concentrations. This causes an unfavourable distribution of field strength along the gradient and alters the focusing time at a given voltage. This latter criticism can also be made against glycerol and Ficoll density gradients.

Ficoll which is a high molecular weight polymer of sucrose and epichlorohydrin could be ranked with sucrose as far as cell electrofocusing is concerned. The viability of cells following focusing in Ficoll gradients was marginally superior than where sucrose had been used. HeLa cells showed plating efficiencies of up to 75% after recovery from electrofocusing using Ficoll density gradients (Sherbet et al., 1972). Ficoll and sucrose are comparable as regards the increase of density of solutions as a function of concentration. Ficoll has low osmolarity at low concentrations, but osmolarity increases exponentially with increase of concentration (Williams *et al.*, 1972; Bach and Brashler, 1970) so much so that an isotonic Ficoll solution requires a concentration of 45% w/w and has a density of $1{\cdot}15\ g\ ml^{-1}$ (Loos and Roos, 1974). A concentration gradient of 50–10%, which would appear appropriate in the light of the above data, results in an unfavourable viscosity distribution, for, according to the data given by Pharmacia, viscosity increases exponentially with increase of concentration from 10 to 50%. As mentioned earlier this phenomenon causes an asymmetric distribution of field strength along the column. In spite of the low density increases, we have used Ficoll gradients of 5–25% (w/v). Although Ficoll solutions of up to 50% (w/v) can be achieved, the preparation of solutions is not as easy with sucrose.

It is probably worthwhile to point out that a mixture of Ficoll and sucrose or other non-ionic material may be adopted with advantage to make supporting density gradients for electrofocusing of cells.

Fredriksson (1975) has used deuterium oxide (D_2O, heavy water) to form density gradients. Heavy water gives a small ($0{\cdot}1$ g cm^{-3}) increment in density which falls a little short of the minimum density increment stipulated as ideal by Haglund (1970), but is still adequate when one employs a miniature focusing column of 36 mm which Fredriksson uses. The unquestionable advantage appears to be the fact that the viscosity of D_2O is about 50% lower than that of an aqueous sucrose solution of equivalent density. Fredriksson (1975) further expatiates on the use of heavy water that being chemically akin to ordinary water it may exert minimum harmful effects on proteins. This probably holds true for cells. Further, heavy water is available in a highly purified form.

The suitability of heavy water in the isoelectric focusing of cells is a matter which requires investigation. Suitable additional components may be desirable in order to obtain a fairly appropriate range of osmolarity. The use of D_2O is also likely to affect the measured pI values of cells, for in D_2O the pK values of protolytic groups is increased with the magnitude of the changes being dependent upon the type of group and on its pK in ordinary water (Li *et al.*, 1961). The discussion in this subsection has been summarised in Table 31.

Isoelectric focusing has been carried out using polyacrylamide gels as stabilising medium for the pH gradient, especially for the separation of biological macromolecules (Wrigley, 1968a,b,c; Fawcett, 1968, 1969; Dale ad Latner, 1968, 1969a,b; Leaback and Rutter, 1968; Leaback *et al.*, 1969). Particle electrofocusing has not been performed to any degree using gels. Rice and Horst (1972) performed some experiments with viruses by polyacrylamide gel isoelectric focusing in columns. Haglund (1970) has listed a number of advantages in using polyacrylamide gel for stabilising pH gradients. One most obvious difficulty is the determination of pH along the column. The method adopted by Rice and Horst (1972) is to run a parallel control gel under identical conditions but without the sample. While the gel which contains the material under investigation is fixed and stained to reveal the focused bands, the control gel is sliced into 2-mm-thick slices, each of which is eluted with 1 ml of water at room temperature for two hours and its pH measured. Rice and Horst (1972) considered 3% polyacrylamide gels as suitable for investigations with small spherical viruses. Lowering of the gel concentration increases the median pore size of the gel but the gels become increasingly more difficult to handle. Although polyacrylamide gel electrophoresis (PAGE) has been used to characterise biological

Table 31 *Suitability of various solutes for density gradient in isoelectric focusing of cells*

Solute	Concentration range (%)	Density range	Osmolarity	Viscosity	Effect on cells
Polyethylene glycol MW 6000	40–10		High, non-linear with concentration		Toxic, caused cellular aggregation and disruption
Glycerol	60–10	1·15–1·02		Too high in lower half of gradient	No effect on cell pI
Sucrose	50–5	1·23–1·01		Too high at high concentrations	Good viability, no effect on cell pI
Ficoll	25–5	1·095–1·015	Low	Low viscosity at 25%	Good viability, no effects on cell pI
Heavy water (not tested in cell IEF)	100–0	1·106–1·0	Low (?)	Less than sucrose gradient of comparable density	Increase of pK of protolytic groups

All data except concerning heavy water from published or unpublished work from author's own laboratory. Data about heavy water only pertains to IEF of β-lactoglobulins (Fredriksson, 1975).

macromolecules, the mechanism by which PAGE produces its resolution is probably not exclusively by its sieving effect. It has been suggested, for instance, that weak interactions may be involved. Polyacrylamide fibres are formed by additive polymerisation through C:C with amide groups projecting from the carbon chain. It is possible that the polar amide groups enter into hydrogen bonding with polar groups in solution or those occurring in the macromolecule being investigated or on the surface of the particle being examined.

ELECTRODE SOLUTIONS

The ampholytes are prevented from oxidation at the anode or reduction at the cathode by means of anode and cathode lock solutions. In our laboratory we employ a 0·5% orthophoric acid solution in distilled water as anode solutions, and a 2% ethanolamine solution in dense gradient solution as cathode solution. These electrode solutions are suitable for use with a 3–10 gradient. It should be pointed out, however, that the electrode solution should be chosen to give a pH not substantially different from the lowest and highest pH of the proposed pH gradient. Even ampholines themselves could be used as electrode solutions. For ampholine ranges applicable to study of isoelectric behaviour of cells, choice of anode and cathode solutions are suggested in Table 32.

Table 32 *Suggested anode and cathode solutions for isoelectric equilibrium studies with cells*

Ampholine range	Anode solution	Cathode solution
pH 3–10	Orthophosphoric acid (0·5%)	Ethanolamine (2%) Sodium hydroxide (6%)
pH 3–6	Orthophosphoric acid	Ampholine 7–10
pH 4–7	Orthophosphoric acid Ampholine 2·5–4·0	Ampholine 7–10
pH 5–8	Ampholine 2·5–4·0	Sodium hydroxide Ampholine 9–11

PREPARATION OF DENSITY GRADIENT

Density gradients could be layered manually which is a laborious and time consuming procedure, or using an appropriate gradient mixer. A gradient mixer not only saves labour but gives a reproducible and an immediately linear density gradient. For LKB and IEF-PPHE focusing columns of volume 110 ml, a gradient mixer such as the LKB-8121 is suitable. For the smaller IEF-M3 column, we use a simple gradient mixer designed and constructed in our laboratory (Fig. 24). The mixer which is

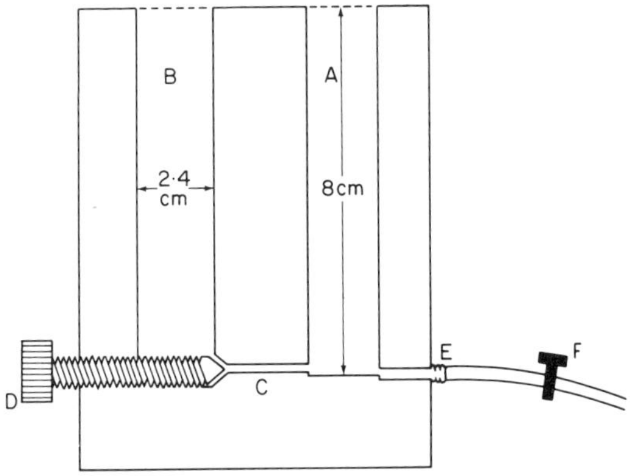

Fig. 24. Gradient mixer to generate linear density gradients of volume 10–30 ml (constructed by L. M. Bowler).

made of perspex consists of two chambers (A and B) of diameter 2·4 cm and height 8 cm. They are interconnected by a narrow channel (C) which can be obliterated by needle valve (D). The chamber A opens to the outside by aperture (E). The dense solution is poured into chamber A and the light solution into chamber B, with the connecting channel in the closed position. Appropriate amounts of ampholines are measured out into the chambers. A magnetic flea is introduced into chamber A and adjusted to rotate at a suitable speed. The whole assembly is placed on a magnetic stirrer. The outlet (E) is connected to a peristaltic pump. The interconnecting channel is now opened and chamber is drained with the aid of the peristaltic pump. A linear gradient of any volume between 10 and 30 ml can be prepared using this simple gradient mixer. The speed of layering the gradient should be no more than about 1 ml min^{-1}. Table 33 gives the volumes of electrode solutions, and the volumes of dense and light solutions and the appropriate quantities of ampholines required for LKB-8101, IEF-PPHE and IEF-M3 columns.

The general procedure adopted in our laboratory is to fill the focusing column with the bottom electrode solution with the teflon valve and gasket of the LKB-8101 or the needle valve of the IEF-M3 column in the open position thus establishing communication between the electrode chamber and the focusing chamber. The density gradient is now layered

Table 33 *Volumes of electrode solutions and density gradient components for isoelectric focusing of cells*

Column	Electrode solution (bottom) (ml)	Density gradient Heavy solution (ml)	Density gradient Light solution (ml)	Ampholines (40%) Heavy solution (ml)	Ampholines (40%) Light solution (ml)
LKB-8101 } IEF-PPHE }	20	55	55	1·9	0·6
IEF-M3	1·0	7·0	8·2	0·25	0·1

and this pushes the electrode solution into its own chamber. When the column is filled, it is topped up with the upper electrode solution in sufficient quantity so as to give contact with the upper electrode.

GENERATION OF THE pH GRADIENT

CHOICE OF POLARITY

This is dictated purely by convenience and there are no theoretical reasons to stipulate the direction of flow of current. Nevertheless, Haglund (1970) recommends that cathodes be located at the top while working with pH gradients below 6 and anode at the top for gradients above pH 6. This is suggested because at pH values above 7, for example, the conductivity increases. When the pH gradient is formed therefore there is greater heat dissipation in pH range 5–7 than in the higher range. In spite of this recommendation, it is worthwhile to confirm pI measurements by changing of polarity.

VOLTAGE REQUIREMENTS

A potential difference is applied to the column using a DC power pack. In LKB-8101 and IEF-PPHE column a current flow of 3 mA is maintained, while in IEF-M3 column the current flow is restricted to 2 mA. The voltage required to achieve this level of current flow depends upon the nature of the density gradient material, the range of the density gradient used, the pH range of the ampholines employed, etc. Table 34 summarises the initial voltage requirements. Although these figures were based on a large number of experiments they may only be considered as a guideline. It is important that the maximum power applied

Table 34 *Initial voltage requirements in isoelectric focusing with different density gradients and pH ranges*

Focusing column	Density Gradient	Range (%)	Ampholine range	Initial voltage (V) (2mA current)	Power dissipation (W)
LKB-8101	Sucrose	50–10	3–10	200	0·4
IEF-PPHE			3–6	350	0·7
			5–8	450	0·9
	Ficoll	25–5	3–10	75	0·15
			3–6	120	0·24
			7–10	140	0·28
	Glycerol	60–10	3–10	350	0·7
			7–10	600	1·2
		75–20	3–10	600	1·2
			All small ranges	>1200	2·4
	PEG	40–10	3–10	150	0·3
IEF-M3	Sucrose	50–10	3–6	280	0·56
	Ficoll	25–5	3–6	140	0·28
	Glycerol	75–20	3–6	320	0·64
			7–10	520	1·04

PEG's polyethylene glycol 6000.

should not exceed 2·0 W in the case of LKB-8101 and IEF-PPHE columns and preferably kept at around 0·5 W in case of the IEF-M3 column. The low power dissipation in the latter is stipulated mainly in the event of focusing being carried out at room temperature.

At the beginning of isoelectric focusing the conductivity of the column is fairly uniform and high. This is because the ampholines are in a charged state. As the molecules become isoelectric their conductivity drops considerably. The procedure adopted in our laboratory is successively to increase the applied voltage, but still keeping power dissipation below 2 W for the LKB-8101 and IEF-PPHE columns and at less than 1 W in the case of the IEF-M3 column. This is essential because with the equilibration of the ampholines at their respective pI the conductivity of different regions of the pH gradient will vary resulting in an unfavourable distribution of power dissipation. Ampholines at both sides of pH 6–7 have greater conductivity. Therefore, if one chooses to place the most acidic or most alkaline end of the pH gradient at the denser end of the supporting density gradient, one is able to compensate for the effects of the variations in the conductivity of the column.

FOCUSING TIME

The time required to produce a pH gradient at equilibrium varies according to the range of the gradient. A pH range of 3 to 10 and gradients in the acidic range require 24–72 hours in an LKB-8101 column while to obtain pH gradients in the alkaline zone is takes between 4 and 6 days. In microanalytical columns the time required to generate a stable pH gradient is considerably shorter. Godson (1970) reckons that the time required to form a 3–10 pH gradient in his J-type apparatus is two hours, while in the microcolumn used by Fredriksson (1975) the time taken is as little as 1·5 hours. With the IEF-M3 column a minimum 4 hours are required to produce a suitable pH gradient of any range. Reduction in electrolytic time has also been achieved using short and broad columns with vertical cooling system (Rilbe and Pettersson, 1975).

LOADING OF SAMPLE

In preparative electrofocusing of proteins the sample is most conveniently mixed with the lighter component used in the density gradient mixer. Alternatively layering of the density gradient could be stopped midway in the column and the sample layered after dissolving it in sucrose solution of matching density. The layering of the density gradient is then continued to completion. This procedure is very useful in the loading of a cell suspension on to the column. Nonetheless the cells must remain in the column during the prolonged electrolysis time. In order to overcome this problem the post-pH equilibrium (PPHE) loading method has been devised (see Fig. 22). In this method the cells are loaded only after a stable pH gradient has formed. Loading is achieved by piercing the silicone rubber septum of the loading port using a size VI serum needle, connected via a silicone tubing to a peristaltic pump. A small quantity of the density gradient (2·0 ml in IEF-PPHE and 0·1 ml in IEF-M3 columns) is withdrawn directly into a centrifuge tube in which the cells have been previously pelleted and the cells are then suspended in it. The direction of flow of the peristaltic pump is now reversed and the sample is returned to the focusing column. The rate of withdrawal and introduction of the sample should be kept as low as possible. The recommended rate for IEF-PPHE column is 0·5 ml min^{-1} and for IEF-M3 not more than 0·1 ml min^{-1}. The cells are equilibrated for 2–3 hours in the IEF-PPHE column and for 20–30 minutes in the IEF-M3 column. Post-loading of cells is not only efficient and reduces time of equilibration of sample but it also improves considerably the viability of cells. The optimum sizes of protein and cell samples are given in Table 35.

Table 35 *Optimum sample sizes in isoelectric focusing*

Column	Sample size	
	Protein (mg)	No. of cells $\times 10^{-6}$
LKB-8101	5–25	30
IEF-PPHE	5–25	30
IEF-M3	0·5–1·0	0·2–0·5
Godson (1970)	0·5–1·0	—
Osterman (1970)	1·5	—
Fredriksson (1975)	<0·1	

ELUTION OF COLUMN, MEASUREMENT OF pH VALUES AND CELL DENSITY

Before draining the column the central electrode chamber of LKB-8101 and IEF-PPHE columns or the corresponding electrode chamber of the IEF-M3 column are shut off from the main focusing chamber so that the electrode solution does not mix with the density gradient. Fractions of suitable size are collected and the pH values of the fractions are read accurately. The cell densities in the different fractions can be measured using a haemocytometer, but for drawing a profile of cell densities, it is sufficient to measure optical densities (at 420 nm) of the various fractions. At low cellular densities the absorbance is proportional to cell densities.

Recently we have mechanised the whole procedure (see Fig. 25). In this assembly the focusing column is connected to a pH flow cell and to an optical density flow cell of an SP 800 spectrophotometer. Suction is applied at the end with a peristaltic pump. The signals from the SP 800 and the pH meter are recorded on a chart recorder. By interposing a device which alternates the signals from the SP 800 with signals from the pH meter, both can be recorded on a single pen chart recorded (see Fig. 26).

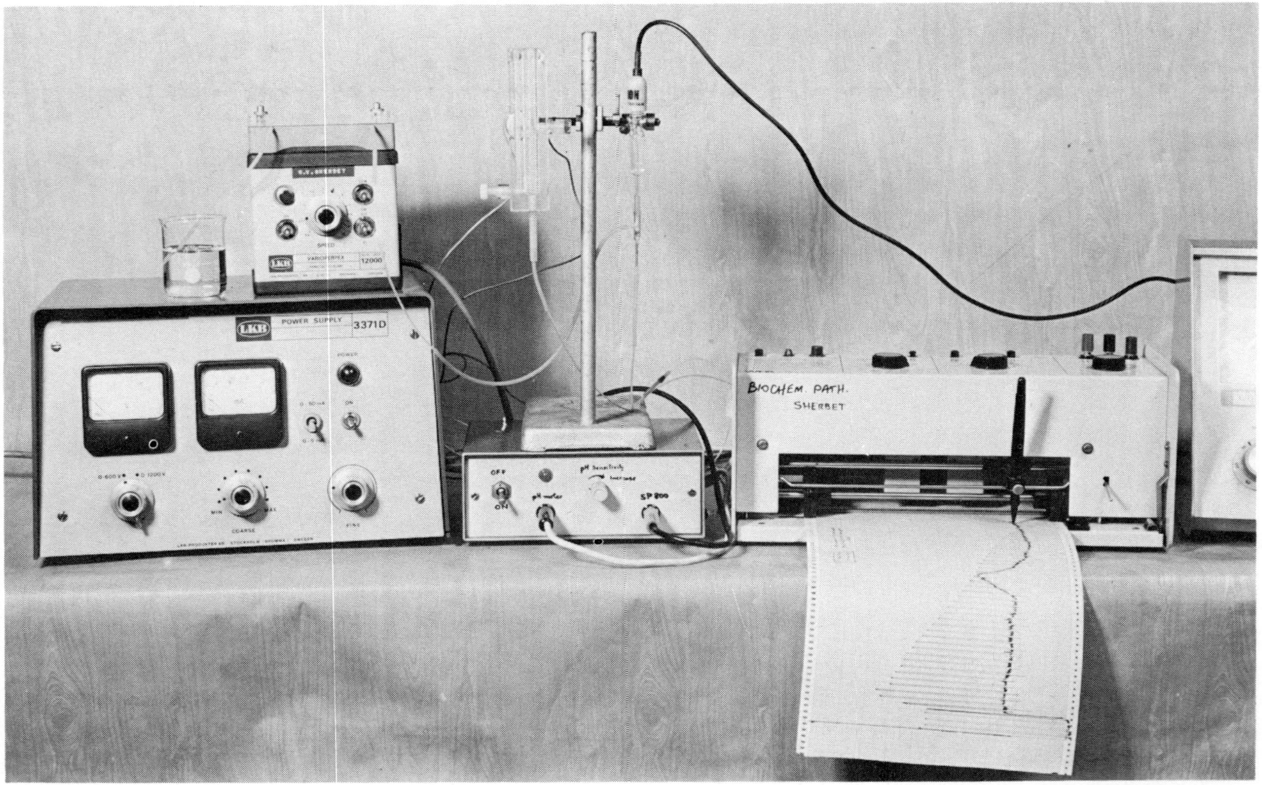

Fig. 25. Isoelectric focusing assembly as used in author's laboratory employing IEF-M3 focusing column connected to a flow-through cell of SP 800 spectrophotometer and then to a pH flow cell. Signals from both pH meter and the spectrophotometer are recorded on a single pen chart recorder by interposing a mechanism to alternate the signals. The recording assembly using a single pen chart recorder was devised by L. M. Bowler (photograph by L. M. Bowler).

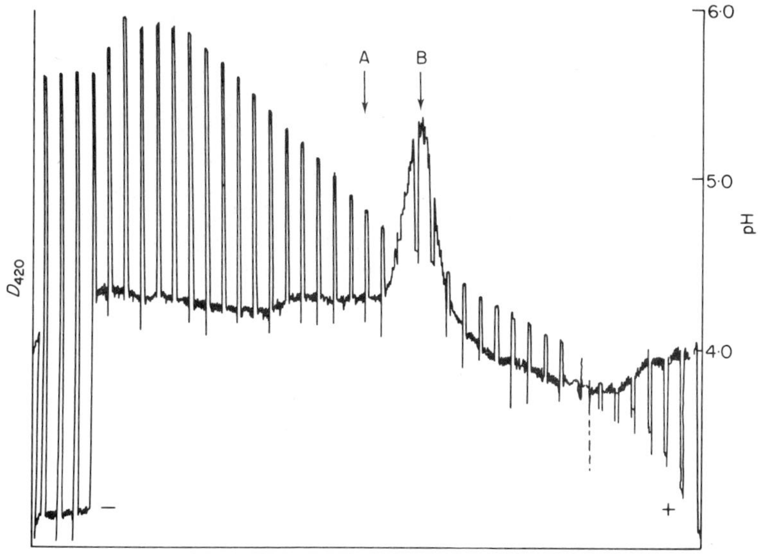

Fig. 26. Isoelectric profile of an experiment with 3T3 cells; A-point of post-pH loading and B-point of equilibrium (Sherbet and Lakshmi, unpublished results).

CELL pI AND EXPERIMENTAL CONDITIONS

Before one can enter into the discussion of the interpretation of isoelectric data, it may be worthwhile to show that the pI is a characteristic feature of the cell surface, and that it is quite independent of factors such as the ampholines, their pH range, the supporting density gradients, etc.

It is fairly certain that ampholines do not form complexes with the protein sample (see pp. 152–153) but as yet there is no direct evidence that ampholines do not form complexes with the cell surface components. If they did, the pI of the cells is bound to change from the true value to an artifactual one. That the pI is not artifactual, however, is indicated by the observation that the same pI value is obtained irrespective of whether a cell is loaded onto the focusing column before the pH gradient is generated or when post-loaded. It could be argued that if pre-loaded i.e. before generation of pH gradient, the cell will have been in contact with the whole range of ampholines used. In post-loading, however, the cell will have been in contact with a much smaller spectrum of ampholine molecules. If the pI were dependent on the ampholines and is determined by the formation of complex between the surface components and ampholine molecules, different values should be obtained for pre- and post-loaded cells. By the same token the observation that polyoma virus-transformed BHK-21 (Py cells) show a pI of 6·4 in both pH 3–10 and 5–8 gradients indicates that the pI is independent of the ampholines used (Table 36) (Sherbet *et al.*, 1972).

Table 36 *Data indicating that the cell pI is independent of ampholines, the density gradient and the buoyant density of the medium at cell pI*[a]

Cell type	Cell pI	pH gradient range	Supporting gradient (%)	Supporting medium at cell pI: Specific gravity	Supporting medium at cell pI: Concentration (%)
Py cells[b]	6·4	3–10	Sucrose 55–10	1·1128	27
	6·4	5–8	Sucrose 55–10	1·081	20
	6·35	3–10	Glycerol 60–10	1·15	59 (=35% sucrose)
	6·4	5–8	Glycerol 60–10	1·10	40 (=24% sucrose)

[a] Data from Sherbet *et al.* (1972).
[b] Py cells are polyoma virus-transformed BHK-21 cells.

The cell pI is also unaffected by the solute used to produce the supporting density gradient. Again Py cells are isoelectric at pH 6·4 irrespective of whether sucrose or glycerol is used to provide the density gradient.

The position which the cells occupy in the column at equilibrium is not determined by the buoyant density of the cells. In other words, the cells do not sediment or float upwards until their buoyant density matches the density of the supporting medium. This independence of cell pI is clearly demonstrated by data given in Table 36. It may be noticed that Py cells have a pI of 6·4 irrespective of the density of the supporting medium in the isoelectric zone. In the same vein, reversal of polarity of current flow makes no difference to the measured pI values. Sherbet *et al.* (1972) describe a further expriment in which the pI of HeLa cells was determined before and after adsorption of γ-globulin molecules on the cell surface. Treatment with γ-globulins produced a statistically significant difference in the pI (see p. 178; Table 38).

The adsorption of γ-globulin can hardly be expected to produce appreciable changes in the buoyant density of the cells. However indirect may all this evidence appear, there is hardly any doubt that the measured pIs are true pI values quite independent of the ampholine range, supporting density gradient solute and other experimental factors.

The pI does indeed depend upon the temperature at which focusing is carried out. It also depends upon the concentration of ampholine used. In isoelectric focusing experiments the ionic atmosphere is provided by the multivalent ampholine molecules, and therefore it may be expected that their concentration influences the surface potential of the particles. Although as mentioned before (p. 157) the concept of ionic concentration and the double electrical layer assumes a different significance so far as isoelectric equilibrium studies are concerned, the assumption that the ionic atmosphere is provided by ampholine molecules appears to be justified in view of the observations made by Godson (1970) that haemoglobin migrates faster at lower concentrations of ampholine.

ISOELECTRIC FOCUSING OF VIRUSES

Isoelectric focusing has recently been used for the characterisation of viruses, their purification, and for the determination of their isoelectric point as a measure of their homogeneity.

An early investigation by Mandel (1971) on Type 1 poliovirus labelled with ^{32}P showed that the virus was isoelectric at two pH values, namely pH 7 and 4·2. The viral activity in both regions was measured by infectivity as well as radioactivity distribution. Mandel also found a very small peak of infectivity in the region of pH 2·9. These three bands were designated as A, B, and C respectively. The specific infectivity (PFU/c.p.m.) of Bands A and B was similar but that of Band C very low indicating either the presence of extraneous radioactivity in Band C or the radioactivity represented disrupted viral RNA. When virus particles that were recovered from Bands A and B were separately re-electrofocused they resolved again in a bimodal distribution. This appears to indicate that while the viral population is homogeneous, each particle can exist with their capsid proteins in two resonating conformational states, since it was possible to alter the isoelectric characteristics of the virus by loading the virus at the alkaline or acid region of the focusing chamber. Mandel (1971) showed that viruses loaded at approximately neutral pH zone, showed the usual bimodal distribution, but those loaded in the alkaline zone showed a pI of 7·0. Particle which had been loaded in the acidic zone were isoelectric at pH 4·5. A similar isoelectric behaviour pattern has also been reported for the foot and mouth disease virus by Talbot (1975). This data shows that polio virus resonates between states A and B, with the state A being favoured at neutral pH values. On heat inactivation or on irradiation with ultraviolet light, the spectrum of isoelectric equilibrium distribution changes completely. Both state A and state B particles lose their infectivity on heating, but while state A particles change to B, B particles remain in state B in spite of loss of infectivity. Ultraviolet irradiation also alters a majority of state A particles into B, but there is also a concomitant increase in B and C particles. A third process of inactivation used by Mandel, namely using neutralising antiserum also converts state A particles into state B. These experiments support the view that not only does the virus resonate between the two states but that the process also has some relevance to the biological activity of the virus (see also p. 181 regarding bimodal distribution of cells in isoelectric focusing) (Table 37 gives a summary of Mandel's experiments).

Table 37 *Transition of polio virus from state A to B on inactivation*

	Isoelectric equilibrium pattern		
Mode of inactivation	pH 7·0 Band A	4·5 Band B	2·9 Band C
Normal	++	++	–
Heat inactivated	+	+++	+
u.v. irradiation for 10 minutes	+	+++	++
Homologous antiserum	–	++++	+

Comparison based on data from Mandel (1971). The specific activities had not been determined for each band in each experiment. The isoelectric equilibrium pattern indicated may be taken only as a qualitative reflection of the spectrum of distribution.

Chlumecka *et al.* (1973) examined the isoelectric behaviour of three variants of the Mengo encephalitis virus. The three variants showed the usual bimodal pattern with a major peak composed of 75–90% of virus input. Isoelectrically they differed considerably from one another. The pI values of the major and minor peaks for the three variants were as follows: L-Mengo 8·1 and 4·6; M-Mengo 4·4 and 6·3; S-Mengo 4·6 and 6·8. These isoelectric characteristics are compatible with their ability to attach to cultured L-cells and their ability to haemagglutinate human type O erythrocytes.

A bimodal distribution in isoelectric equilibrium was reported by Herschman and Helinski (1967) in a study of colicin E_2 and E_3. While colicin E_3 was homogeneous, E_2 showed a bimodal distribution with the peaks showing a pI difference of 0·2 pH units. This isoelectric behaviour of the bacteriocins may be analogous to the conformational changes seen in viral particles.

Although these investigations have not been pursued further, there is little doubt that isoelectric equilibrium analysis has considerable potential use in the physicochemical characterisation of capsid proteins, and probably also in the study of the interaction of viruses with cells. Joklik and Darnell (1961) suggested that the inactivation of viruses following interaction with host cells, may be accompanied by alterations in viral capsid. Mandel (1971) has indeed found that polio virus eluted after interaction with HeLa cells was mainly in state B and there also occurred a small increase in particles isoelectric at pH 2·3.

More recently Rice and Horst (1972) have examined the isoelectric behaviour of viruses in polyacrylamide gel columns (see Table 38). Of the five viruses investigated four appeared to have low pI values. The Brome mosaic virus, on the other hand, had a pI of 6·8. These viruses also appeared to be homogeneous, with the exception of $Q\beta$ bacteriophage. But Rice and Horst (1972) did find that the homogeneity or otherwise of the virus sample depended upon loading. Both the Brome mosaic virus and cowpea chlorotic mottle virus appeared heterogeneous when loaded from the alkaline end. It is probable that this reflects the same phenomenon of molecular transitions of capsid proteins suggested by the work of Mandel (1971).

Electrophoresis has also brought to light such heterogeneity in viruses at pH values above their isoelectric points (Bockstahler and Kaesberg, 1962; Bancroft *et al.*, 1968). Agarwal (1964) described two electrophoretic forms of cowpea mottle virus, namely the slow and fast migrating forms. The slow form may be converted *in vivo* into the fast form and this can be achieved *in vitro* by using proteolytic enzymes (Niblett and Semancik, 1969). Presumably therefore the two forms differ in their capsid component. Silver *et al.* (1976) were able to resolve these two forms into two further zones each. They were also able to isolate a mutant form which could be converted more rapidly from the slow form to the fast form. Silver *et al.* (1976) have suggested that this process of conversion could be due to a change in the amino acid sequence in the capsid protein, which makes the component susceptible to the action of proteolytic enzymes while also altering the electrokinetic properties of the particle.

Table 38 *Isoelectric points of animal cells, bacteria and viruses determined by isoelectric equilibrium method*

Type	pI	Reference
HeLa (Human carcinoma of the cervix)	6·85	Sherbet *et al.* (1972)[a]
γ-Globulin-treated HeLa	6·36	
Yoshida ascites sarcoma	6·35	
Ehrlich ascites sarcoma	5·6	
Py cells	6·4	
Normal liver cells	6·5	
Butter yellow-induced rat hepatoma	6·75	Sherbet and Lakshmi (unpublished)[a]
3T3 mouse fibroblasts	4·6	Sherbet and Lakshmi (unpublished)[b]
SV-3T3 cells	4·8	Sherbet and Lakshmi (1976)[b]
SV-CHK	4·66	
SV-TRK	4·6	
Py-3T3	4·78	
Human astrocytoma Grade I	5·15	Sherbet and Lakshmi (1974)[b]
Grade II	5·0	
Grades III and IV	4·48	
Human meningioma	4·73	
Human fetal brain cells	4·38	
Escherichia coli	5·6	Sherbet and Lakshmi (1973, 1975)[a]
Human erythrocytes	5·6	Just *et al.* (1975)
Mouse erythrocytes	5·8	
Rabbit erythrocytes	6·0	
Boar spermatozoa	6·5	Moore and Hibbit (1975)
Turnip yellow mosaic virus	4·0	Rice and Horst (1972)[c]
Cowpea chlorotic mottle virus	4·1	
$Q\beta$ bacteriophage	4·1	
Satellite of tobacco necrosis virus	4·3	
Brome mosaic virus	6·8	
Poxviruses	4·0	Hill *et al.* (1972)[d]

[a] pI determined at 25°C using 0·008 M ampholine.
[b] pI determined in 0·009 M ampholine.
[c] pI measured at 2·0% ampholine.
[d] pI determined by electrophoresis (= pIE).

The capsid proteins of the cowpea mottle virus are of two kinds and are of molecular weight 42 000 and 22 000, in the fast form (Wu and Bruening, 1971); the smaller component of the slow form has a slightly higher molecular weight of 25 000 (Geelen *et al.*, 1972). Since the number of polypeptides in the viral capsid is limited, changes in the composition should be easily detectable by means of isoelectric focusing and it should also be possible to test these possibilities by examining the isoelectric states of the slow and fast moving forms and also the mechanisms by which proteolytic treatments might bring about the conversion of the slow into the fast form.

ISOELECTRIC CHARACTERISTICS OF CELLS

The isoelectric behaviour of animal cells, normal and neoplastic, has been actively investigated in our laboratory since 1968. A preliminary report on some of the early work on Yoshida ascites sarcoma, Ehrlich ascites sarcoma and HeLa cells was published by Sherbet and Lakshmi (1969a) and Sherbet *et al.* (1970). The potential use of the technique of isoelectric focusing in the characterisation and study of surface properties of cells and in the separation of cell types from a mixed population, was no doubt appreciated early (Sherbet and Lakshmi, 1969a). Much of our work, in fact, has taken this course and we have been able to achieve at least in part the aims which we set ourselves. That a fair degree of characterisation of the cell surface is possible will be apparent in the rest of this chapter. But isoelectric focusing does not appear to be suitable for separation of cell types, as we had hoped, due to drawbacks inherent in the technique.

The isoelectric points of a variety of cell types are given in Table 38. No definite pattern emerges from this data. HeLa cells which are derived from a tumour of the cervical epithelium have the highest pI. These cells produce an epithelial pattern even in tissue culture. Normal liver cells also showed a high pI value. It may be that a high pI value is associated with cellular component derived from organised structures. This statement is consistent also with lower pI values given by the ascitic tumours and even lower pIs possessed by normal and virus transformed fibroblasts. Human fetal brain cells had the lowest pI, of the variety of animal cells examined in our laboratory. Rice and Horst (1972) obtained even lower pI values for some viruses.

An interesting feature of early isoelectric equilibrium studies of cells using the LKB-8101 column and the IEF-PPHE column was the bimodal distribution of cells (Sherbet and Lakshmi, 1969a; Sherbet *et al.*, 1972)

(Fig. 27). As a general rule the major peak (I) with higher isoelectric points was mainly composed of dye-excluding or viable cells, while the minor peak (II) was composed mainly of non-viable cells. Although the pI of viable cells ranged from 5·6 to 6·85 the non-viable cells showed a pI around 5·0 irrespective of their pI in the viable state (see Table 39). Often the minor peak (II) was a diffuse broad band with a spread of pI values, so that cells in fractions which were closer to the major peak, had intermediate pI values. In these cells only the nucleus appeared to be stained with trypan blue—in contradistinction to the overall cytoplasmic and nuclear staining of the non-viable peak II cells. This seems to suggest that a gradual change occurs in the surface of the cells which, on the one hand, results in an increase in the permeability of the cell membrane to the dyes, and on the other, also results in an increase in the net surface charge borne by the cells. Both these changes might result from changes in the conformation of membrane constituents. An analogous phenomenon was subsequently described in the polio virus by Mandel (1971), where the virus resonated between two states, and where inactivation of the virus converted virus in state A to state B which has a

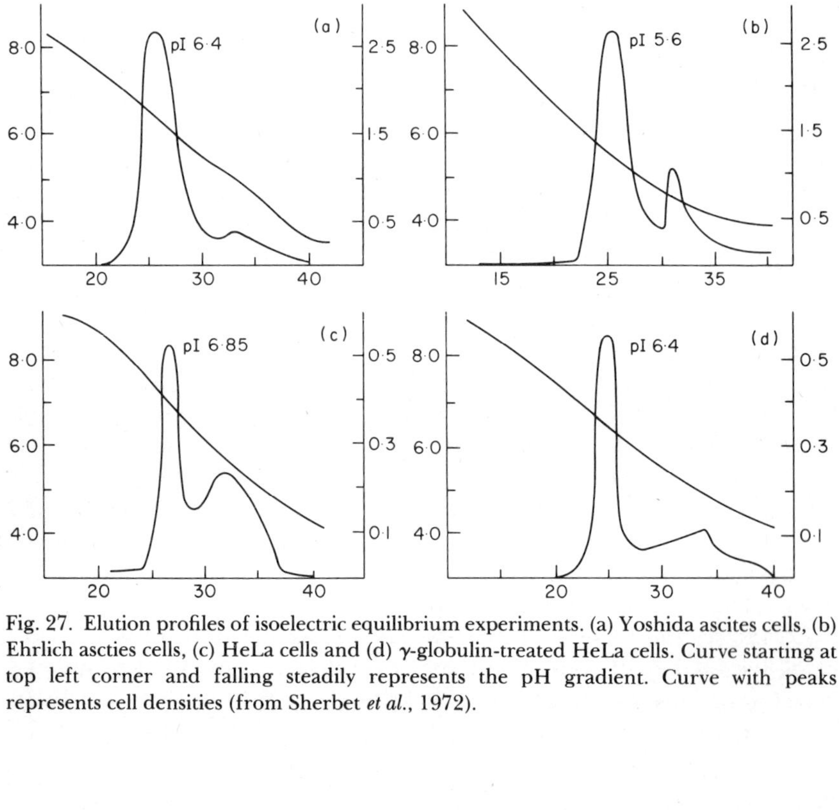

Fig. 27. Elution profiles of isoelectric equilibrium experiments. (a) Yoshida ascites cells, (b) Ehrlich ascties cells, (c) HeLa cells and (d) γ-globulin-treated HeLa cells. Curve starting at top left corner and falling steadily represents the pH gradient. Curve with peaks represents cell densities (from Sherbet *et al.*, 1972).

Table 39 *Bimodal distribution of cells in isoelectric focusing*

	Isoelectric point	
Cell type	Peak I	Peak II
Yoshida ascites cells	6·35±0·19	4·79±0·3
Ehrlich ascites sarcoma	5·6	4·73±0·17
Py-cells	6·4±0·01	5·11±0·02
HeLa	6·85±0·11	5·32±0·11
Normal liver cells	6·5	5·05

From Sherbet *et al.* (1972).

lower isoelectric point (see pp. 175–177). The bimodal distribution of cells differs from the situation obtaining in viruses in that cells of peak I were viable and those of peak II non-viable but in the case of polio virus viral activity is detected in both states A and B. Nonetheless, in inactivation state A becomes altered to B reducing the specific activity of Band B. The analogy cannot be carried further, for in the case of viruses the bimodal distribution and the transition from A to B probably has a biological significance.

The bimodal distribution was not seen in experiments on normal and transformed 3T3 fibroblasts and human astrocytoma and meningioma cell lines. This may be because these experiments were performed with IEF-M3 column with no more than $0{\cdot}5\times10^6$ cells. Even if the bimodal pattern had been present, probably it would not have been detected because of the insensitivity of cell density monitoring and recording equipment used.

CELL pI AND IONISABLE GROUPS OF THE CELL SURFACE

It was recognised early that isoelectric point was a determinate physical characteristic of cells. As referred to earlier (p. 147), it was shown that erythrocytes from different species had different isoelectric points; so did rough and smooth variants of *Shigella flexneri* and *Salmonella typhii* and different strains of *Diplococcus*. But the value of isoelectric data as a potential source of information on the physicochemical make up of the cell surface does not appear to have been recognised. The rest of this chapter is devoted to such an interpretation of isoelectric data and it is hoped that this discussion would highlight the enormous use that can be made of such data.

CHARACTERISATION OF IONISABLE GROUPS

Since the isoelectric and electrophoretic properties of cells are determined by the protein and glycoprotein components, the cell surface can be treated as a multivalent ion containing negatively and positively charged groups which ionise as a function of pH. At a certain pH, namely the pI, an equivalent dissociation is achieved by the cationic and anionic groups and the surface will bear no net electric charge.

Considering a small molecule such as an amino acid, the dissociation constants of the carboxyl and amino groups and the hydrogen ion concentration will determine whether the carboxyl or the amino groups are ionised. At a certain hydrogen ion concentration the amino acid may be present almost exclusively in the zwitterionic form. This pH value represents the isoelectric point of the amino acid, and this can be calculated from the dissociation constants of the ionisable groups as follows:

$$pI = \frac{pK_a + pK_b}{2} \tag{44}$$

where pK_a and pK_b are the negative logarithmic values of the dissociation constants of the acidic and basic groups.

For complex ampholytes such as proteins with a variety of ionisable groups of different dissociation constants, the pI is given by the equation:

$$pI = -\log\left[\frac{K_{a1} + K_{a2} + K_{a3} \ldots \times K_w}{K_{b1} + K_{b2} + K_{b3} \ldots}\right]^{\frac{1}{2}} \tag{45}$$

where K_{a1}, K_{a2}, etc. are the dissociation constants of the anionic groups; K_{b1}, K_{b2}, etc. are dissociation constants of the cationic groups and K_w the dissociation constant of water.

Since the dissociation constants of the weaker anionic and cationic groups are negligible in comparison with those of the strongest dissociating groups, the dissociation constants of the weaker groups can be neglected and eqn (45) can be rewritten as follows:

$$pI = -\log\left[\frac{\sum K_a \times K_w}{\sum K_b}\right]^{\frac{1}{2}} \tag{46}$$

where K_a and K_b are the dissociation constants of the strongest dissociating anionic and cationic groups. For simplicity of calculation the following relationship may be used:

$$pI = \tfrac{1}{2}(pK_a - pK_b + pK_w) \tag{47}$$

where pK_a and pK_b are the dissociation constants of the strongest anionic and cationic groups respectively and pK_w that of water.

Equation (46) thus gives us a device to interpret the isoelectric points of cell surface and deduce the dissociation constants of the strongly dissociable groups and by comparing these values with known values of dissociation of ionogenic groups in proteins and glycoproteins, it is possible to provide a degree of characterisation of the ionogenic groups of the cell surface.

Such a characterisation has been attempted by Sherbet *et al.* (1972) for Py-cells, Yoshida and Ehrlich ascites and HeLa cells (Table 40), using known pK values of the ionisable groups in amino acid residues in proteins, the terminal carboxyl groups of proteins and sialic acid moieties of proteins and the terminal amino groups of protein molecules (Table 41). Table 40 shows that the pI values of the tumour cells are compatible with the presence of sialic acid and α-carboxyl anionic groups with pK ~3·0 and with α-amino and imidazolyl groups. The high pI values also exclude the presence of anionic groups such as phosphate or sulphate with pK values very much below 3. The low pI values (~5·0) of the peak II cells, on the other hand, strongly indicates the presence of phosphate groups on the surface of these cells. This seems further to

Table 40 *Probable ionogenic groups on the surface of some tumour cells*

	Ionogenic groups		pI[a]	
Cell type	Present	Absent	Calculated	Observed
Py-cells	Anionic pK ~ 3·0			6·4 (6·1)
Yoshida and Ehrlich	Sialic acid α-COOH			5·6 (5·4)
ascites, and HeLa	cationic ~5·5	pK < 3·0	5·85	6·85 (6·6)
cells	Imidazolyl and			
	α-NH_3^+	pK > 6·0(?)		
Peak II	Anionic ~2·0			
	Phosphate, α-COOH	pK > 6·0	5·25	5·0 (4·8)
	sialic acid -COOH			
	Imidazolyl and			
	α-NH_3^+			

[a] pI values were calculated from surface pK of ionogenic groups corrected using the Hartley-Roe (1940) equation. The values given within brackets are the observed values corrected after Hartley-Roe.

From Sherbet *et al.* (1972).

Table 41 *pK_a values of some ionogenic groups in proteins*

Ionisable group	pK_a (at 25°C)
Phosphate	2·12
Sialic acid carboxyl	2·6
α-COOH	3·0–3·2
β-COOH (of aspartic acid residue)	3·0–4·7
γ-COOH (of glutamic acid residue)	4·5
Imidazolyl (of histidine residue)	5·6–7·0
α-amino	7·6–8·4
ε-amino (of lysine residue)	9·4–10·6
Thiol (of cysteine residue)	9·1–10·8
Phenolic hydroxyl (tyrosine residue)	9·8–10·4
Guanidinyl (arginine residue)	11·6–12·6

Data from Dawes (1972) and Svennerholm (1956).

confirm the suggestion made earlier that the appearance of the peak II cells is caused by conformational changes produced in the membrane components of the normal cells.

HARTLEY–ROE TREATMENT OF ISOELECTRIC POINTS

The calculated pI values of Py-cells, Yoshida ascites and HeLa cells differ by 9–17% from the observed values. This may conceivably be due to the fact that the bulk pH usually differs from the surface pH. This is due to several factors such as differences in the dielectric constants of the bulk phase and the interface, the interference in ionisation due to a close approximation of the carboxyl groups, and the presence of Donnan equilibrium between the interface and the bulk phase (see Danielli, 1936). These factors result in a lower surface pH than the bulk pH.

Hartley and Roe (1940) calculated the concentration of ions in the interface from the ζ potential. According to Hartley and Roe, the hydrogen ion concentration near the particle surface is given by the equation:

$$[H^+]_i = [H^+]_b \times e^{-D\zeta/kT} \tag{48}$$

where $[H^+]_i$ is the hydrogen ion concentration at the interface, $[H^+]_b$ that in the bulk phase, e electronic charge, D the dielectric constant of water, ζ represents the ζ potential of the particle, k the Boltzmann's constant and T the absolute temperature. The dissociation constant K_s is related to the bulk dissociation constant K_b as given in the following equation:

$$K_s = K_b \times e^{-D\zeta/kT}. \tag{49}$$

At 25°C, the relationship between the pH at the surface of the particle (pH_s) and the pH of the bulk phase (pH_b) can be written as:

$$pH_s = pH_b + \zeta/60. \tag{50}$$

Since ζ potentials of cell surfaces have negative values pH_s is lower than pH_b.

Now since

$$\zeta = \frac{4\pi\eta v}{D} \tag{14}$$

at 25°C, η the viscosity of water is 89·4 poises and D its dielectric constant is 78·54, eqn (50) assumes the form:

$$pH_s = pH_b + 0{\cdot}238v \tag{50}$$

where v is the electrophoretic mobility of the particles.

In Table 40 the pK values used for the calculation of isoelectric points were corrected using eqn (50), so also have the values of observed isoelectric points of cells been corrected. The calculation of values of v or ζ is described in a later section. It may appear incongruous that the pI is corrected using an equation which contains the quantity v which is itself based on the value of the pI. But in fact, although v is *based* on the pI it is *not derived* from the pI. This will become apparent when the calculation of v from pI is discussed.

CHARACTERISATION OF IONISABLE GROUPS BY CHEMICAL MODIFICATION

The considerable ambiguity in the interpretation of the isoelectric points in terms of the probable ionogenic groups occurring on the surface of cells has prompted further attempts at their characterisation. These attempts have involved chemical modification of specific ionogenic groups which is followed by the determination of the pI of the modified surface. This procedure has the merit that the pK values of individual groups can be determined in isolation, and has also the added advantage of allowing the characterisation of the weaker ionogenic groups. The main disadvantage associated with this approach is that the chemical treatments are often too harsh to be suitable for the study of animal cells, not to mention the liability of such experiments to be subjected to criticism on grounds that the information obtained from such experiments may not have much relevance to the original unsullied state of the cell membrane. On account of some of these difficulties, much of the work to be described here (Sherbet and Lakshmi, 1973) was performed on *E. coli* cells, which can be grown in quantities large enough for chemical modification.

Electron microscopy of *E. coli* and other Gram negative bacterial cells has revealed the presence of three distinct layers in the cell wall (Kellenberger and Ryter, 1958; Murray *et al.*, 1965; Ogura, 1963). The innermost layer is the cell membrane which is surrounded by the rigid layer of peptidoglycans. The outermost layer contains lipopolysaccharides, phospholipid and lipoprotein. It has been suggested that the lipopolysaccharide is present in the outer layer in the form of a complex with phospholipid and protein, the protein in turn being covalently bound to the rigid peptidoglycan layer (Tomasz, 1971; Wright and Kanegasaki, 1971). Physicochemical considerations discussed elsewhere in this chapter suggest that the outermost layer is the subject of this isoelectric equilibrium investigation.

The isoelectric point of unmodified *E. coli* cells has been determined to be 5·6. The cells were subjected to a number of chemical treatments, such as with ethyleneimine (Gittens and James, 1963a) which reacts with carboxyl groups to form β-aminoethyl esters. Treatment of the cells with depolymerised paraformaldehyde has been used as a means to block the amino groups of the cell surface by Heard and Seaman (1961) and Ward and Ambrose (1969). This method was found to be suitable, although rather slow in bringing about the blocking of carboxyl groups. The presence of thiol groups was detected by employing a specific reagent, 6,6′-dithiodinicotinic acid also known as carboxypyridine disulphide (CPDS). The effects of these various treatments are shown in Fig. 28 which are optical density profiles of typical experiments.

The surfaces of *E. coli* and *A. aerogenes* have been shown to contain predominantly carboxyl and amino groups (Gittens and James, 1963a). The intrinsic pK of carboxyl groups of glucuronic acid component of the bacterial surface is 3·2 at 20°C (Gittens and James, 1963), while that of α-amino groups in proteins varies between 7·0 and 8·4 (Dawes, 1972; Tanford, 1962). The pI for *E. coli* cells calculated on the assumption that these are the only predominant groups is 5·38, which differs from the observed values by only 4% (see Table 42). The high isoelectric point of the unmodified surface also excludes strong anionic groups such as phosphate groups. This inference is supported by the fact that the pI of the purely carboxyl surface of formaldehyde-treated cells (HCHO-cells) is as high as 3·85. If phosphate groups had been present, the pI of

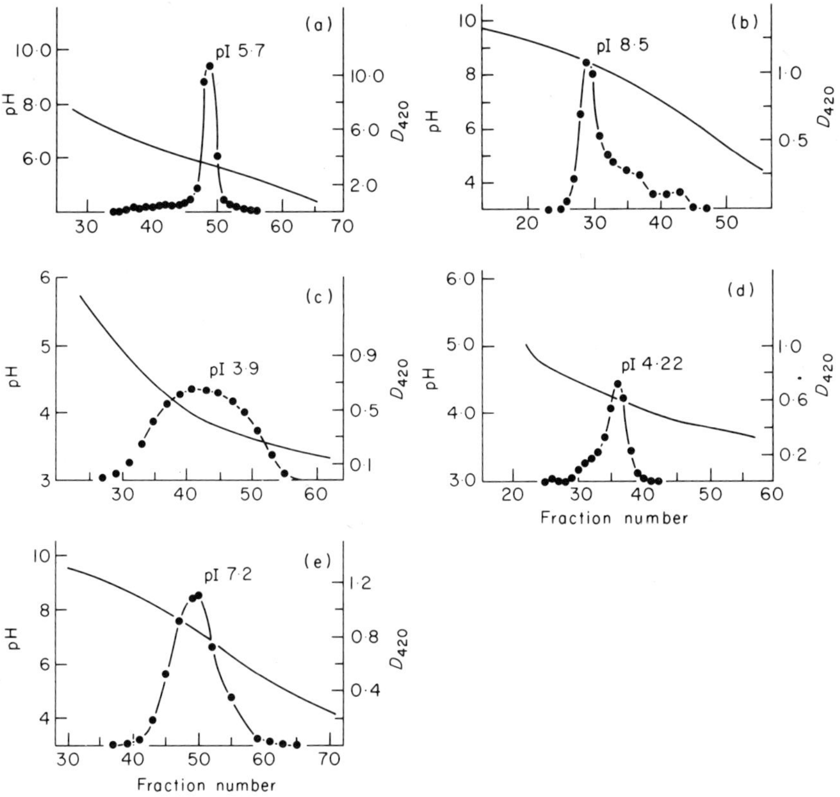

Fig. 28. Elution profiles of isoelectric equilibrium experiments with *E. coli* cells. (a) Unmodified cells, (b) ethyleneimine-treated acid washed cells (AW-EI), (c) formaldehyde-treated *E. coli* (HCHO) cells, (d) *E. coli* cells treated with 6,6′-dithiodinicotinic acid (*E. coli*-CPDS cells) and (e) AW-EI cells subsequently treated with CPDS. Curve starting at left top corner and falling steadily represents the pH gradient (from Sherbet and Lakshmi, 1973).

HCHO-cells would have been below 3·0 and that of the unmodified cells as low as about 4·4. The pI of HCHO-cells also suggests that the surface contains other species of carboxyl groups such as the β- and γ-carboxyl groups in addition to the α-carboxyl groups. On the other hand, the isoelectric behaviour of ethyleneimine-treated cells suggests the possibility that the surface possesses weakly acidic groups such as thiols in addition to α-amino groups. A calculation based on this assumption gives a pI of 8·5 which agrees extremely well with the observed pI of 8·55 of the ethyleneimine-(AW-EI) treated cells. This assumption has been shown to be true by treating the ethyleneimine-treated cells with carboxypyridine disulphide (CPDS) which is a specific thiol reagent.

Table 42 *Ionogenic groups on the surface of* Escherichia coli *cells*[a]

Cells	pI	Nature of ionogenic groups in the isoelectric zone[b] — Present	Excluded	Calculated[c] pI
Escherichia coli unmodified	5·6	carboxyl; amino group pK ~ 7·5	Acidic groups pK < 3·2; ε-amino groups	5·38
Ethyleneimine (EI)—treated *E. coli* cells	8·55	Amino groups pK ~ 7·5 Thiols	Amino groups of phospholipids; phenolic hydroxyl and guanidyl groups	8·5
Formaldehyde-treated *E. coli* cells	3·85	Carboxyl groups of glucuronic and neuraminic acids; β- and γ-carboxyls	Acidic groups pK < 3·2	3·5

[a] From Sherbet and Lakshmi (1973).
[b] Isoelectric zone is estimated to extend to a depth of 60 Å below the cell surface.
[c] $pI = \frac{1}{2}(pK_a - pK_b + pK_w)$ where pK_a, pK_b and pK_w are ionisation constants of the strongest anionic and cationic groups and the ionisation constant of water, respectively.

CPDS reacts with the thiol groups and introduces a carboxyl group (see p. 203 for discussion of its mode of action). Thus a treatment of ethyleneimine-treated cells with the thiol reagent causes a reduction of the pI from 8·55 to 7·2 (Fig. 28) unequivocally demonstrating the presence of thiol groups. The unmodified *E. coli* cells also show a reduction in pI from 5·6 to 4·2 (Fig. 28) following treatment with CPDS, which provides additional proof of the presence of thiol groups.

A similar isoelectric equilibrium analysis of 3T3 mouse fibroblasts and 3T3 cells transformed by simian virus 40 (SV-40) (SV-3T3 cells), though at a preliminary stage, has yielded interesting comparisons. The analysis of the isoelectric behaviour given in Tables 43 and 44 shows that the same kind and number of anionogenic groups are present on the surface of both normal and transformed cells. However the basic groups present on SV-40-transformed cells appear to be much weaker than those present on normal fibroblasts. The isoelectric points of both untreated and ethyleneimine-treated SV-3T3 cells suggest a pK of ~6·5–7·0 for these basic groups, while a pK of 7·5 presumably corresponding to $\alpha\text{-}NH_3^+$ groups is indicated in the untransformed cells. The difference in the composition of the basic groups was also reflected

Table 43 *Ionogenic groups on the surface of 3T3- and SV-40 transformed 3T3 fibroblasts*

	3T3 fibroblasts		SV-40-transformed 3T3	
Treatment	pI	Probable ionogenic groups	pI	Probable ionogenic groups
None (control)	4·6	Sialic acid carboxyl; α, β, γ-carboxyls; α-NH_3^+ pK ~ 7·5	4·8	Sialic acid carboxyl; α, β, γ-carboxyl; cationic group pK ~ 7·0 (imidazolyl?)
Formaldehyde-treated cells	4·05	Carboxyl groups only	4·1	Carboxyl groups only
Ethyleneimine-treated cells	8·2	Amino groups pK ~ 7·5 Thiol	8·58	Cationic group pK ~ 6·5 (imidazolyl?); thiols

Sherbet *et al.* (unpublished work).

in the time course curve for the modification of amino groups by formaldehyde. There was a linear increase in the net negative charge (due to modification of cationic groups) in the first two days of treatment of untransformed 3T3 cells with formaldehyde. But there was no change in the isoelectric behaviour of SV-3T3 cells for four weeks when an effect seemed apparent and this effect reached a maximum at eight weeks. Thus the kinetics of modification by formaldehyde are completely different, which supports the observation made earlier. In addition, SV-3T3 cells appear to possess about 30% more basic groups than do the untransformed cells (Table 44). If one may venture a suggestion, however premature, these investigations reveal the appearance of a new component(s) at the cell surface, which appears to be basic in nature. However, whether this basic material is directly associated with the process of viral transformation is yet unclear. Although less significantly than the amino groups, the acidic groups increased by about 5%. In other words, it is possible that acidic components such as glycoproteins may have also been acquired. Sherbet and Lakshmi (1976) have indeed shown the appearance of a surface antigen(s) on transformed cells which appear to bind concanavalin A which has affinity for α-D-mannopyranose and α-D-glucopyranose moieties. Lin *et al.* (1977) have recently demonstrated the appearance of two glycoprotein surface antigens with pI values of 4·5 and 4·7 on the plasma membrane of lymphocytes transformed by SV-40.

On the other hand, the work of Weiss *et al.* (1975) may be interpreted as indicating a loss of RNAase- and NANase-susceptible material from

Table 44 *Composition of ionogenic groups on the surface of 3T3- and SV-40-transformed 3T3 cells*

	3T3 cells			SV-40 transformed 3T3 cells		
Treatment[a]	pI	Corrected[b] pI	Charge density $\times 10^{-13}$ cm^{-2}	pI	Corrected[b] pI	Charge density $\times 10^{-13}$ cm^{-2}
Control	4·6	4·22	−2·212	4·8	4·45	−2·029
Formaldehyde-treated cells	4·05	3·6	−2·706	4·1	3·65	−2·666
Ethyleneimine-treated cells	8·2	8·39	+1·106	8·58	8·82	+1·448

From Sherbet *et al.* (unpublished work).
[a] Cells treated as described by Sherbet and Lakshmi (1973).
[b] pI corrected as described on p. 184.

the surface of SV-40-transformed 3T3 fibroblasts presumably as a result of the process of transformation. In two transformed cell lines showing revertant characteristics these or similar components may have been reacquired. Transformation of chick embryo fibroblasts by a temperature-sensitive mutant of Rous sarcoma virus is reported to result in the loss of expression of a protein component on the cell surface, although the protein may still be synthesised by the cell and released into the medium (Critchley *et al.*, 1976). In revertant cells, the protein is said to appear rapidly on the surface (Hynes and Wyke, 1975).

A biochemical investigation reported earlier by Sheinin and Onodera (1972) in fact revealed both qualitative and quantitative differences between the membrane components of normal and simian virus and polyoma virus-transformed 3T3 fibroblasts. Electrophoretic analyses in polyacrylamide gels of both glycoprotein and protein components have indeed revealed the presence of additional bands representing some weakly basic and other highly basic components.

CALCULATION OF SURFACE CHARGE FROM ISOELECTRIC DATA

EQUATION FOR SURFACE CHARGE OF A LARGE PARTICLE

In the following section a relationship is established between the isoelectric point of a cell and its surface charge, and an equation applicable to large spherical particles is derived. Animal and bacterial cells may be considered as large particles because, assuming a spherical shape for them, their radii are very much larger than the thickness of the electrical double layer, $d(1/K)$. In the derivation of this relationship the assumptions have been made that: the particles are spherical in shape and large (with $r \gg d$); the electrical charges are distributed symmetrically on the surface of the sphere and the potential P on the surface of the cells due to the electric charges expressed on the surface is equivalent to the electrochemical potential difference between a solution at pH 7 and a solution isoelectric with the cell surface. The basis for this assumption and the measurements flowing from this will be discussed later.

Consider a spherical particle of radius r bearing negative charge Q on its surface. The potential P_1 due to charge Q which is the work performed to bring a charge of unlike sign from infinity to a given distance r equal to the radius of the particle assuming a symmetrical surface distribution of the charge, is given by the following equation:

$$P_1 = \frac{-Q}{Dr} \tag{51}$$

where D is the dielectric constant of water.

The particle bounded by the electrical double layer can be envisaged as two concentric spheres (Fig. 16). Now the potential P_2 due to the outer sphere of charges will be:

$$P_2 = \frac{Q}{D(r+d)} \tag{52}$$

where d is the thickness of the electrical double layer.

A summation of P_1 and P_2 gives the resultant potential:

$$P = \frac{Q}{D(r+d)} - \frac{Q}{Dr} \tag{53}$$

or

$$P = \frac{-Q}{Dr} \times \frac{d}{(r+d)}. \tag{54}$$

Since $d \ll r$, eqn (54) can be rewritten as:

$$P = \frac{-Qd}{Dr^2}. \tag{55}$$

Substituting $1/K$ for d:

$$P = \frac{-Q}{Dr^2 K}$$

i.e.

$$-Q = PDr^2 K. \qquad (56)$$

THE EVALUATION OF POTENTIAL *P* FROM ISOELECTRIC DATA

The electrometric determination of pH is made by means of measurements of potentials of certain electrodes. If two zinc electrodes are immersed in 1·0 M and 0·1 M solutions of Zn^{2+} as salts, respectively, and connected by a copper wire and a salt bridge to complete the circuit, the potential of each of these electrodes is determined by the strength of Zn^{2+} (as salts). For according to the law of mass action the rate at which Zn^{2+} combine with No. 1 electrode (i.e. immersed in 1 M Zn^{2+} salts) is far greater than the rate at which such combination occurs at the No. 2 electrode (i.e. that immersed in 0·1 M Zn^{2+} salts), and hence there is higher electron pressure at the No. 2 electrode than at No. 1 electrode and a consequent flow of electrons from No. 2 electrode to No. 1 electrode. In essence the potential of an electrode cannot be measured in isolation as an absolute potential but only as a potential difference between a given electrode and a standard electrode.

A glass electrode cell may be represented as follows:

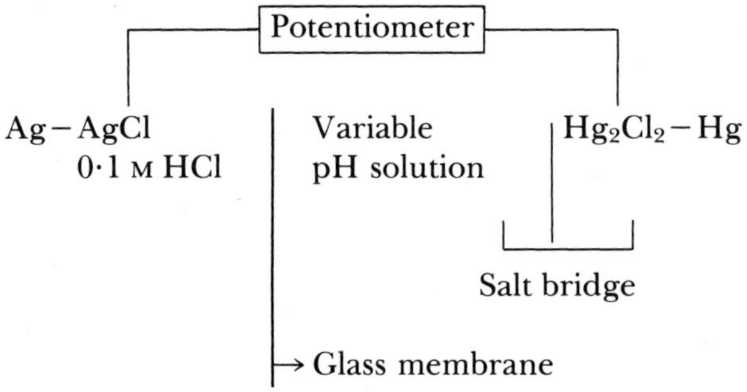

with silver–silver chloride half-cells with known $[H^+]$ provided by 0·1 M HCl separated by a glass membrane from the variable hydrogen ion activity and a saturated calomel half-cell. Now a potential difference develops across the glass membrane due to differences in the hydrogen ion activity. The difference in potential (emf) is given by:

$$\text{emf} = \frac{RT}{F} \log_e \frac{[H^+]_1}{[H^+]_2} \tag{57}$$

where R is the molar gas constant 8·314 joules °C^{-1}, T the absolute temperature and F the Faraday 96 500 coulombs mol^{-1}. Substituting these values in eqn (57):

$$\text{emf} = 0{\cdot}0001984T \times \log_{10} \frac{[H^+]_1}{[H^+]_2} \tag{58}$$

but since $[H^+]_1 = 1$

$$\text{emf} = 0{\cdot}0001984T \times \log_{10} \frac{1}{[H^+]_2}. \tag{59}$$

Now since

$$\log_{10} \frac{1}{[H^+]_2} = -\log_{10} [H^+]_2$$

we can write eqn (59) as

$$\text{emf} = 0{\cdot}0001984T \times \text{pH}. \tag{60}$$

At 25°C, $T = 298$K

$$\text{emf} = 0{\cdot}0591 \times \text{pH} \tag{61}$$

The cell surface can be considered as a multivalent ion, and by analogy with multivalent ions such as proteins, the isoelectric point of a cell surface may be defined as the pH at which it bears equal number of negative and positive charges i.e. it bears no net electric charge and therefore electrically neutral. Now, if one can postulate the occurrence of a hypothetical cell or a particle which is isoelectric at pH 7, the hydrogen ion concentration in equilibrium with the surface is 10^{-7} moles litre^{-1}. A cell or particle isoelectric at pH 5, has a much higher (×100) hydrogen ion concentration in equilibrium. When this particle is suspended in a pH 7 buffer is will carry net negative charges. The potential due to these charges could then be postulated to be equal to the difference in the electrochemical potential between a solution at pH 7 and another at pH 5. According to this postulate potential P on the surface of the cell is given by:

$$P = 0{\cdot}0591(\text{pI} - 7). \tag{62}$$

This value of P can be substituted into eqn (56) to obtain:

$$-Q=\frac{(\mathrm{pI}-7)0{\cdot}0591\times Dr^2K\times 3{\cdot}33\times 10^{-3}}{e} \qquad (63)$$

since P is measured in volts (V) and $\mathrm{V}=3{\cdot}33\times 10^{-3}$ e.s.u. and e is the electronic charge $4{\cdot}8\times 10^{-10}$ e.s.u.

CALCULATION OF ELECTROPHORETIC MOBILITY FROM ISOELECTRIC DATA

RELATIONSHIP BETWEEN EPM AND CHARGE DENSITY DERIVED FROM ISOELECTRIC DATA

The negative charge densities of three cell types, namely HeLa, liver cells and Ehrlich ascites cells, calculated using eqn (63) showed a linear relationship with their EPMs measured by means of free electrophoresis (Table 45). This relationship was so significant that it suggested the possibility of predicting EPMs of cells from their isoelectric data. From Fig. 29 it may be predicted that a net charge density of 10^{13} cm^{-2} would produce a mobility of 1 TU at 25°C and $I=0{\cdot}145$ M. The actual mobility of such a cell under these electrophoretic conditions is obtained by adding a factor of 9·7 units to the value obtained from the relationship in Fig. 29. In other words:

Table 45. *Data on the density of electronic charges and EPMs of HeLa, Liver and Ehrlich ascites cells*

Cell type	No. of electrons cell^{-1} $\times 10^{-6}$	Average surface area $\times 10^{8}\ \mathrm{cm}^2$	σ (No. of electrons cm^{-2}) $\times 10^{-13}$	Observed EPM TU
HeLa	3·9	308·0	0·13	−9·8
Liver cells	27·9	659·0	0·42	−10·09
Ehrlich ascites	100·7	850·0	1·115	−10·65

From Sherbet *et al* (1972).

$$v=9{\cdot}7+\frac{\sigma}{10^{13}} \qquad (64)$$

where σ is the electronic charges cm^{-2}.

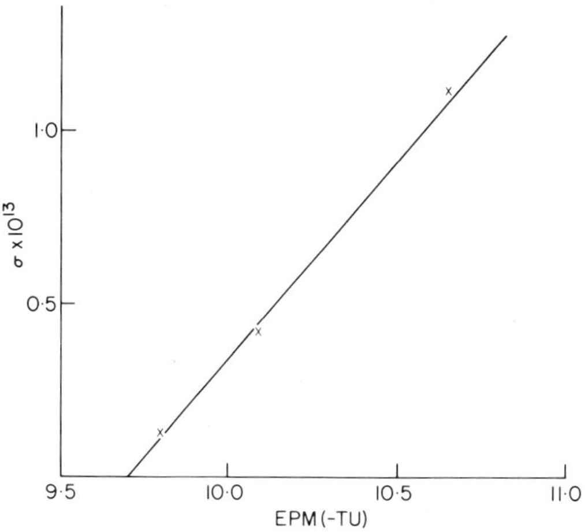

Fig. 29. Relation between σ (charge density calculated from isoelectric data) and electrophoretic mobility (in Tiselius units) (based on data from Sherbet *et al.*, 1972).

Using eqn (64) the EPMs were calculated for a number of cell types (Table 46). In 9 out of 14 cases the calculated mobilities deviated by 10% or less from the observed mobilities. In the remaining cases where the deviation was 20–28%, the observed mobilities themselves appear to have some experimental error incorporated in them, since these mobilities are not compatible with their respective charge density values. Nevertheless, it is obvious that eqn (64) has two major drawbacks, namely it can be used to predict mobilities accurately only where the surface charge density is high, or speaking in electrophoretic terms where EPMs of cells exceed 9·7 TU and the relationship as given in Fig. 29 implies an anodic mobility of cells in all cases. This apparent incompatibility is resolved in the following section.

CALCULATION OF EPM FROM ISOELECTRIC DATA

Isoelectric focusing and electrophoresis are two facets of electrokinetic behaviour of charged particles. The ionic conditions can be accurately defined and rigidly adhered to in electrophoresis, but as far as isoelectric focusing is concerned, however, the ionic atmosphere is provided by ampholines which are polyvalent ions. The assumptions made in describing the ionic conditions makes it imperative that we emphasise the differences between the expression of surface charge of a particle in the two different systems, and that there is no apparent or intrinsic relationship between isoelectric and electrophoretic data.

Table 46 *EPMs of some cell types calculated from charge density-mobility relationship*

Cell type	σ $\times 10^{-13}$	EPM (−TU)	Calculated[a] EPM (−TU)	Deviation (%)	Reference
Erythrocyte					
Human	0·77	10·8	10·5	−3·0	Cook *et al.* (1961)
Human	0·6	9·7	10·3	−6·0	Lipman *et al.* (1966)
Human	0·8	11·3	10·5	−8·0	Abramson *et al.* (1942)
Human	0·73	11·6	10·4	−9·7	Eylar *et al.* (1962)
Pig	0·5	10·7	10·2	−4·0	Abramson *et al.* (1942)
Guinea-pig	0·56	12·1	10·3	−14·0	Abramson *et al.* (1942)
Toad	0·5	8·1	10·1	+20·0	Lipman *et al.* (1966)
Rat	0·75	14·5	10·5	−28·0	Abramson *et al.* (1942)
Horse	0·88	13·9	10·1	−23·0	Eylar *et al.* (1962)
Lamb	0·75	11·9	10·5	−11·0	Eylar *et al.* (1962)
Calf	0·64	10·2	10·3	+1·8	Eylar *et al.* (1962)
Chicken	0·57	9·0	10·3	+12·6	Eylar *et al.* (1962)
Epithelium (Toad)	0·36	11·3	10·1	−10·5	Lipman *et al.* (1966)
Kidney tumour (Hamster)	1·0	11·5	10·7	−6·0	Ambrose (1967)

[a] Calculated using eq (64) and Fig. 30, p. 195.
From Sherbet *et al.* (1972).

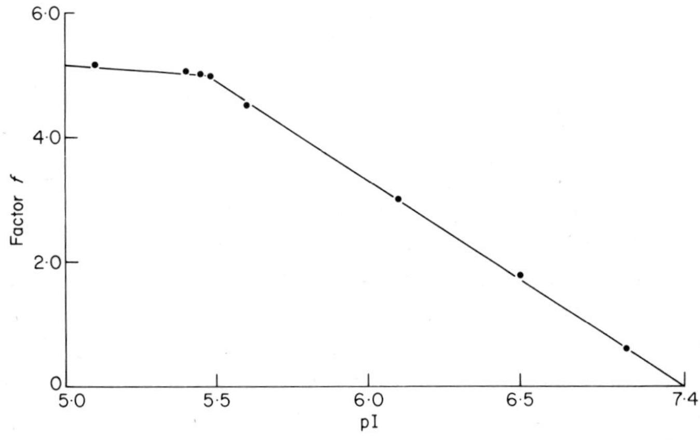

Fig. 30. Variation of $f = v_c/v_0$ as a function of pI (see Table 47) (Sherbet and Lakshmi, unpublished results).

The electrophoretic mobility of a particle is given by:

$$v = \frac{\sigma e}{K\eta} \tag{65}$$

where σ is the number of charges cm^{-2}, and $K = 0{\cdot}327 \times 10^8 I^{\frac{1}{2}}$ where I is the ionic strength. If σ values obtained using eqn (63) were used in the calculation of v, the values of v deviated from the observed values as a function of the respective isoelectric point (Table 47), by a factor, the pH compensation factor, $f = v_c/v_0$ where v_c is the EPM calculated from isoelectric data using equation (65) and v_0 is EPM in free electrophoresis. The variability of factor f as a function of the isoelectric point may be seen in Fig. 30. A linear increase in the value of f is seen with decreasing pI until a pI of 5·47 is reached. Below pH 5·47 the value of f decreased only slightly. This relationship can be enumerated as given below.

Table 47 *Relationship between observed EPM and EPM calculated from isoelectric data*

Cell type	pI	EPM (−TU) Calculated v_c	EPM (−TU) Observed v_0	$f = \frac{v_c}{v_0}$
HeLa	6·85	5·99	9·8	0·571
Liver cells	6·5	18·08	10·09	1·793
Escherichia coli (AW-EI)[a]	6·1	30·8	10·3	3·0
Ehrlich ascites	5·6	47·87	10·65	4·494
Escherichia coli (HCHO)[a]	5·5	51·38	10·6	4·87
Escherichia coli (HCHO)	5·48	52·07	10·52	4·95
Escherichia coli (HCHO)	5·45	53·1	10·62	5·0
Escherichia coli (HCHO)	5·4	54·81	10·75	5·1
Escherichia coli (HCHO)	5·3	58·23	11·6	5·0
Escherichia coli (HCHO)	5·1	65·1	12·6	5·15

[a] *Escherichia coli* (AW-EI) and *E. coli* (HCHO): *E. coli* cells treated with ethyleneimine and formaldehyde respectively (Sherbet and Lakshmi, unpublished work).

For cells whose pI is between pH 7·04 and 5·47,

$$f = (7{\cdot}04 - \text{pI})3{\cdot}1. \tag{66}$$

When the cells are positively charged as a result of experimental modification of surface groups:

$$f = [7{\cdot}04 - (\text{pK}_w - \text{pI})] \times 3{\cdot}1 \tag{67}$$

provided that the value of $\text{pK}_w - \text{pI} > 5{\cdot}47$, where pK_w is the ionisation constant of water. If however, the pI of the cell is below 5·47, or if the value of $\text{pK}_w - \text{pI}$ is less than 5·47, then

$$f = (5{\cdot}47 - \text{pI})0{\cdot}2 + 5{\cdot}0$$

or

$$f = [7{\cdot}04 - (\text{pK}_w - \text{pI})]0{\cdot}2 + 5{\cdot}0 \tag{68}$$

in the case of positively charged surfaces.

Now, introducing the pH compensation factor f into equation (65), the equation for v from isoelectric data can be written as follows:

$$v = \frac{\sigma e}{K\eta f} \tag{69}$$

where v is obtained in units of cm $\text{sec}^{-1}\ V^{-1}$cm.

The employment of the pH compensation factor allows EPMs of cells to be calculated with reasonable accuracy, as may be seen in Table 48. Since EPM is proportional to the ζ potential, the pH compensation should in theory apply to the calculation of ζ potential as well. In Table 48 calculated ζ potential values for a variety of cell types are also given, and these agree reasonably well with values obtained using cell electrophoresis.

SURFACE CHARGE DENSITIES OF SOME CELL LINES CALCULATED FROM ISOELECTRIC DATA

The estimation of net charge of macromolecules, though not on the surface of cells, from isoelectric data has been attempted before. Pauling *et al.* (1949) observed that a difference of 0·22 pH units existed between the isoelectric points of normal and sickle cell haemoglobin (HbS). This observation was confirmed by Svensson (1962) using isoelectric focusing. Pauling *et al.* (1949) in fact went further and estimated this difference in

Table 48 *Isoelectric and electrophoretic data for various cell types*

Cell type	pI	f	EPM (TU) v_c	EPM (TU) v_0	Reference for v_0	ζ^a (mV)
Py-BHK 21	6·4	1·984	−10·4	−11·9	Forrester *et al.* (1964)	−17·9
3T3	4·6	5·174	−15·9	−13·0	Adam and Adam (1975)	−27·5
SV-3T3	4·8	5·134	−14·7	−17·8	Adam and Adam (1975)	−25·4
Yoshida ascites sarcoma	6·41	1·953	−10·4	−12·7	Maekawa *et al.* (1971)	−17·9
Escherichia coli	5·6	4·464	−10·7	−11·1	Gittens and James (1963)	−18·56 (−16·6)
Escherichia coli (AW-EI)	8·55	5·0	+10·6			+18·31 (+16·4)
Escherichia coli (HCHO)	3·85	5·324	−20·3			−35·0 (−31·5)
HeLa cells	6·85	0·589	−8·7	−9·8		−15·1 (−12·7)
Ehrlich ascites	5·6	4·464	−10·7	−10·7		−17·9 (−15·5)
Rat liver cells	6·5	1·674	−10·2	−10·1		−17·7 (−17·0)

[a] $\zeta = P \times 0{\cdot}0592/f$ values given within bracket are ζ potential obtained by cell electrophoresis.

the pIs to represent 2–3 extra net positive charges on the HbS molecule and this is known to be due to the substitution of glutamic acid residue by valine in the β-chain of the haemoglobin molecule.

The charge densities of a variety of cell lines are listed in Table 49. Ehrlich ascites, Yoshida ascites and Py-BHK 21 cells have pI in the range of 5·6 to 6·4 which is attributable, as discussed earlier, to the presence of phosphate, carboxyl and thiol groups. In spite of this similarity in the nature of their cell surface, it will be noticed that the negative charge density on Ehrlich ascites cells is nearly twice that of Yoshida ascites or Py-BHK21 cells. HeLa cells appear to have rather low charge density. From free electrophoresis, however, a charge density of $0{\cdot}7 \times 10^{13}\ \text{cm}^{-2}$ has been calculated. At the other end, 3T3 and SV-3T3 cells have a charge density around $2{\cdot}0 \times 10^{13}$ which is somewhat higher than calculated from their ζ potential (=27·5 mV, Table 48). This essentially means that eqn (63) underestimates charge densities of cells which have pI values between 6 and 7 and gives higher σ values for cells with low pI,

Table 49 *Surface charge densities of some cell lines*

Cell type	Isoelectric point[a] Observed	Isoelectric point[a] Corrected	Charge density[b] $\sigma \times 10^{-13}$ cm^{-2}
Rat liver cells	6·5	6·26	0·589
Ehrlich ascites	5·6	5·35	1·313
Yoshida ascites	6·41	6·16	0·669
3T3 cells	4·6	4·22	2·212
SV-3T3 cells	4·8	4·45	2·029
Py-3T3 cells	4·78	4·36	2·101
Py-BHK 21 cells	6·4	5·15	0·676
HeLa cells	6·85	6·62	0·302
HeLa: γ-globulin treated	6·36	6·11	0·708

[a] pI values corrected using eqn (50).
[b] Calculated using eqn (63).

as compared with charge densities derived on the basis of free electrophoresis.

This deviation is due to either or both of the following factors: state of conductance, and localised increased temperature in the pH gradient. Gittens and James (1963b) have cautioned about the interpretation of electrokinetic measurements, as surface conductance plays a very important role, and they also showed that the major proportion of the observed surface conductance seems to arise in the Stern layer. It is known that surface conductance decreases with higher pH. Besides, the conductivity of ampholytes is the lowest for pH values between 6 and 7 (Davies, 1970). In addition changes in viscosity along the pH gradient also affect conductance, the conductance being lower in more viscous solutions. It appears possible, therefore, that a correction factor for conductance may be required.

The differences in the conductivity along the pH gradient cause alterations in the distribution of field strength. In the low conductance zone this will result in greater power dissipation and consequent localised increases in temperature. Since the isoelectric point is also dependent on temperature this again may influence the outcome and interpretation of data.

In trying to explain the deviation of charge densities calculated from isoelectric data from those based on electrophoresis, lies the implicit assumption that the values obtained from electrophoresis are more accurate and reliable. This is not necessarily true. In fact, Pastushenko and Donath (1976a) state that electrophoretic mobility is not proportional to the surface potential and that the concept of ζ potential may not be applicable in cases where the glycoprotein layer is anything up to 100 Å thick. The surface charge of erythrocytes, for example, when calculated from electrochemical data, gives a value of 2×10^8 electrons $cell^{-1}$. If it is assumed that these charges are distributed in a homogeneous manner, a surface potential (ψ) of approximately 20 mV can be estimated. At ionic strengths of 0·1 M this is equal to a potential of a particle bearing 10^7 electrons $cell^{-1}$ (Pastushenko and Donath, 1976b). In the interpretation of surface charge densities from ζ potentials one is indeed treading on shady uncharted area.

ESTIMATION OF IONISABLE GROUPS ON THE SURFACE OF *Escherichia coli* CELLS

ESTIMATION OF CARBOXYL AND AMINO GROUPS

Equation (63) also provides a device for the quantitation of specific ionogenic groups, if it is used in conjunction with chemical modification or enzymic excision of specific groups. Experiments were described earlier where *E. coli* cells were subjected to treatment with formaldehyde, ethyleneimine and 6,6′-dithiodinicotinic acid in order to achieve a degree of characterisation of these various groups. The same experimental data can also be used for quantitation of the different species of ionisable groups. Such quantitative data are presented in Table 50.

Unmodified *E. coli* cells bear a net negative charge density of $1{\cdot}3159\times10^{13}\ cm^{-2}$. Ethyleneimine (EI) reacts with carboxyl groups to form β-aminoethyl esters and the cells so treated (designated AW-EI cells) can be expected to have a purely amino surface. The corrected pI of these cells therefore gives the amino group density of $1{\cdot}4334\times10^{13}\ cm^{-2}$. Similarly, when amino groups are blocked using formaldehyde, one obtains *E. coli* (HCHO-cells) which have a purely carboxyl surface. From the pI of those cells it is estimated that the surface has a carboxyl group density of $2{\cdot}8896\times10^{13}\ cm^{-2}$. The effective net negative charge density would therefore be $1{\cdot}4562\times10^{13}\ cm^{-2}$. This is about 11% higher than the net negative charge density on the unmodified cell surface. The discrepancy is not large, nevertheless, at least a part of this is attributable to the presence of thiol groups on the surface of the cells.

Table 50 *Estimated ionisable groups on the surface of* Escherichia coli *cells*

Cell designation	Isoelectric point		No. of effective charges cm^{-2} $\times 10^{-13}$	No. of thiol groups cm^{-2} $\times 10^{-13}$	Total charges cm^{-2} $\times 10^{-13}$	Net negative charges cm^{-2} $\times 10^{-13}$
	Observed	Corrected				
Escherichia coli (unmodified)	5·6	5·35	−1·3159		−1·3159	−1·3159
Escherichia coli (AW-EI) cells	8·55	8·8	+1·4334	0·0367[a]	+1·4701	−1·4195
Escherichia coli (HCHO) cells	3·85	3·4	−2·8896		−2·8896	
Escherichia coli (CPDS) cells	4·28	3·86	−2·5002	1·1843[a]		

[a] Number of thiols expected to dissociate at cell pI are approximately 3·1% of the total number of thiol groups. Data from Sherbet and Lakshmi (1973) except that the pI values were corrected by Hartley-Roe treatment and electrical charges are expressed cm^{-2}.

ESTIMATION OF THIOL GROUPS

The isoelectric behaviour of ethyleneimine-treated *E. coli* cells suggested the possibility that the cell surface possessed weakly acidic groups such as thiols. A number of thiol reagents of varying suitability are known. 6,6′-dithiodinicotinic acid (carboxypyridine disulphide, CPDS) is a

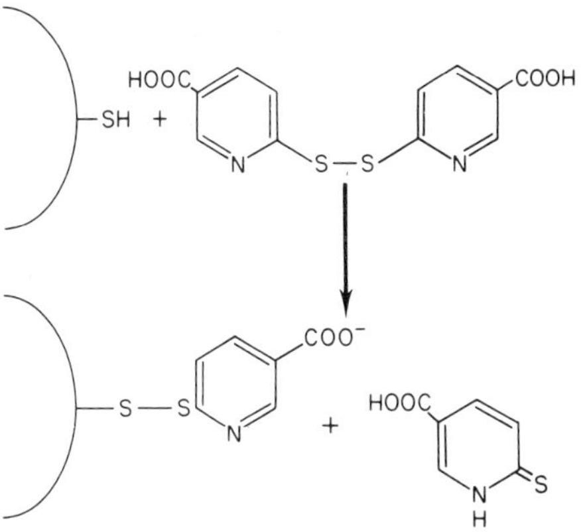

Fig. 31. Binding of 6,6′-dithiodinicotinic acid (CPDS) to the cell surface. The disulphide bond of the CPDS molecule splits up and one-half forms a disulphide bridge with the cell surface thiol group which results in the introduction of one negative charge for each thiol group reacted.

specific and reversible thiol reactant which does not enter the cell or damage the cell membrane (Grasetti *et al.*, 1969; Mehrishi and Grasetti, 1969). The suitability of this compound is attributable to the fact that interaction of this compound with thiol groups results in the splitting up of the disulphide bond of the CPDS molecule, one-half forms a disulphide bridge with the cell surface thiol and the other half is released in the stable thione form. The reaction is schematically represented in Fig. 31. In other words, for each thiol group reacted with a CPDS molecule, one carboxyl group is introduced into the cell surface. As may be seen in Table 50, when *E. coli* cells are treated with CPDS the pI is lowered. The difference in the pI of *E. coli* and CPDS-treated *E. coli* cells is equivalent to the introduction of $1{\cdot}184\times10^{13}\ \mathrm{cm}^{-2}$ carboxyl groups. In other words, the density of thiol groups on *E. coli* cell is $1{\cdot}1843\times10^{13}\ \mathrm{cm}^{-2}$. At the pI of 8·55 of AW-EI cells about 3·1% of these groups may be expected to be ionised thus neutralising some of the positive charges on AW-EI cells. The total amino group density should therefore be amended to $1{\cdot}4701\times10^{13}\ \mathrm{cm}^{-2}$. When this correction is made, a net negative charge density value of $1{\cdot}4195 \times 10^{13}\ \mathrm{cm}^{-2}$ is obtained which agrees very well with the net negative charge density estimated on the surface of unmodified *E. coli* cells.

Although such an approach to the quantitation of ionisable groups on the cell surface can be considered only provisional, this series of experiments does underline the potential value of isoelectric equilibrium analysis in such quantitative investigations.

ASSAY OF ANTIGEN–ANTIBODY INTERACTION AT THE CELL SURFACE

Several methods are available for the investigation of antigen–antibody interactions. In *in vitro* systems these interactions can be visualised as precipitin lines formed in gels in Ouchterlony plates, by immunodiffusion, immunoelectrophoresis, etc. Quantitation of antibody–antigen binding can be achieved by ammonium sulphate precipitation method (see Minden and Farr, 1973). Antibody–antigen interaction can also be measured by the changes in intensity or polarisation of fluorescence when fluorescent-labelled antibodies are employed (see Parker, 1973).

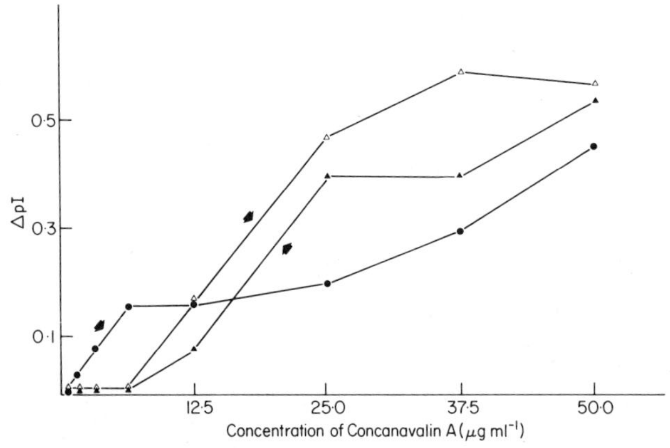

Fig. 32. Kinetics of binding of Con A by 3T3, trypsin-treated 3T3- and SV-40-transformed 3T3 mouse fibroblasts. The phase of binding indicated by arrow suggests that Con A binds the same receptors or receptors with equivalent affinity for Con A, in the three cellular systems. The point at which the kinetics of binding changes is considered as the point of saturation of the high affinity receptors (see pp. 213–222).

Interaction of antibody with cell surface antigen can be demonstrated using fluorescent-labelled antibodies. This method also allows a study of the distribution of antibodies on the surface. Among other qualitative methods used are the agglutination reaction among cells which is brought about by cross linking of multivalent antigens of the cell surface by antibody molecules and complement-mediated cytotoxicity testing which is monitored using dye inclusion or ^{53}Cr release as an indicator of loss of viability of cells consequent upon treatment with antisera. In Chapter 4, the interaction of antibodies with cells and the consequence of such interaction to the electrokinetic properties of the cells was discussed (pp. 93–99). The purpose of this section would be to show that the method of isoelectric focusing can be employed for the demonstration of the binding of antibodies to the cell surface, and in the characterisation and quantitation of cell surface antigens.

EFFECTS OF ANTIBODY BINDING ON CELL pI

As discussed in Chapter 4 cell electrophoresis has shown that binding of antibodies to the cell surface causes changes in the electrophoretic mobility of the cells. This may be attributed to changes in net surface charge densities produced by the introduction of antibody molecules on the cell surface. It may be expected therefore that the binding of immunoglobulin molecules to surface antigens will be reflected by corresponding changes in the pI of cells. Early experiments by Sherbet *et al.* (1972) showed that non-specific adsorption of γ-globulin molecules on HeLa cells caused a significant reduction in their isoelectric point. Similar reductions in the pI of simian virus-transformed 3T3 (SV-3T3) cells was noticed when they were incubated with antisera from rabbits that had been immunised with SV-3T3 cells (Table 51) (Sherbet and

Table 51 *Effects of antibody binding on cell pI*

Target cell	Treatment	Isoelectric point	
		Control cells	Treated cells
HeLa[a]	γ-Globulin (Cohn Fraction IV)	6·85±0·11	6·36±0·14
SV-3T3[b]	Anti-SV-3T3 serum (1:2 diluted)	4·72±0·01	4·4±0·01
SV-3T3	Anti-SV-3T3 serum (absorbed with untransformed 3T3 cells) (1:2 diluted)	4·72±0·1	4·325±0·01

[a] From Sherbet *et al.* (1972).
[b] From Sherbet *et al.* (1974) and Sherbet and Lakshmi (1976).

Lakshmi, 1976). In order to show that the effect was not due to non-specific adsorption of serum components or non-specific antibodies, not only control cells were treated with serum from non-immunised animals, but the antiserum designated S1 (see Table 52) was absorbed with untransformed 3T3 cells. The resulting absorbed serum (S2) also produced reductions in the pI of SV-3T3 cells. These experiments not only demonstrate the binding of antibodies to the cell surface but indicate also that the antiserum S2 contains antibodies specific for antigen(s) associated with SV-3T3 cells. In order to investigate this aspect in greater detail, a quantitation of the binding of antibodies to the cells was undertaken.

Table 52 *Antisera against SV-3T3*

Antiserum	Inactivation	Absorptions	50% cytotoxicity titre
S1	56°C, for 30 minutes	None	1:128
S2	56°C, for 30 minutes	Equal volume of packed untransformed 3T3 cells, 4×45 minutes at 37°C.	1:32
S3	56°C, for 30 minutes	Equal volume of packed untransformed trypsin-treated 3T3 cells, 4×45 minutes at 37°C	

From Sherbet and Lakshmi (1976).

ESTIMATION OF ANTIBODY BINDING

Estimations of antibody binding to the cell surface are made by an "antiserum dilution" method which consists of measuring the effects on cell pI of two or more successive dilutions of the antisera and estimating the change in pI that would be produced when the receptor sites are saturated by the antibodies. This estimated change in pI is termed as "receptor saturation pI differential", the RSPD value. From the RSPD values the numbers of antigenic sites are calculated using equation (63).

Two basic assumptions are made in these calculations, that stoichiometry exists in antibody binding to antigenic sites and that the result of antibody binding is to introduce one net electric charge for each molecule bound. While these assumptions simplify the calculations, it should be borne in mind that the values so calculated become relative and not absolute values. Nonetheless, they allow valid comparisons to be made of the ability of a number of different cell types to bind antibodies present in a given antiserum.

SHARED ANTIGENS ON SV-3T3 AND POLYOMA VIRUS-TRANSFORMED 3T3 (Py-3T3) CELLS AND THEIR CHARACTERISATION

A variety of surface changes accompany viral transformation of cells. Of these, alterations in the expression of surface antigens have been the subject of considerable study. These may involve the appearance of new antigens, loss of normal cellular antigen or unmasking of cellular antigens normally not detected on the cell surface(see Easty, 1974). Two classes of new antigens have been detected in transformed cells. The first class of antigens has been shown to be virus specific and coded for by the virus. The T-antigens described by Pope an Rowe (1964), Rapp *et al.* (1964) and Sabin and Koch (1964) and the U-antigens (Lewis *et al.*, 1969) which have been shown to be present in the nuclei and perinuclear regions of cells transformed by papovaviruses and the transplantation rejection antigens located on the cell surface (Habel, 1961; Sjögren *et al.*, 1961; Defendi, 1963; Koch and Sabin, 1963; Sjögren and Hellström, 1967; Tevethia *et al.*, 1965; Irlin, 1967; Lhérisson *et al.*, 1967; Malmgren *et al.*, 1968) are examples of virus-specific and virus-coded antigens. The second class of antigens are not specific for the transforming virus but appear to be host cell coded. The carcinoembryonic antigens associated mainly with malignancies of the gastrointestinal tract (Laurence and Neville, 1972) and on SV-40 transformed hamster cells (Duff and Rapp, 1970; Coggin *et al.*, 1970) and transformation-associated surface antigens (TASA) may be examples of this second class of antigens.

The purpose of this subsection is to discuss the work described by Sherbet *et al.* (1974) and Sherbet and Lakshmi (1976) where isoelectric focusing was used to demonstrate and partially define certain antigen(s) occurring on the murine 3T3 cell line transformed by simian virus-40.

Sherbet and Lakshmi (1976) examined the effects of the antisera described in Table 52 on the pI of 3T3 cells transformed by SV-40 and polyoma virus (SV-3T3 and Py-3T3) and two other cell lines, Chinese hamster kidney and rabbit kidney cells, both transformed by SV-40 (SV-CHK and SV-TRK, respectively). The antisera produced a marked reduction in the pI of SV-3T3 cells, but neither S1 nor S2 affected the pI of SV-CHK or SV-TRK cells. On the other hand, antiserum S1 produced a reduction in the pI of Py-3T3 cells, but the reduction was much smaller than in the case of SV-3T3 (Table 53). One may conclude from these experiments that SV-3T3 and Py-3T3 cells share certain surface antigens which are not detected on SV-CHK and SV-TRK and on untransformed 3T3 cells i.e. the antigens detected are transformation-associated surface antigens (TASA), but not specific for the transforming virus.

Table 53 *Effects of anti-SV-3T3 serum on pI of some cell lines transformed by SV-40 and polyoma virus*

Treatment	Isoelectric point SV-3T3	Py-3T3	SV-CHK	SV-TRK
None	4·8±0·01	4·78±0·01	4·66±0·01	4·6±0·01
Control serum	4·72±0·01	4·76±0·003	4·7±0·01	4·74±0·04
S1 (1:2 diluted)	4·4±0·01	4·69±0·01	4·73±0·03	4·78±0·02
(1:4 diluted)	4·46±0·003	4·73±0·01		
S2 (1:2 diluted)	4·33±0·005			4·72±0·01

From Sherbet and Lakshmi (1976).

The numbers of TASA sites were estimated using both S2 (the fully absorbed serum) and S1, the unabsorbed serum (Table 52). From either method an estimate of $8{\cdot}6\times10^{11}$ cm^{-2} TASA sites was obtained (Tables 54 and 55). SV-3T3 and Py-3T3 not only shared the antigens but their expression was comparable (Table 56).

The presence of virus-induced antigens on SV-40-transformed cells was demonstrated by Häyry and Defendi (1968, 1969). Subsequently (1970) these authors suggested that these antigens are components of

Table 54 *TASA sites on SV-3T3 cells*

Serum dilution	Isoelectric point Control	Antiserum (S2)	ΔpI	RSPD value[a]	Density of TASA sites $\times10^{-11}$ cm^{-2}
1:2	4·72±0·01	4·325±0·01	0·395	0·108	8·6
1:4	4·76±0·02	4·392±0·02	0·368		

From Sherbet and Lakshmi (1976).

[a] RSPD value: receptor saturation pI differential value (see p. 206).

Table 55 *3T3 specific and TASA sites on SV-3T3 cells*

Cell type	Antiserum S1	Density of sites $\times 10^{-11}$ cm^{-2}: 3T3 specific	TASA	Total sites calculated	Total sites observed	Deviation of observed sites (%)
			8·61			
SV-3T3	1:2	21·21	4·31	25·52	25·52	
	1:4	21·21	2·16	23·37	23·92	2·39
	1:8	21·21	1·08	22·29	21·33	0·18
	1:16	10·61	0·54	11·51	10·37	6·98

From Sherbet and Lakshmi (1976).

Table 56 *3T3 specific and TASA sites on Py-3T3 cells*

Antiserum (S1) dilution	Density of sites $\times 10^{-11}$ cm^{-2}: 3T3-specific	TASA
		6·36
1:2	7·98	3·18
1:4	7·98	1·59

From Sherbet and Lakshmi (1976).

the normal cell membrane and that they can be unmasked by a mild treatment of the cells with trypsin or papain. They also put forward the hypothesis that during transformation the oncogenic virus uncovers specific sites on the cell surface. The net result of this process would then be the expression on the surface of the transformed cell of apparently virus-induced but host cell-coded antigens. The TASA detected on SV-3T3 cells do not appear to be antigens uncovered in this way, for antiserum S2 and antiserum S1 fully absorbed with trypsin-treated untransformed 3T3 cells, bound SV-3T3 cells to the same degree. It is possible therefore that we are dealing with a species of transformation-associated surface antigen(s) that are host cell coded, yet whose expression is controlled by the virus. The virus may not only control expression of TASA, but also the expression of 3T3-specific antigens, as the following discussion would show.

The unabsorbed antiserum S1 may be expected to contain not only antibodies against TASA as shown earlier but it will also contain antibodies against 3T3-specific antigen. Serial dilution experiments using the unabsorbed antiserum provides a device to assess also the expression of these 3T3-specific antigens. Since 3T3-specific antigens form a major antigenic component as compared with the TASA components, it may be expected that in lower dilutions, antiserum S1 will continue to saturate the antigenic sites. On the other hand, since the titre of anti-TASA antibodies may be expected to be low, it may be assumed that the number of bound TASA sites is also halved at each dilution. Using such a device we have been able to show that SV-3T3 cells have about $21 \times 10^{11}\ \mathrm{cm}^{-2}$ 3T3-specific sites (Table 55). But in Py-3T3 cells only $8 \times 10^{11}\ \mathrm{cm}^{-2}$ 3T3-specific sites were detected (Table 56). This may indicate that the two viruses affect different features of the transformed phenotype or reflect the divergent evolution of a cell line after transformation by different viral agents.

In Table 57 the various characteristics which can be used to define TASA are presented. We consider that isoelectric equilibrium method can go even further in elucidating the nature of these antigens. Preliminary experiments, for example, show that concanavalin A partially inhibits the binding of anti-TASA antibodies to the surface antigens. It is estimated (see Table 58) that antibodies and concanavalin A share about $12{\cdot}7 \times 10^{11}\ \mathrm{cm}^{-2}$ receptor sites, or *pro rata* $9{\cdot}8 \times 10^{11}\ \mathrm{cm}^{-2}$ sites are shared by TASA and concanavalin A. This figure agrees so well with the estimate for TASA sites themselves that one may suggest that the TASA antigens contain α-D-mannopyranose or α-D-glucopyranose moieties towards which concanavalin A has predominant specificity (see Lis and Sharon, 1973). In SV-40-transformed lymphocytes Lin *et al.* (1977) have

Table 57 *Characteristics of TASA detected on SV-40 and polyoma virus-transformed 3T3 fibroblasts*

Antiserum	SV-3T3	Py-3T3	SV-CHK	SV-TRK	Nature of antigen
S1	++	+	0	0	Species specific
S2	++	+	0	0	Transformation associated and host cell coded
S3	++		Not tested		Not cryptic normal antigen

Sherbet and Lakshmi (1976).

Table 58 *Shared receptors between concanavalin A and anti-TASA antibodies*

S1 dilution	ΔpI of cells S1 alone	 S1+25 μg Con A		Con A receptors bound by antiserum $\times 10^{-11}$ cm^{-2}	No. of shared receptors $\times 10^{-11}$ cm^{-2}
1:8	0·26	0·5	0·24	12·7	9·8
1:16	0·14	0·54	0·4		

From Sherbet and Lakshmi (1976).

indeed described the appearance of surface antigenic components which are described as being glycoprotein in nature with pI values of 4·5 and 4·7, although admittedly it is difficult to say whether the same class of antigen is involved. The methods used in estimating concanavalin A binding to cell surfaces has been described in the following section and it is easy to see how experiments could be divised to characterise the TASA antigens using lectins of different specificities.

ESTIMATION OF CONCANAVALIN A RECEPTORS

Plant lectins have found a variety of applications. Lectins have been used for blood typing, investigations into structures of blood group substances, as mitogens, and also in recent years as a probe in investigations into the structure and organisation of the cell surface. Concanavalin A (Con A) which is a lectin derived from the jack bean *Canavalia enisformis* has been used extensively. This lectin has high specificity for interaction with polysaccharides which contain multiple terminal non-reducing α-D-glucopyranosyl and α-D-mannopyranosyl residues (Goldstein *et al.*, 1965). A number of other lectins are known which have different carbohydrate specificities, such as, for instance, wheat germ allgutinin (WGA) with specificity for *N*-acetylglucosamine (Le Vine *et al.*, 1972), *Ricinus* lectin for D-galactose (Nicolson and Blaustein, 1972), *L. culinaris* lectin for D-glucose and D-mannose (Entlicher *et al.*, 1969; Young *et al.*, 1971), soya bean lectin for *N*-acetyl-D-galactosamine (Lis *et al.*, 1970). Because of such specificity, plant lectins have been used to explore membrane structure and organisation. Thus differences have been demonstrated in the lectin agglutinability between normal and transformed cells (Inbar and Sachs, 1969a; Aub *et al.*, 1965; Burger, 1969; Burger and Goldberg, 1967; Burger and Martin, 1972;

Sela *et al.*, 1970; Ben-Bassat *et al.*, 1970; Inbar *et al.*, 1971a, 1972a,b), in the distribution of Con A receptors (Inbar *et al.*, 1972a,b; Ben-Bassat *et al.*, 1971; Nicolson, 1972a; Shoham and Sachs, 1972; Martinez-Palomo *et al.*, 1972; Bretten *et al.*, 1972; Unanue *et al.*, 1972; Inbar and Sachs, 1973) and in Con A-induced cytotoxicity (Shoham *et al.*, 1970; Inbar *et al.*, 1972c). It is also known that differences exist in amino acid and carbohydrate transport sites (Inbar *et al.*, 1971b). Agglutination of cells mediated by lectins involves the movement of the receptor–lectin complex into localised patches (Nicolson, 1971, 1972a; Inbar *et al.*, 1973b; De Petris *et al.*, 1973; Noonan and Burger, 1973a). The mobility of the receptors also appears to be a temperature-dependent phenomenon (Nicolson, 1973). The phenomenon of cap formation following interaction of antibody with surface antigen has also been described (Taylor *et al.*, 1971; Yahara and Edelman, 1972, Edidin and Weiss, 1972; Loor *et al.*, 1972). These observations support the view that the cell membrane is a dynamic fluid structure (Singer and Nicolson, 1972). The differences in lectin-mediated agglutination of normal and transformed cells may be a consequence of the differences in the fluidity of the membrane (Nicolson, 1973). Lymphocytes from patients with chronic lymphocytic leukemia and malignant lymphoma show reduced mobility of Con A receptors (Ben-Bassat and Goldblum, 1975; Ben-Bassat *et al.*, 1974, 1976a,b; Mintz and Sachs, 1975).

The interaction of Con A with the cell membrane components can be detected by immunofluorescence (Malucci, 1971; Nicolson, 1973), by employing fluorescence-labelled Con A molecules (Shoham and Sachs, 1972; Inbar *et al.*, 1973a) or by electron microscopy using electron-dense ferritin-conjugated Con A (Singer, 1959; Roos and Temmink, 1975) or coupled with cytochemical method (Barrat and Avramaes, 1973; Rowlatt *et al.*, 1973; Garrido *et al.*, 1974; Garrido, 1975). These methods are adequate for the demonstration of qualitative differences in Con A binding to the cell surface, although Roos and Temmink (1975) have attempted a semi-quantitation using ferritin-conjugated Con A. Fluorescence-labelled Con A has been used for a comparative quantitation of binding with embryonic cells (Neri *et al.*, 1975). Fernandez and Berlin (1976) have recently described a new method for studying lectin distribution by determining resonance energy transfer between donor and acceptor chromophores conjugated separately to lectin molecules. An aggregation or capping of the lectin receptors will be related to the distribution of the donor and acceptor chromophores and therefore to the degree of resonance energy transfer.

Quantitative determinations of Con A binding have been made using Con A labelled with radioactive ^{63}Ni, ^{125}I or with tritium (Inbar and

Sachs, 1969; Inbar *et al.*, 1972; Ozanne and Sambrook, 1971; Noonan *et al.*, 1973; Noonan and Burger, 1973b; Inbar and Sachs, 1973; Collard and Temmink, 1975). The use of radioactively labelled Con A samples is associated with some disadvantages. Ozanne and Sambrook (1971) found ^{63}Ni-labelled Con A unsatisfactory. They obtained varying results from batch to batch of Con A and also found that only up to 30–40% of ^{63}Ni counts bound to the cells competed with the corresponding hapten and it was difficult to saturate the receptors with radioactive ^{63}Ni. Most investigators have not taken into account the possibility of labelled lectin entering into the cells. In spite of the use of corresponding haptens to demonstrate specificity of recognition of receptors by the lectin molecules, counting cell-bound activity may produce erroneous results.

The electrokinetic properties of a cell are determined purely by the state of ionisation of the chemical groupings of the membrane components occurring in the electrokinetic zone of shear, or in the isoelectric zone. While the electrokinetic behaviour can be altered by modifying the charged state of the cell surface by excision of surface components or by the adsorption of exogenous molecules, the occurrence or the introduction of exogenous charged molecules intracellularly is not expected to alter the electrokinetic behaviour or the isoelectric properties of the cell.

EFFECTS OF CON A BINDING ON CELL SURFACE NEGATIVITY

Yamada and Yamada (1973) used cell electrophoresis to examine binding of Con A to rat ascites hepatocarcinoma cells. They observed that anodic EPM of cells increased with treatment in concentration of up to $10\,\mu g\,ml^{-1}$ i.e. the net negative surface charge density is raised as a result of the binding. This successful electrokinetic investigation prompted an examination of such interaction by the isoelectric equilibrium method. In agreement with the observations of Yamada and Yamada (1973) treatment of cells with Con A decreased their pI, i.e. increased their electronegativity (Table 59; Fig. 32, p. 204). This observation is compatible with the isoelectric profiles of Con A preparation. In our laboratory Con A (Sigma) was found to be isoelectrically heterogeneous containing four components with pI values of 6·53, 6·4, 6·25 and 6·07 (Sherbet and Lakshmi, unpublished work). Entlicher *et al.* (1971) earlier described the appearance of four bands in isoelectric focusing with pI values 5·7, 5·32, 4·95 and 4·74. Although these values are lower than the ones we obtained, which may be attributable merely to experimental conditions, both these studies indicate that Con A

Table 59 *Concanavalin A receptors on untransformed and SV-40-transformed 3T3 fibroblasts*

	Concentration of Con A $\mu g\ ml^{-1}/\Delta pI$									Concanavalin A receptor density $\times 10^{-11} cm^{-2}$			
										Estimates from			
Cell type	0·781	1·562	3·125	6·25	12·5	25·0	37·5	50·0	RSPD[a] value	RSPD value[b]	Molecular weight[c]	Other sources	Reference
Untransformed 3T3	—	0·03	0·08	0·16	0·16	0·2	0·3	0·47	0·16	12·7	12·6	11·4	Burger and Goldberg (1967)
												10·3	Inbar and Sachs (1969)
												12·5	Ozanne and Sambrook (1971)
												5·6	Phillip *et al.* (1974)
												6·2	Collard and Temmink (1975)
Trypsin-treated 3T3 cells	—	—	—	—	0·08	0·4	0·4	0·49	0·4	36·6	35·9	46·0	Inbar and Sachs (1969)

SV-3T3	0·02	—	—	—	0·17	0·47	0·59	0·52	0·56	43·8	49·5	39·5	Burger and Goldberg (1967)
												58·0	Inbar and Sachs (1969)
												53·8	Noonan and Burger (1973)
												19·8	Phillip *et al.* (1974)
												39·0	Collard and Temmink (1975)

[a] Change in pI produced when receptors are saturated (receptor saturation pI differential, RSPD value).
[b] Using $\sigma = \text{PDK} \times 3{\cdot}3 \times 10^{-3}/4\pi e$ (see eqn 63) where σ = no. of receptors cm^{-2} and $P = \text{RSPD} \times 2{\cdot}303RT/F$.
[c] From saturation concentration $[C]_s$ of Con A for 3T3 5·41 μg; trypsin-treated 3T3 11·44 μg and SV-3T3 15·25 μg 10^{-6} cells; MW Con A is assumed to be 110 000 (see p. 218 for method of calculation).

molecules are negatively charged at physiological pH values, which would account for the increase in the net electronegativity of the cell surface after interaction with Con A.

Although Con A has strong affinity and a high specificity for α-D-mannopyranosyl and α-D-glucopyranosyl residues, it is also known to interact with lesser affinity with β-D-fructofuranosides and D-arabinofuranosides (Lewis *et al.*, 1967; So and Goldstein, 1969; Goldstein and Misaki, 1970). In addition there is also a difference in its affinity to manno- and glucopyranosyl residues. So and Goldstein (1968) measured binding constants of (Ka) $1{\cdot}4\times10^4$ litre $mole^{-1}$ and $0{\cdot}3\times10^4$ litre $mole^{-1}$ to α-D-mannopyranoside and α-D-glucopyranoside respectively. In view of this, it is customary to test the specificity by examining the binding in the presence of an appropriate hapten (α-D-mannopyranoside). If Con A–receptor interaction is specific, binding of Con A will be abolished by the presence of the competing hapten molecules. In addition, one may use the kinetics of Con A binding to infer the specificity of interaction.

In the investigation of Con A binding by normal 3T3, simian virus transformed 3T3 (SV-3T3) cells and trypsin-treated normal 3T3 cells, both these methods were employed to check the specificity of binding. In untransformed cells a linear reduction in pI was observed with increasing Con A concentration until it reached $6{\cdot}25\ \mu$ gm ml^{-1}. Above this point no change in pI occurred until Con A concentration reached $37{\cdot}5\ \mu$ gm ml^{-1} when further reductions in pI occurred. The reduction in pI occurring in the initial linear phase is abolished by subsequent treatment of the cells with methyl α-D-mannopyranoside (α-MMP). It may be considered therefore that the phase of Con A binding during which the decrease in pI is linear represents Con A-specific binding. It may also be noticed in Fig. 32 (p. 204) that the rate at which the pI decreases is virtually the same for 3T3, SV-3T3 and for trypsin-treated normal 3T3 cells, for this particular phase of Con A binding. This may indicate that either the same receptors or receptors with equivalent affinity are bound in the three cellular systems. The point at which the change in kinetics of binding occurs is considered as the point of maximal pI differential (RSPD) caused by the saturation of receptors which have high affinity for Con A. In subsequent phases of binding receptors which have much lower affinity for Con A appear to be bound and this causes changes in cell pI at a much slower rate. Ultimately, at high concentrations the three cell systems appear to bind the same amount of Con A. The RSPD values are utilised in quantifying the binding of the lectin. The concentration of Con A required for saturating receptors on 10^6 cells are $5{\cdot}41\ \mu g$, $11{\cdot}44\ \mu g$ and $15{\cdot}29\ \mu g$ respectively for untransformed 3T3, trypsin-treated untransformed 3T3 and SV-3T3 cells. These values may appear low compared with data published by Shoham and Sachs (1972) and Inbar and Sachs (1973), although these authors did not set out to determine saturation concentrations. Comparison of the saturation values obtained in our investigations with Con A concentrations required for half-maximal agglutination may also be misleading, for cell agglutinability changes from day to day after subculturing (Inbar *et al.*, 1972) and agglutinability is also temperature sensitive (Inbar *et al.*, 1971a). Values of 60–120 $\mu g\,ml^{-1}$ Con A have been reported for half-maximal agglutination of 3T3 and SV-3T3 cells (Inbar and Sachs, 1969a; Nicolson, 1971, 1972b). These values doubtless provide valuable comparison between the two cell lines. The wide differences between concentrations of Con A required to saturate high affinity receptors and to cause agglutinability may indicate that agglutinability does not result immediately on saturation of the high affinity receptors but binding of Con A to receptors with less affinity may also be required to achieve agglutination.

From the estimates obtained in the author's laboratory, high affinity binding sites of 3T3 cells account for about 36% of total lectin binding. In rabbit thymocytes less than 25% of the total lectin bound is accounted for by high affinity receptors (Schmidt-Ullrich *et al.*, 1976). Schmidt-Ullrich *et al.* have also pointed out that cell agglutination by Con A may involve adsorbed lectin which may not be bound to specific receptors.

Vlodavsky and Sachs (1975) have pointed out that cell agglutination by lectins is a multi-step and complex process and that lectin receptors have functional heterogeneity. The efficiency of agglutination depends not only upon the degree of affinity between the receptor and the lectin but also on the distribution of the receptors on the membrane and on the degree of their mobility. Ratishauser and Sachs (1974) found that aggregation of lectin receptors is not necessary and in fact inhibits lectin-mediated aggregation. The inhibition of lateral mobility and patching and capping of immunoglobulin receptors seems to require high lectin concentrations (Yahara and Edelman, 1972). It appears therefore that a realistic comparison cannot be made between saturation concentrations of lectin determined by isoelectric equilibrium method and the concentrations required to achieve half-maximal agglutination of cells.

QUANTITATION OF CON A BINDING BY ISOELECTRIC EQUILIBRIUM METHOD

The advantage of using isoelectric equilibrium method over free electrophoresis is that the former allows a quantitation to be made of Con A binding to the surface and thus provides the much needed corroboration of the quantitation of binding made using radioactively labelled Con A.

The number of Con A receptors can be estimated from the maximum pI difference obtained when the receptors are saturated, termed the receptor saturation pI differential (RSPD) value, on the assumption that one net negative charge is introduced into the cell surface for each Con A molecule bound. This assumption may be considered reasonable for the following reasons: the major component of Con A has a pI of 6·53 from which it is estimated that each molecule possesses no more than five net negative charges and it is also known that each receptor site may involve two saccharide residues (Kalb and Levitskyi, 1968; So and Goldstein, 1968). The difference in net charge densities brought about by the treatments were calculated using eqn (63).

QUANTITATION OF CON A BINDING FROM SATURATION CONCENTRATION AND MOLECULAR WEIGHT OF CON A

The density of receptors can also be calculated from the saturation concentration values and molecular weight of Con A using the following relationship:

$$\sigma = \frac{[C]_s \times N}{\mathrm{M}nA} \tag{70}$$

where σ is receptor density (i.e. number of receptors cm^{-2}), $[C]_s$ in g is the concentration of Con A required to saturate receptors, N is Avagadro's number ($6{\cdot}023 \times 10^{23}$), M is the gram mole (110 000 in the case of Con A), n is the number of cells ml^{-1} and A is the surface area of the cell.

For the 3T3, SV-3T3 system saturation concentration $[C]_s$ is calculated as follows:

$$[C]_s = \frac{\mathrm{RSPD} \times 50}{1{\cdot}3} - [C]_m \tag{71}$$

where $[C]_m$ is the minimum concentration of Con A required to produce a detectable pI change. This relationship, not applicable to other systems, is based on the observation (Fig. 32) that rate of change of pI is 1·3 pH units 50 μg^{-1} Con A at a cell concentration of 10^6 ml^{-1}.

CON A RECEPTORS ON 3T3 AND SV-3T3 CELLS

From the isoelectric data given in Table 59, it is estimated that untransformed 3T3 cells possess $12{\cdot}7 \times 10^3$ Con A receptors μm^{-2}. The density of receptors calculated from saturation concentrations agrees very well with this estimate. Inbar and Sachs (1969b) obtained a value of $10{\cdot}3 \times 10^3$ sites μm^{-2} while Burger and Goldberg (1967) and Ozanne and Sambrook (1971) obtained values of $11{\cdot}4$ and $12{\cdot}5 \times 10^3$ μm^{-2} respectively. Phillips *et al.* (1974) and Collard and Temmink (1975) have, however, obtained considerably lower values.

In SV-3T3 cells there appears to be a three-fold increase in receptor density as compared with their untransformed counterparts. A similar increase was also described by Inbar and Sachs (1969b), Burger and Goldberg (1967), Ozanne and Sambrook (1971) and Phillips *et al.* (1974). Noonan and Burger (1973) found that polyoma-transformed 3T3 cells also bound 3–5 times more Con A than untransformed 3T3 cells. Márquez (1976) demonstrated that hamster embryo fibroblasts transformed by herpes virus bound more Con A than did the corresponding normal cells. A number of other investigators have claimed that normal and transformed cells bind similar amounts of Con A (Cline and Livingstone, 1971; Arndt-Jovin, 1971; Ben-Bassat *et al.*, 1971; Inbar *et al.*, 1971a; Kraemer *et al.*, 1972; Trowbridge and Hilborn, 1974; Rottman *et al.*, 1974). It should be borne in mind, however, that the alleged similarity may be because surface areas of the cells may not have been accurately determined. Collard and Temmink (1975) found that 3T3 and SV-3T3 cells bound similar amounts of Con A. But they found that the surface area of untransformed cells was nearly seven times larger than that of the transformed cells. Therefore the density of binding is effectively seven times greater in transformed cells than in untransformed cells. Such an inaccuracy is not expected to enter into calculations of isoelectric data because the equations evolved for calculation of surface charge density provide an inherently direct relationship between charge density and change of pI. In addition, isoelectric data relate only to Con A adsorbed on to the surface and any Con A that may have entered the cell does not exert any electrokinetic effect. When radioactively labelled Con A is used, it is difficult, if not impossible, to ensure that one is measuring only the radioactivity bound to the membrane and not that present intracellularly. The increased receptor densities on SV-3T3 cells are compatible with the findings that untransformed cells require 3–4 times higher Con A concentrations to bring about half-maximal agglutination than do the transformed cells (Noonan and Burger, 1973b; Nicolson, 1972b, Arndt-Jovin and Berg, 1971).

Trypsin and other proteolytic enzymes are known to alter the surface components by removing sialomucopeptides and acidic mucopolysaccharides. Trypsin is known to alter cellular agglutinability (Inbar and Sachs, 1969a; Burger and Goldberg, 1967; Burger, 1966; Weber, 1973) and also alter the distribution of lectin-binding sites (see p. 114 for references). Inbar and Sachs (1969) and Márquez (1976) observed that a treatment with trypsin resulted in the exposure of cryptic Con A receptor on hamster cells. But subsequent work (Vlodavsky *et al.*, 1973; Shoham and Sachs, 1974) has failed to confirm these findings. There are also a number of other investigations in which no differences have been detected in lectin binding following trypsin treatment (Nicolson, 1974). Interestingly enough, electrokinetic experiments (summarised in Table 59, Fig. 32) described earlier in this section, and those of Yamada and Yamada (1973) seem to resurrect this disputed aspect of trypsin treatment of cells. Isoelectric equilibrium experiments indeed show that Con A binding by untransformed cells alters completely on treatment with trypsin, and assumes features characteristic of transformed cells (Sherbet and Lakshmi, 1978). Borysenko *et al.* (1977) have shown that untransformed 3T3 cells do not show a Con A-induced clustering of the lectin receptors, unlike transformed 3T3 cells. Such clustering can be induced in normal cells if they are pretreated briefly with trypsin.

Gahmberg and Hakomori (1976) do not appear to subscribe to the view that trypsin causes conformational changes in the membrane, but they do believe that trypsin confers on normal cells features which are characteristic of transformed cells. They have reported that normal NIL and BHK cells contain two components, galactoprotein *a* and *b*. The transformed counterparts are said to lack the component *a* but possess galactoprotein *b*. Con A and *Ricinus* agglutinin are said to bind component *a* of normal cells but component *b* of transformed cells. The preferential binding to component *a* in normal cells may be a reflection of a higher affinity of the lectins to that component. According to Gahmberg and Hakomori, trypsin removes component *a* and therefore binding characteristics of normal and transformed cells come to resemble each other. These observations are compatible with the behaviour of the fibroblast surface antigen described by Vaheri and Ruoslahti (1974) and Keski-Oja *et al.* (1976). This antigen is lost in viral transformation of the cells and is removed from the cell surface by trypsin treatment. It may be pointed out, however, that work in the author's laboratory has suggested that Con A binds receptor or receptors of different class in normal 3T3, trypsin-treated 3T3 and in SV-40-transformed cells, as indicated by the kinetics of binding (Fig. 32, p. 204).

The kinetics of binding (shown in Fig. 32) were analysed using a Hill plot (logit Y/Y_{max} vs. log z, where Y is ΔpI and Y_{max} the RSPD value and z the concentration of the lectin). The following values were obtained: normal 3T3 – slope $s = 12{\cdot}34$, intercept $a = -0{\cdot}45$; trypsin-treated 3T3 cells $s = 11{\cdot}69$, $a = -10{\cdot}5$; SV-40-transformed cells $s = 10{\cdot}15$, $a = -8{\cdot}23$. These data indicate that the receptors occurring in normal 3T3 cells and SV-40-transformed cells belong to two distinct classes but probably possess similar affinity to the lectin, although it could also be argued that the difference in the Hill coefficient is large enough to suggest a higher affinity in case of the normal cells. A third class of receptors are also detected on the untransformed cells, and these have much lower affinity to Con A than the other two classes.

On the basis of our observations, we would postulate that receptors of different class bind Con A in the different phases of interaction of the lectin with the cell surface. Three phases of interaction may be distinguished, namely Phases A, B and C. It may be postulated that two, possibly three, classes of receptor, K1, K2 and K3 are involved in the interaction in the three phases respectively. The untransformed 3T3 cells contain receptor K1 and K3, of which K1 appear to bind first. This is followed by the binding of K3 receptors at high concentrations of the lectin. This binding may be non-specific because α-MMP does not compete with the bound Con A. In trypsin-treated normal cells and SV-3T3 cells, binding to K1 receptors is probably not detected because these receptors are buried in deeper zones not amenable to detection by the isoelectric equilibrium method. This is followed by the binding of K2 receptors which do not appear to be present in untransformed 3T3 cells. This pattern of binding is summarised in Table 60. The divergence of views about the binding pattern could be due to unsuspecting

Table 60 *Concanavalin A binding pattern of normal 3T3, trypsin-treated normal 3T3 and SV-40-transformed 3T3 murine fibroblasts*

Binding phase	A	B	C	Total Con A receptors ($\times 10^{-3}$ μm^{-2})
Receptor class	K1	K2 ($\times 10^{-3}$ μm^{-2})	K3	
3T3 cells	12·7	—	24·7+	37·4+
Trypsin-treated 3T3 cells	14·6	36·6	?	53·6
SV-3T3 cells	20·3	44·6	?	67·3

From Sherbet and Lakshmi (1978).

comparisons being made of densities of receptors belonging to different classes. This analysis also suggests that K2 receptors appear on the cell surface as a result of trypsin treatment and as a result of viral transformation, possibly indicating a conformational reorganisation of the membrane components (Sherbet an Lakshmi, 1978).

Clear differences in the kinetics of Con A binding between differentiated and undifferentiated neuroblastoma cells have also been described. Rosenberg and Charalampous (1977) have reported a higher Hill coefficient for the binding of the lectin to the differentiated cells than to the undifferentiated cells, again indicating possible differences in the structure and organisation of the receptors. The processes of differentiation as well as transformation produce distinct alterations also in the social behaviour of cells. Whether the altered expression of lectin receptors is merely related to this change in the behaviour or whether it is an occurrence intrinsically associated with the processes of differentiation and transformation, is still unclear.

CELL SURFACE CHANGES AND TUMOUR PROGRESSION

During malignant transformation the cell surface undergoes many alterations such as, for example, in surface components, membrane permeability and susceptibility to agglutination by plant lectins. Alterations in the density of net negative electric charge on the surface has also been described by several investigators (Ambrose *et al.*, 1956; Forrester, 1965; Fuhrman, 1965; Ruhenstroth-Bauer, 1965; Weiss and Sinks, 1970). The surface charge characteristics of normal and neoplastic cells as understood by electrophoretic measurements have been discussed earlier (pp. 134–137). Since malignancy is a progressive characteristic of a neoplasm, it may reasonably be expected that the cell surface also will change with tumour progression.

Attempts have been made in the past to correlate EPM of tumour cells with their malignancy. Purdom *et al.* (1958) examined the electrophoretic mobilities of a number of sublines derived from MCIM sarcoma of mice possessing various degrees of malignancy. They concluded from this study that the more malignant a cell line the greater was its EPM. In a number of strains of rat ascites hepatoma cells, an increasing ability of the cells to metastasise could be correlated with a progressive increase in electrophoretic mobility (Sakai, 1967). But Weiss and Hauschka (1970) found no correlation between EPM of TA3 ascites tumour cells and the survival times of a variety of host strains of mice in which the tumour cell line could be carried. However, malignancy assessed in terms of survival times, is always with respect to the particular conditions of the host itself. Therefore a comparison of the survival times of a number of host strains carrying a given tumour does not provide adequate grounds for refuting a correlation between malignancy and electrophoretic mobility. The experiments of Weiss and Hauschka (1970) are unsatisfactory since they may have been performed under a false premise.

The malignancy of a tumour is also reflected by its degree of differentiation and this characteristic has often been used in determining malignancy and prognosis (Bloom, 1950a,b, 1965; Bloom and Richardson, 1957). Sherbet and Lakshmi (1974b,c) have recently evolved an embryological assay for malignancy. This assay system is based on the interaction between embryonic and tumour cells (Sherbet and Lakshmi, 1970; Sherbet *et al.*, 1970; Lakshmi and Sherbet, 1974). The tumours are graded for malignancy depending upon the nature, frequency and intensity of the embryonic responses induced by the tumours. In a recent investigation Sherbet and Lakshmi (1974a) examined the isoelectric characteristics of human astrocytomas in culture in relation to their malignancy as assessed not only by the survival times of the patients but also using histological and embryological grading of their malignancy. These experiments have indicated a clear correlation not only between malignancy and the isoelectric point of the surface but also between the rate of growth of the tumours in culture and their surface charge densities (Table 61). It may be seen from Table 61 that a greater degree of malignancy of tumour is associated with a higher charge on the cells. It may also be noticed that the rate of tumour growth is also correlated with the net negative charge density on the surface of the cells. It is difficult to conclude from this data whether the higher charge density is characteristic of malignancy or merely associated with a higher growth rate. Non-malignant meningiomas and normal fetal brain cells too possessed surface charge densities equal to or even higher than some very malignant astrocytic tumours. Thus charge densities seem to vary irrespective of malignancy. There is evidence that regenerating liver cells have higher electrophoretic mobilities than non-regenerating liver cells (Ben-Or *et al.*, 1960; Eisenberg *et al.*, 1962; Chaudhuri and Lieberman, 1965). Several other investigators (Simon-Reuss *et al.*, 1964; Mayhew *et al.*, 1968) have also argued that differences in surface charge densities reflect differences in rates of growth rather than malignancy. Neither is surface charge related to the state of differentiation, according to Elul *et al.* (1975). They found that while the EPM of differentiated and undifferentiated neuroblastoma cells differed considerably in their EPM

Table 61 *Isoelectric characteristics of some human intracraneal tumour cell lines*

	Malignancy rating			Isoelectric point				
Tumour	Average[a] survival time (months)	Histological grade	Embryological grade	Mechanically harvested	Versenated cells	Trypinised cells	Net negative charges $\times 10^{-4}\ \mu m^{-2}$	Mean[b] generation time (h)
Astrocytoma	2–6	III, IV	10–12	4·7 ± 0·04	4·6 ± 0·04	4·48 ± 0·04	20·05	165 ± 18
	7–10	II	~7	4·75 ± 0·05	5·05 ± 0·08	5·0 ± 0·08	15·91	249 ± 24
	>10	I	~5			5·15 ± 0·09	14·72	
Meningioma	Non-malignant			4·7 ± 0·05	4·78 ± 0·07	4·73 ± 0·08	18·06	194 ± 32
Fetal brain cells	Normal cell line			4·65 + 0·05	4·45 + 0·05	4·38 + 0·07	20·84	66·0

[a] Average survival time of patients.
[b] Mean generation time of cells in culture.
From Sherbet and Lakshmi (1974a).

in normal medium, in serum-deprived medium both had similar electrophoretic mobilities. Elul *et al.* therefore suggested that the differences in their mobilities in normal culture medium is due to the high rate of their division. Elul *et al.* argue that in serum-free medium neuroblastoma cells stop dividing and differentiate (Seeds *et al.*, 1970; Kates *et al.*, 1971; Booher *et al.*, 1973). However, the EPMs in serum-free medium are at the same level as undifferentiated cells in normal medium, while differentiated cells show much reduced mobilities. If the presence of serum stimulated cell division, as it is known to do, one would expect similar EPM values for differentiated cells in normal medium and serum-free medium, but considerably higher EPMs in serum-stimulated dividing (undifferentiated) cells.

The higher EPMs of undifferentiated (dividing) cells are compatible with the observations of Mayhew and O'Grady (1965) and Brent and Forrester (1967) who found the highest EPMs in peak mitotic phase. But recently Adam and Adam (1975) observed that density-inhibited cultures of 3T3 cells actually showed a reduction in EPM when subjected to growth stimulation by serum of trypsin. Adam and Adam (1975) have further shown that SV-40-transformed 3T3 cells which are not liable to density-dependent growth inhibition, do not show any variation in EPM on being subjected to the same stimuli of cell division. This investigation not only contradicts the view that dividing cells possess higher surface charge densities but also introduces a new element of the involvement of surface charge in the regulation of cell division.

Evidence has grown considerably in the past few years which suggests the active involvement of the cell membrane in the regulation of cell division. The triggering of cell division by plant lectins, proteases and serum may involve surface architecture of cells (Burger, 1973; Burger *et al.*, 1972). The higher sialic acid content reported for fucose-containing glycoproteins of and the higher sialyltransferase activity associated with mitotic cell membrane fractions (see pp. 139–140) indicates the possible involvement of sialic acids in growth regulation.

Although the question of surface charge, growth rate and malignancy, must be deemed as unresolved, one particular aspect of the work of Sherbet and Lakshmi (1974a) does appear to suggest the treatment of surface charge and malignancy relationship still as a possibility quite distinct from the well-known and unequivocal relationship between surface charge and growth rate. The isoelectric data given in Table 61 shows that the pIs of highly malignant Grade III and IV astrocytomas are reduced by trypsinisation. On the other hand, the less malignant Grade II astrocytomas and the non-malignant meningiomas, and the normal fetal brain cells are unaffected by trypsin. For reasons stated earlier (pp. 113–114) the reductions in pI on trypsinisation could not have been due to the adsorption of trypsin molecules on the cell surface. The enzyme is known to strip material off the surface of erythrocytes (Ponder, 1951; Pondman and Mastenboek, 1954; Seaman and Heard, 1960). More recently Huggins *et al.* (1976) found that trypsin treatment of sarcoma 180 tumour cells caused a loss of surface glycoprotein and a high molecular weight polypeptide suggested to play a role in membrane stabilisation. The trypsin susceptibility of Grade III and IV astrocytomas indicates that these tumours may possess some additional trypsin-labile surface coat material associated with their malignancy. In a line of BHK21 cells and their polyoma virus-transformed counterparts also an analogous situation appears to exist (Latner and Turner, personal communication). A short treatment with 0·04% trypsin appears to increase the EPM of the virus-transformed cells but it has no effect on the mobility of the untransformed cells. Latner and Turner have suggested the possibility that the enzyme removes some basic substances associated with the surface of the transformed cells. It may be worthwhile recalling here that Sherbet and Lakshmi (unpublished work, see pp. 188–190 and Tables 43 and 44) found that SV-40-transformed 3T3 cells possessed 30% more cationic groups at the surface than the corresponding normal cells.

It has been reported that cells transformed by oncogenic viruses have thicker surface coats than do corresponding normal cells (Martinez-Palomo, 1970; Poste, 1970). It appears reasonable to suggest that the appearance of trypsin susceptibility in malignant cells is associated with appearance of a new surface component(s). Electrophoretic analyses of membrane components of normal and virus-transformed 3T3 cells have revealed the presence of a new protein and glycoprotein components in the transformed cells (Sheinin and Onodera, 1972). Sherbet and Lakshmi (1976) demonstrated the appearance of a surface antigen on the surface of SV-40- and polyoma-transformed cells. Recently Lin *et al.* (1977) detected two antigenic proteins on SV-40-transformed lymphocytes which were not detectable on the membrane of normal lymphocytes. In other studies on virus-transformed 3T3 cells, Gahmberg and Hakomori (1973) and Gahmberg *et al.* (1974) found that in these cells components of molecular weight 30 000–200 000 had been deleted with the appearance of other components of molecular weight 85 000–105 000. On the other hand, there are several papers in which decrease in or disappearance of high molecular weight surface components have been described (Hynes, 1973; Wickus and Robbins, 1973; Stone *et al.*, 1974; Bussel and Robinson, 1973; Hynes and Wyke, 1975; Gahmberg and Hakomori, 1976; Vaheri an Ruoslahti, 1974; Keski-Oja et al. 1976, see also p. 139).

MONITORING OF LABORATORY CELL LINES BY ISOELECTRIC FOCUSING

It is almost universally acknowledged that growth of cells over prolonged periods in tissue culture conditions subjects the cells to spatial, toxicological and nutritional constraints. Over prolonged periods in culture adaptive selection is liable to occur. By a process of alternate *in vitro* and *in vivo* growth (Laskov and Scharff, 1970) it is possible not only to adapt some murine plasmacytomas for growth in tissue culture, but it is also possible to select variant plasmacytoma cells with altered growth properties and altered immunoglobulin phenotype (Hausman and Bosma, 1975). It is not unknown for laboratory cell lines to undergo spontaneous transformation or to become contaminated with other lines being handled in the same laboratory. In present day medical research, tissue culture is being used extensively and it is imperative that cell lines are monitored routinely using one of several criteria such as growth kinetics, surface enzyme or other antigenic markers, etc. In this context, the isoelectric properties of cells may provide a new additional criterion for monitoring of cell lines or clones. This method has in fact been used in our laboratory to monitor murine 3T3 cells maintained over prolonged periods and human astrocytoma cell lines kept in culture for 90–120 days. The data provided in Table 62 indicate that the 3T3 cells used in various biochemical investigations showed the same pI. This means that the surface characteristics of these cells are fairly stable. H. S.

Table 62 *Cell pI in relation to duration of growth in tissue culture*

Cell type	Transfer number	Duration in culture (days)	Isoelectric point
Balb c/3T3[a]	27	?	5·35[a]
	54	?	5·25
Human astrocytoma[b]	Primary	7	4·65
Grade IV	3	80	4·55
	3	90	4·5
	4	120	4·4
Human fetal brain cells[b]	Primary	7	4·4
	5	90	4·5
Astrocytoma[b]	Primary	40	4·9
Grade II	1	70	5·05

[a] Sherbet and Lakshmi (unpublished work). The pI values were measured at 20°C, I= 0·01 M.

[b] From Sherbet and Lakshmi (1974). pI values were measured at 25°C, I=0·009 M.

Smith *et al.* (1976) have examined several surface-mediated cultural features such as growth properties, saturation density, morphology and tumorigenicity of SV-3T3 cells and found that these were stable over 50 passages. These observations support the stability of the isoelectric characteristics of 3T3 fibroblasts noticed in our laboratory.

Non-established human astrocytoma Grade IV cells seem to have undergone some surface alterations in relation to the period in culture of not more than 120 days, roughly equivalent to four transfers. On the other hand, the less malignant Grade II astrocytomas and normal fetal brain cell lines showed comparatively more stable isoelectric properties. H. S. Smith *et al.* (1976) found that subtle changes had occurred in the expression of the viral genome of the transformed 3T3 cells. It appears therefore that isoelectric characteristics are a reasonably sensitive means which can be usefully employed in periodical monitoring of laboratory cell lines.

INTERACTION OF DRUGS WITH CELL SURFACE

One of the important features of the concept of chemotherapy proposed by Ehrlich (1957) is the recognition that a chemotherapeutic agent should possess chemical groups which can mediate the binding of the agent to receptor groups on the cell surface and a "toxophoric" group which brings about the destruction of the target cell. In order that the target cells can be destroyed selectively, it is necessary to ensure that a potential chemotherapeutic agent has affinity for receptors occurring only on the target cell.

Another way of destroying a target cell is by interfering with its metabolic processes with as much selectivity as one can achieve in designing the chemotherapeutic agent. There are several anti-metabolites which have been used with varying degrees of success in controlling the growth of neoplasms (see Drews, 1976). Nonetheless, achieving the desirable selectivity of interaction has always been difficult. Unger (1976) has suggested that a search for anti-metabolites which affect the process of biosynthesis and assembly of surface components could be promising, especially in view of the numerous observations related to the differences in the surface components, properties and altered growth and immunological properties of the malignant cells.

Two important parameters to be examined in the early stages of designing of chemotherapeutic agents are the binding of the agent at the cell surface and whether the agent produces alterations in the nature of the cell surface. Both these parameters lend themselves to investigation by cell electrophoresis, but probably more efficiently by the isoelectric equilibrium method. Although admittedly neither method has been used extensively in the investigation of drug interaction with the cell surface, in this section the available work will be discussed with a view to demonstrating the utility of the method, especially the isoelectric equilibrium method, in this and related fields.

EFFECTS OF POLYIONIC COMPOUNDS AND CHONDROITIN SULPHATE ON SURFACE CHARGE

Mehrishi (1970c) examined the effects of poly I:C (polymer of ionosinyl and cytosinyl residues) which is said to be tumour inhibitory, and of chondroitin sulphate (said to be a tumour promoting agent) on the surface charge of Ehrlich ascites tumour cells. Although the compounds are supposed to produce antagonistic results, both produced increases in their EPM. The only difference was the degree of increase in EPM. While poly I:C produced a 300% increase, chondroitin sulphate raised the EPM by a mere 24%. It may, however, be assumed that the binding of these compounds, especially of poly I:C, is non-specific, and the differences in the EPM produced may not be of any significance. Mehrishi (1970c) found only a 34% increase in the EPM of human blood platelets, consequent upon their treatment with poly I:C. Admittedly there is considerable variation in the binding of poly I:C by different cell types. In the absence of a proper control, interpretation of the difference in the binding is extremely difficult, if not impossible.

EFFECTS OF HISTONE ON BEHAVIOUR OF HAMSTER KIDNEY CELLS

Histones which are basic chromosomal proteins are known to produce a wide spectrum of effects *in vitro* and *in vivo.* They are toxic to viruses (Fischer and Wagner, 1954) and bacteria (Hirsch, 1958), inhibit template activity of chromatin (see De Reuck and Knight, 1964; Allfrey *et al.*, 1965; Bonner and Huang, 1963; Georgiev *et al.*, 1966; Paul and Gilmour, 1966, 1968; Iwai, 1969). Histones are also known to inhibit morphogenesis (Brachet, 1964; Sherbet, 1966a,b; Sherbet and Lakshmi, 1967b, 1969b). Vorobyev and Bresler (1963) claimed that histones controlled tumour growth *in vivo* and also altered tumour histology. *In vitro,* on the other hand, histone appears to alter the morphology and cultural behaviour of BHK-21 cells and these changes are similar to those observed with viral transformation (Latner and Longstaff, 1971). Latner *et al.* (1973, 1974) not only confirmed this, but also showed that histones were transforming the cells and not merely selecting cells which had changed cultural behaviour. In electrophoretic studies of a control group of cells derived from a non-invasive tumour formed by BHK-21 cells, and cells derived from invasive tumours produced by histone-transformed cells, they showed the emergence of a bimodal electrophoretic pattern in the histone-treated group. This group appeared to contain two distinct types, the EPM of one type was comparable with the EPM of control cells. The second type possessed approximately 20% higher EPMs than did the cells of the control series.

Histones are cationic proteins and may be expected to stick to the cell surface which bears a net negative charge. In the experiments described by Latner and Longstaff (1971) and Latner *et al.* (1973) the transforming effect of histone is not due to surface alterations produced by adsorption of histone molecules to the surface. The adsorption of the cationic proteins on the surface would have decreased the EPM of the cells rather than increase it. It seems therefore that the histone effect is an intracellular effect. The histones pesumably affect the functional biochemistry of the cells and this causes an alteration, qualitative or quantitative, of the surface components. These changes are reflected in the altered behaviour of the histone-treated cells (see also pp. 127–128 regarding the effects of cationic local anaesthetics on cell surface associated phenomena such as mobility, adhesion to substratum, cell fusion, rounding, etc. It has been argued that local anaesthetics may cause surface alterations which may be transduced into alterations in the cytoskeletal elements).While discussing the possible mode of action of the histones, Latner *et al.* (1974) suggest that it is possible that the exogenous histones modify genetic expression. Kolodny (1973) found that subconfluent mouse fibroblast cultures showed higher saturation densities and also a disordered cultural morphology on treatment with histones. These effects did not appear when confluent cultures were treated with histones. Histones are also capable of altering the lactic dehydrogenase isozyme pattern (Latner and Longstaff, 1969). Earlier Sherbet (1966a,b) showed that calf thymus histones produced inhibitory effects on the development of chick embryos. These effects appeared to depend upon the stage chosen for treatment, which suggested that the exogenous histones may be inhibiting the mRNAs being synthesised at the time of treatment. Histones also inhibit the induction of neural tissue differentiation by the primary organiser (Sherbet and Lakshmi, 1967b). Most of the observations tend to suggest a direct effect of the histones on the expression of genetic information as suggested several years ago by Sherbet (1966c). The transforming effect of histones may indeed be a result of a similar direct effect. However, Latner *et al.* (1974) do not exclude other possibilities, or the possibility that the effects may be caused by non-histone proteins present as an impurity in the histone preparations.

Another interesting aspect of this investigation is the observation made by Latner *et al.* (1974) that the cells "transformed" by histones are less tumorigenic than the control cells. They point out that this may be because the cells become more antigenic, presumably by the acquisition of new surface antigens. The transformation of Balb/3T12 cells by polyoma virus was also said to have been accompanied by a reduced tumorigenicity (Aaronson and Todaro, 1968).

Cell electrophoresis has also been used in monitoring the effects of ultrasound X irradiation and u.v. irradiation on tumour cells. This topic has been treated elsewhere (pp. 141–142).

EFFECTS OF 4-HYDROXYANISOLE ON CELL BEHAVIOUR

An investigation similar to the work of Latner (pp. 229–231) has been conducted in this laboratory on the effects of the anti-oxidant 4-hydroxyanisole (4-OHA) (P. A. Riley and Sherbet, unpublished work).

It has been reported that para substituted phenolic compounds induce invasive behaviour in the basal cells of epithelium when applied locally to guinea-pig skin or hamster cheek pouch epithelium (Riley and Seal, 1968; Seal *et al.*, 1969; Wood and Smith, 1969a,b). This phenomenon of "microinvasion" has not only been observed in early spontaneous epithelial tumours (Sugár, 1968; Frithiof, 1969; Cohen *et al.*, 1971; Olsen *et al.*, 1968; Frasca *et al.*, 1968) but also in psoriatic skin (Cox, 1969) and in healing wounds (Odland and Ross, 1968; Ross and Odland, 1968). Microinvasion does not occur in the application of benzene or turpentine (Tarin, 1968). It appears therefore that the substituted phenolic compounds alter cellular behaviour by producing alterations at the cell membrane.

The effect of 4-OHA on 3T3 fibroblasts were examined using isoelectric equilibrium method. It was argued that the adsorbed molecule would be orientated in such a way that the hydroxyl groups would contribute to the surface charge, and since phenolic hydroxyl group has a high pK value (see p. 184) the pI of the cell surface would increase. Paradoxically, 4-OHA treatment caused a reduction in the pI. In addition, time course experiments over a six week period during which 4-OHA (10 μM) was included in the culture medium showed no detectable change in pI for one week, after which an increasing effect was noticed. These experiments seem to suggest that 4-OHA has an indirect action on the cell surface. Treatment of 3T3 cells with the non-phenolic anti-oxidant sodium ascorbate, also produces reduction in pI. However, an examination of formaldehyde titration curves (neutralising amino groups as described on pp. 186–188) indicates differences in the acid components present on 4-OHA and ascorbate-treated cells. Obviously, this is due to differences in their modes of action.

The discussion in this subsection indicates the potential use of the isoelectric equilibrium method in the investigation of interaction of drugs with membranes. The method will not only enable a rapid screening of potential drugs, but may also provide useful guidelines in designing experiments for the elucidation of the mode of action of drugs.

ISOELECTRIC BEHAVIOUR OF SUBCELLULAR ORGANELLES

MITOCHONDRIA AND LYSOSOMES

The application of isoelectric focusing in the characterisation of cell surfaces in the author's laboratory has also recently prompted investigations into the surface characteristics of subcellular organelles. Just *et al.* (1975) have reported that mitochondria and lysosomes derived from rat liver have pI values of 7·4 and 7·2 respectively. This observation is of considerable interest, and suggests a lack of acidic substances which are so characteristic of the plasma membrane, and presumably also of membranes of the endoplasmic reticulum and the nuclear membrane which form a continuum. The characterisation of the internal membrane system or the membranes of subcellular organelles has received little attention so far. The isoelectric equilibrium technique may provide an extremely useful means of investigation.

RIBOSOMES

Preliminary investigations into the isoelectric behaviour of ribosomes (Sherbet, Lakshmi and F. Cajone, unpublished work) has suggested two further important areas of application of the method. When examined by post-loading isoelectric equilibrium method, polysome preparations from rat liver have shown a characteristic distribution pattern, with a part of the material focusing in the pH range 5·0–6·25 and the remainder as a well defined peak at pH 4·66. Since the focusing conditions are not conducive to the maintenance of the integrity of the polysomes, it appeared probable that the pI 4·66 material could be ribosomal subunits. This was confirmed by recovering this material and examining its sedimentation pattern. Purified 50*S* and 30*S* subunits have pI values of 5·1 and 4·66 respectively which may indicate that the pI 4·66 material may be composed predominantly of 30*S* subunit. It appears likely that the 50*S* subunit co-focuses with the material appearing in pH 5·0–6·25 range. Much of this latter material may have been derived by degradation to varying degrees of the messenger RNA into mostly 10*S* and larger fragments, as indicated by the isoelectric points of SV-40 DNA and 10*S* and 4*S* RNA preparations used as markers (Table 63).

Table 63 *Isoelectric behaviour of rat liver polysomes*

Sample	pI[a]
Polysomes	5·0 – 6·25 4·66 ± 0·08
50*S* subunit	5·1
3O*S* subunit	4·68
SV-40 DNA	4·36
10*S* RNA	5·4 – 5·5
4*S* RNA	6·0 – 6·7

[a] Determined at 0·01 M ampholine concentration, by post-loading method and equilibration for 20 min in LKB-PPHE column.
From Sherbet, Lakshmi and Cajone (unpublished work).

These experiments, although at a preliminary stage, strongly suggest the possibility that the isoelectric point may be a useful indicator of the degree of tertiary structure achieved by the RNA molecules. Larger molecules are more likely to assume a tertiary structure dictated by complementary base sequences than shorter molecules, and greater the degree of tertiary organisation achieved lower will be the isoelectric point, since such a configuration will offer predominantly the negatively charged groups for isoelectric detection.

Finally, in addition to providing a characterisation and serving as an index of the degree of tertiary organisation, isoelectric focusing may also serve as a reliable means to isolate and purify subcellular organelles, and also in the study of the mechanisms involved in the association of ribosomes with membranes and ribonucleic acids.

THE ISOELECTRIC ZONE

Isoelectric focusing and free electrophoresis are manifestations of the electrical properties of surfaces. The difference between the two methods exists mainly in the description and definition of the electrokinetic phenomenon. Thus free electrophoresis describes the velocity of particles under given ionic conditions and pH. Variation of the ambient pH alters the ionised state of the particle and therefore alters the particle velocity. In electrophoresis one can alter the ambient pH such that the particle is immobile on account of being electrically neutral with equal numbers of positive and negatively charged groups being ionised as a function of a given pH. This is the isoelectrophoretic point of the particle. On the other hand, isoelectric focusing while making use of the electrokinetic behaviour of the particle, does not allow the particle to electrophorese freely. The focusing column is made up of a gradient of pH from the acidic end to the most basic end. As a result from the moment a particle becomes electrokinetic, the ionised character of the particle is continually changing until the particle has reached a point where it bears no net electric charge and therefore it comes to rest. This is the isoelectric point.

In the preceding text a distinction has been made between these two isoelectric points, as the isoelectrophoretic point (pIE) and the isoelectric point (pI). Bull (1964) observed that in electrophoresis of proteins, the point of zero charge determined by titration methods does not correspond to the isoelectrophoretic point. A significant difference between pI and pIE of bacterial cells was noticed by Sherbet and Lakshmi (1973). It appears that this may be due merely to the ionic conditions under which the determinations are made. At physiological ionic strengths the thickness of the electrical double layer is approximately 8 Å. If the particle surface is smooth the ionisable groups in this zone will produce the electrokinetic activity. If the particle surface is rugged only those groups located on the ridges or "hills" would exert electrokinetic effects. Now, lowering the ionic concentration would increase the electrokinetically effective zone i.e. detect more and probably also new groups which were beyond detection at higher ionic concentrations, which in turn may be expected to alter the pI of the particle surface. Bull (1964) has in fact cited evidence that lowering the ionic concentration has the effect of increasing the isoelectric point of proteins. Most of the isoelectric equilibrium studies with cells were performed with ampholines at a concentration of approximately 0·009 M and consequently pI values have tended to be much higher than pIE values of corresponding cell types.

No distinct relationship between pI and the ampholine concentrations is possible for two important reasons. Firstly, the ampholines are a mixture of molecules of diverse molecular weights. The data on their molecular weight and other physical characteristics are being continually revised. Secondly, in the mathematical approximation not only has it been assumed that the ampholines provide the ionic environment but it is also assumed that the ionic atmosphere is analogous to that provided by point charges. Given that these assumptions constitute a reasonable approximation to and definition of the conditions of isoelectric focusing, one may suggest that this method is able to probe the cell surface to a greater depth than does free electrophoresis. The region which is amenable to characterisation by isoelectric focusing has been termed as the isoelectric zone (IEZ) to distinguish it from the electrokinetic zone (EKZ) that can be probed by free electrophoresis. The differences in the dimensions of these zones is brought home quite clearly by the calculations of carboxyl group densities in *E. coli* cells by free cell electrophoresis and isoelectric equilibrium experiments. Sherbet and Lakshmi (1973) found that in formaldehyde-modified *E. coli* cells the density of carboxyl groups was $2{\cdot}8896 \times 10^{13}$ cm^{-2} (see Table 50) i.e. one carboxyl group every 350 Å^2. The corresponding value obtained by Gittens and James (1963) from electrophoresis of *E. coli* at the ionic concentration of 0·05 M is one carboxyl group every 1730 Å^2. In other words, the density of carboxyl group revealed by the isoelectric method is five times greater than that indicated by free cell electrophoresis. At the ionic strength of 0·05 M, cell electrophoresis can probe the surface to a depth of 13·7 Å. It is suggested therefore that the isoelectric zone extends to a depth of approximately 60–70 Å below the surface. The dimensions of the zones point to the unequivocal advantage of the isoelectric method over cell electrophoresis. To probe the cell membrane to a depth equivalent to the IEZ, cell electrophoresis has to be carried out at ionic strengths as low as 0·002 M. More direct experimental evidence is needed to confirm the dimensions of the isoelectric zone.

6. Partition of Cells in Aqueous Two Phase Systems

Tiselius (1960) stated that the methods devised to isolate and purify biological macromolecules also play a role in the determination of their structure. For they take advantage of one of the characteristic properties of the molecule or the particle. In centrifugation, for example, the sedimentation coefficient is proportional to size; in gel electrophoresis mobility it is related to the molecular weight; isoelectric focusing uses the isoelectric point of macromolecules as a criterion for separation and so on. These various methods not only allow the separation and purification but they also help to characterise the material being separated. Partition of biological macromolecules and particles in aqueous two phase systems, similarly, takes advantage of certain characteristics of the component being separated, and therefore can be used to provide a limited characterisation of the component under investigation. In this chapter we shall discuss the principles and the molecular and cellular parameters which are related to partition with a view to assessing the utility of this method in the characterisation of cell surfaces.

PRINCIPLE OF PHASE PARTITION

A mixture of immiscible aqueous solutions of two unlike polymers results in a separation into phases over a period of time. If a substance is present in solution, it tends to be distributed between the two phases, and such partition is characterised by the partition coefficient which is the ratio of concentrations of the substance in the top and bottom phases:

$$K = \frac{[C_t]}{[C_b]} \qquad (72)$$

where $[C_t]$ and $[C_b]$ are concentration in moles of the partitioned substance in the top and bottom phases respectively. The properties of the two aqueous phases essentially determine the distribution of soluble substances. But in the partition of particles the interface between the phases has some capacity for absorbing the particles and must therefore be taken into consideration. The interfacial tension between aqueous–aqueous phases is very low and therefore suitable for partitioning of biological particles and cells.

Albertsson (1972) has described the mechanism governing phase separation in terms of interactions between the suspended particle and the molecules constituting the phase. These interactions are of a complex nature and may involve hydrogen bonds, ionic and hydrophobic bonds. If the net effect of these interactions are different with respect to the two polymers, a partitioning of the particle or macromolecule takes place as a function of its affinity to the two phases. Albertsson (1972) has attempted to express these interactions in the following equations.

If the energy required to move a particle from one phase to another is ΔE, at equilibrium the following relationship will subsist between ΔE and the partition coefficient K:

$$K = \exp\left[\frac{\Delta E}{kT}\right] \tag{73}$$

where k is the Boltzmann's constant and T the absolute temperature.

The interaction between the particle or macromolecule is greater, the greater the size of the molecule, and therefore Bronsted (1931) deduced the following expression:

$$K = \exp\left[\frac{M\lambda}{kT}\right] \tag{74}$$

where M is the molecular weight of the macromolecule and λ a constant factor, other than size, which is characteristic of phase–particle interaction. In the case of spherical particles, the quantity M in eqn (74) can be replaced by the surface area, A, of the particle, and the equation rewritten as:

$$K = \exp\left[\frac{A\lambda}{kT}\right]. \tag{75}$$

This exponential λ factor includes properties such as surface free energy and surface charge-related properties of the particle. In the partitioning of biological macromolecules, cells and subcellular particles, the aqueous phases can be buffered and made isotonic. When ions are present in the phases, they are more likely than not, to be distributed unequally between the phases as dictated by their affinities to the two phases (Johansson, 1970a). Although the polymers themselves are non-ionic this results in the generation of an electrostatic potential difference between the two phases (see Albertsson, 1972; Reithermann *et al.*, 1973) which has been termed the partition potential (Baird *et al.*, 1961). The surface charge of the cells being characterised may interact with the potential and thereby provide insights into the charge composition of the surface (Walter, 1975). In addition, the net charge Q on the particle surface is expressed as a function of the pH. Taking these various interactions into consideration, the relationship between K and the various properties of the particle can be expressed as follows:

$$K = \exp\left[\frac{\lambda_1 A + Q(P_1 - P_2)}{kT}\right] \tag{76}$$

where λ_1 is a constant factor involving properties of particles other than the surface area, A, and the net charge, Q. $(P_1 - P_2)$ represents the potential difference between the phases.

The most interesting feature of Bronsted's partition theory is the exponential relationship between K and the particle attributes A and Q. It follows from this theory that even for small changes in charge density and/or surface area, very significant changes occur in the partition characteristics. In other words, the partition characteristics of a given species of particles supplies considerable information about certain surface properties of the particles, such as surface charge.

Partition behaviour of cells can also be related to the lipid content of the membrane provided, of course, that partition is examined in a phase system which has no partition potential (Singer, 1974). This has been demonstrated by introducing changes in the concentration of the polymers. In a two phase system containing a salt which partitions uniformly, cells will not show partition. In such a Dextran-PEG system, changing the concentration of polymers from Dextran 5%-PEG 4% to Dextran 5%-PEG 3·5% to approach critical concentrations i.e. below which phase separation does not occur but a homogeneous solution will result, causes an increase in the partition coefficients of the cells. In the case of erythrocytes this is said to correlate extremely well with the poly/monounsaturated fatty acid content of the cell membrane. The increase in partition here has been shown to be independent of surface charge, at least that contributed by sialic acids (Walter *et al.*, 1976a).

A more recent development in the use of partition behaviour for the characterisation of cell membrane is the technique of "affinity partition". In this technique a small quantity of a polymer ligand is included in the phase system which has no electrostatic potential occurring between the phases. Only the cells which can bind the ligand are attracted into the phase which is preferred by the ligand (Flanagan *et al.*, 1975; Walter and Krob, 1976; Eriksson *et al.*, 1976).

A number of phase polymer derivatives have been used as ligands, for example DEAE-dextran (Walter and Selby, 1967; Walter *et al.*, 1968), trimethylamino-PEG or PEG-sulphonate (Johansson, 1970b). These ligands aid phase separation based on surface charge. Triton X-100 has been used by Albertsson (1973) to take advantage of its hydrophobic interactions in the separation of hydrophobic proteins. Flanagan and Barondes (1975) used α, ω-bis-4-trimethyl-ammonium(phenyl-amino)-PEG to enrich membranes containing receptors for acetyl-choline. PEG-palmitate is another ligand which has found considerable use in the extraction of serum albumin (Shanbhag and Johansson, 1974) and in studies involving erythrocyte membrane (Walter and Krob, 1976; Walter *et al.*, 1976b; Eriksson *et al.*, 1976).

COUNTER CURRENT DISTRIBUTION

The discriminatory ability of a two phase partition system can be increased by introducing multiple stage partition in the two phase system. For the separation of two substances which have partition coefficients in a given two phase system too close to each other, counter current distribution (CCD) will provide a much more efficient separation than a single stage operation. The principle of this method has been discussed in several reviews (Craig, 1960; Craig and Craig, 1956; King and Craig, 1962; Albertsson, 1972). Counter current distribution consists mainly in perfoming repeated extraction of separated phases in equilibrium with fresh bottom and top phase followed by remixing and reseparation, as given in Fig. 33 which depicts a scheme for CCD of two hypothetical substances with partition coefficients of 2·3 (A) and 4 (B). When these substances are partitioned in a two phase system they will be distributed 70, 30 (A) and 80, 20 (B) arbitrary units in the top and lower phases respectively. In Stage I transfer the top phase of the original tube (No. 0) is mixed with fresh lower phase polymer, and its lower phase is mixed with fresh top phase polymer. When this transfer is completed, the tubes are shaken to mix the contents and then allowed to separate. Several transfers are carried out on the same lines (as indicated by arrows in Fig. 33) with the addition of one tube at each transfer. In Fig. 33 seven transfers have been shown with the concentrations of substances A and B expected at partition in each of the tubes of the

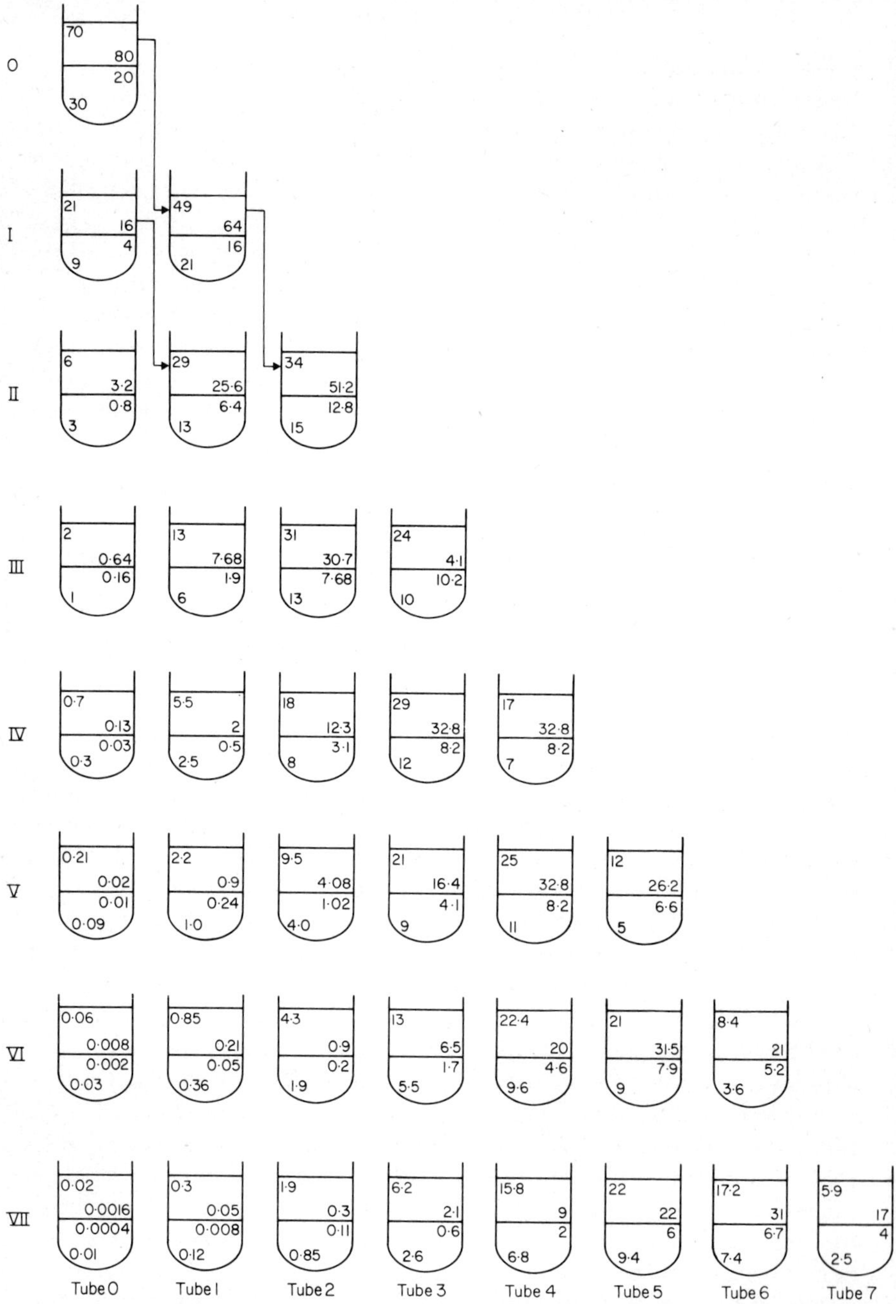

Fig. 33. A scheme for counter current distribution of two hypothetical substances with partition coefficients of 2·3 and 4. When several transfers are carried out as indicated a bimodal distribution emerges at VII.

transfer series. It may be seen from Fig. 34 that at the end of Stage VII transfer a bimodal distribution is evident. If this process is carried out over a large number of transfers, substance A will separate out from B.

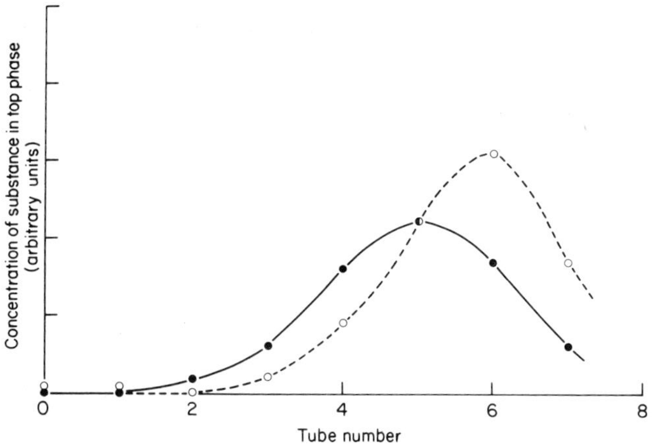

Fig. 34. A scheme for counter current distribution of two hyothetical substances with partition coefficients of 2·3 and 4. Continuous line represents substance with $K=2{\cdot}3$ and broken line $K=4$.

This scheme of CCD describes the partition of soluble components between the two phases, with little or no adsorption of the substances at the interface. Distribution of this type has been termed liquid–liquid counter current distribution, where the distribution ratio G may be written as the product of K and the ratio of volumes of the top and bottom phases:

$$G = K \times \frac{V_t}{V_b} \tag{77}$$

where V_t is the volume of the top polymer phase and V_b that of the lower phase. Since the partition coefficient K is the ratio of the concentrations of the substance in the top and bottom phases (eqn 72), eqn (77) can be written as:

$$G = \frac{[C_t]}{[C_b]} \times \frac{[V_t]}{[V_b]}. \tag{78}$$

In the case of particles, however, the distribution occurs mainly between one bulk phase and the interface. When phase transfers are performed, the bottom phase, the interface and a small portion of the top phase is kept stationary and the rest of the top phase is transferred (see Fig. 35). This is liquid–interface distribution, where distribution ratio, G_i, is given by the following equation (Albertsson and Baird, 1962):

$$G_i = \frac{[C_t](V_t - V_s)}{[C_i] + [C_t]V_s} \tag{79}$$

where V_s is the volume of the part of the top phase which is kept stationary, $[C_i]$ is the concentration of particles in V_s and the interface.

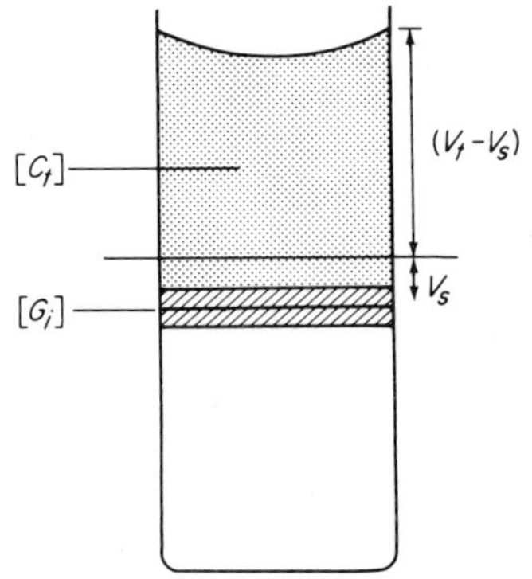

Fig. 35. A schematic representation of liquid interface counter current distribution. $(V_t - V_s)$ is the volume of the transferred portion of the top phase whose volume is V_t. V_s is the part of the top phase which remains stationary. The distribution ratio (G_i) is given by the product of the ratio of concentration of particles in the moving phase and that in the interface plus those present in the part of the top phase which is kept stationary and ratio of the corresponding volumes (see eqn 79).

Several versions of CCD apparatus have been described. Most prominent among these are the all glass apparatus of Craig (see Craig, 1960; Craig and Craig, 1956) and the thin-layer CCD apparatus (see Albertsson, 1971). It is not proposed to describe these two versions except to point out that the thin-layer version has obvious advantages in that the phase layers formed are about 2 mm deep and equilibrium of the separated phases requires a very short time.

SEPARATION OF BIOPOLYMERS AND CELL PARTICLES BY CCD

Counter current distribution has been used extensively in the study of nucleic acids, strain differences of viruses and bacteria, in the separation of chloroplasts and mitochondria, etc. The method has also been used in the demonstration of antigen–antibody binding, binding of antisera to viral and bacterial particles. The topics have been examined in considerable detail by Albertsson (1971). The reader is referred to this review for this aspect of CCD application.

Counter current distribution has also been used effectively for examining the patterns of messenger RNA synthesis, and in investigating possible alterations in these patterns induced by hormones and carcinogenic agents. Alterations in messenger RNA patterns have indeed been reported as effects of hormonal treatments and of carcinogenic agents (Kidson and Kirby, 1964, 1965; Parish and Kirby, 1966). Burczyk (1975) and Burczyk *et al.* (1976) showed that CCD can be employed to separate *Chlorella pyrenoidosa* A24 and *Chlorella* sp.366 both of which are sensitive to infective lytic agent (lysate-containing algophage in the presence of *Caulobacter vibrioides*), from *Chlorella vulgaris* which is resistant to the lytic agent. Presumably the separation of these strains is due not merely to differences in their size, but also to membrane receptors which have affinity to the active particles contained in the infective lytic agent.

As far as animal cells are concerned CCD has been used in the separation of erythrocytes and granulocytes from lymphocytes (Walter *et al.*, 1969a,b). Earlier Walter *et al.* (1965) showed that erythrocytes derived from different species showed differences in behaviour, in CCD, and that erythrocytes of different ages could be separated from leukocytes (Walter *et al.*, 1969b) Injection of phenylhydrazine releases into the blood stream populations of reticulocytes, which can be demonstrated by CCD (Walter and Selby, 1966; Walter and Albertsson, 1966). Despite the considerable importance of these investigations, no attempt will be made to discuss them in detail. The prime objective set for this chapter is to highlight the use of CCD in the study of surface characteristics of cells in relation to biological features such as growth, cell division, differentiation and metabolic state of cells.

PARTITION BEHAVIOUR AND ISOELECTRIC POINT

The partitioning of macromolecules or biological particles in aqueous two phase systems takes place as a function of their affinity to the phases, which is probably determined by the complex interactions between the substance or the particle and molecules of the phase system. The net electric charge borne by the macromolecule or the particle affects the partition behaviour (pp. 237–238). As discussed in Chapter 5 (pp. 181–183), the isoelectric point of a cell or macromolecules is determined not only by the nature of the ionisable groups but also by their number. A relationship between pI and partition behaviour may therefore be expected and has, in fact, been shown to exist. Albertsson (1971) found that the partition coefficients of lysozyme, RNAase and serum albumin could be correlated with their isoelectric points (Table 64). In dextran-PEG phase system, the more basic the substance, the smaller was the partition coefficient. In this experiment the coefficients were determined at pH 7 where lysozyme and RNAase will be positively charged. Both these have K values below 1. Serum albumin and pepsin, on the other hand, will be negatively charged at this pH on account of their relatively low pI, and have K values near or above 1.

Table 64 *Correlation between pI and* K *in dextran-PEG system*

Substance	pI	K^a	Charge at pH 7
Lysozyme	10·5–11	0·5	+ve
RNAase	8·2–9·7	0·6	+ve
Serum albumin	4·7	0·7	−ve
Pepsin	2·5–2·8[b]	6·5	−ve

[a] K in dextran 500 7% (w/w)—PEG 6000 4·4% (w/w) phase system pH 7.
[b] From Haglund (1975).
From Albertsson (1971).

The degree of ionisation of the various groups depends upon the pH of the solution or the medium in which particles are suspended. Consistent with this is the alteration in the partition behaviour of a number of proteins and enzymes which occurs with changes in pH (see Albertsson, 1971). The ionic concentration is also known to affect the isoelectric point (see pp. 157 and 175). A correlation between partition behaviour and ionic concentration is too complex to analyse (Albert and Nyns, 1961) and appears to be related to the nature of interaction between the ions and the polymer constituents of the phase systems (Johansson, 1970).

PARTITION BEHAVIOUR OF ERYTHROCYTES

A number of morphological and physical alterations occur in the course of ageing of erythrocytes (Jandl, 1961; Danon and Marikovsky, 1961;

Prankerd, 1958). Walter and Selby (1966) labelled erythrocytes with ^{59}Fe-ferrous citrate, exsanguinated the animals at various times from 2 to 65 days after injection of the label, and examined the cells by counter current distribution. They observed not only that reticulocytes could be separated from mature young erythrocytes and young erythrocytes from old ones, but also that even a difference of a few days in age produced changes in their distribution pattern. As a result of these and subsequent investigations (Walter *et al.*, 1971) a pattern of CCD distribution of erythrocytes has emerged, which is related to the maturation and ageing of erythrocytes. Young reticulocytes have a low partition coefficient in dextran-PEG system. Partition increases during the period of maturation of reticulocytes until the highest partition is obtained for erythrocytes in circulation. On ageing, there is a diminution of partition and the oldest cells resemble young reticulocytes in their partition behaviour. Consistent with this pattern is the appearance of cells with low partition coefficients in the blood of rats that have been administered phenylhydrazine which is known to produce and release large numbers of reticulocytes into the blood stream.

The differences in the partition behaviour between young and old erythrocytes are compatible with the observation that young erythrocytes have up to 30% more *N*-acetyl neuraminic acid residues than old cells (Marikovsky and Danon, 1969) and they also bind soy bean agglutinin, which has affinity for galactose-containing saccharide moieties (Lis *et al.*, 1970; Gordon *et al.*, 1972), to a greater extent than do old erythrocytes (Marikovsky *et al.*, 1976). A significant decrease in the sialic acid content (and also of galactosamine) of the membrane glycoproteins in the process of maturation and ageing has also been reported by Balduini *et al.* (1974). It may be that all these differences add up to a considerably higher surface charge density on young cells which apparently results in their higher partition efficiency.

Subsequent work reported by Walter *et al.* (1965, 1967) has brought to light a correlation between the distribution pattern of erythrocytes of a variety of donor species and their EPMs. The distribution shifted more to the right of the CCD extraction train with higher EPM values. In other words, the greater the partition coefficient the higher is the electrophoretic mobility.

The EPM of erythrocytes is known to be due mainly to the sialic acids occurring on their surface. Walter *et al.* (1965) however found that the amount of sialic acids extractable from the stroma of cells of different ages was constant, but the electrophoretic and CCD distributions do show much variation. This, the authors suggest, may be due to a diminution of surface area of the erythrocytes with age which will in effect alter the net negative charge densities. But Walter and Coyle subsequently (1968) showed that removal of sialic acid with NANase which is known to result in large reductions in their EPM also reduced their partition coefficients. Differences in the distribution pattern have also been detected between fresh erythrocytes and stored erythrocytes. Brooks *et al.* (1971) have shown that these differences could be due to surface charge-associated properties of the cells. They found that stored erythrocytes have generally higher partition coefficients than fresh cells. As regards their EPM, the following situation existed. Fresh cells unexposed to phase polymers have the same EPM as fresh cells exposed to the polymers, washed and then suspended in standard saline. But stored cells recovered from the CCD peak have greater EPM than stored cells unexposed to the phase polymers. Obviously, the increase in EPM is a result of adsorption on the surface of the cells of the polymer molecules which are easily washed off from the surface of fresh cells but not from the surface of stored cells. Thus the differences in CCD patterns appear to be mainly due to surface charge differences and also charge-related properties determining the less readily reversible binding of the phase polymer molecules to the surface. Comparable experiments designed to relate the CCD pattern with possible alterations in the surface charge of cells as a function of their maturation and ageing have not been performed, but, as suggested earlier, these changes may be due to diminution of the surface area and consequent increase in surface charge densities. On the other hand, the difference in the partition behaviour of erythrocytes of different species may reflect quantitative differences in the sialic acids occurring on their surfaces. Walter *et al.* (1972) found that bovine erythrocytes are of three types as distinguished by their partition behaviour. In these three types, they have noticed that the higher the partition, the greater is the amount of sialic acids released by NANase (Table 65).

Erythrocytes also seem to show species-specific partition when examined in Dextran-PEG systems containing PEG-palmitate as a ligand (Walter *et al.*, 1976a; Eriksson *et al.*, 1976). Walter *et al.* (1976) also showed that partition in the presence of the ligand was independent of the surface charge borne by the cells and that the species-specific partition may have depended upon the ratio of poly/monounsaturated fatty acids in the membranes.

SURFACE CHANGES IN CELL CYCLE

The cell membrane appears to be actively involved in the regulation of cell division. Evidence from a variety of sources which indicates the

Table 65 *Partition behaviour and sialic acid content of bovine erythrocytes*

Type of cell	Partition in phase system[a] I	II	Sialic acid released (μg)
(1)	37 ± 13	13 ± 2	88 ± 8
(2)	62 ± 9	17 ± 3	91 ± 8
(3)	90 ± 11	27 ± 3	137 ± 8

[a] Partition in dextran-PEG system. Expressed as quantity of cells in top phase (% total cells added).
[b] Sialic acid released by 10^{10} cells.
From Walter *et al.* (1972).

association of surface charge variations in cell division has been discussed in Chapter 5 (pp. 223–226). A diminution in the adhesive ability of metaphase cells has been described (Axelrod and McCulloch, 1958). It is also known that changes also occur in the cell size and density in the course of the cell cycle (Mitchison, 1971; Sitz *et al.*, 1970). These features make partition and CCD methods the most effective means of studying the variations in surface properties during the cell cycle. If the cell cycle involves alterations in two features, namely the net charge and size of the cells both of which will positively influence their partition characteristics, partition distribution and CCD will provide the maximum discrimination of cell particles at different stages of the cycle.

Walter *et al.* (1973) examined the CCD distribution of the alga, *Chlorella pyrenoidosa,* in dextran-PEG system. They observed that synchronised cells which have just divided show the lowest partition. Partition increases with time after division and the highest values are obtained just prior to cell division. This pattern may be due not only to increase in surface charge but may also be attributed to the size of the cells. Thus, Walter *et al.* (1973) observed that the newly divided cells were also the smallest in the cell population and that the size and rate of their sedimentation increases with time after cell division. However, there are no means by which one can divorce these two parameters and examine the possible surface changes alone in relation to partition behaviour. Increase in size of the cells will mean a reduction in net charge densities, unless this change is compensated by increases in the acidic component of the surface.

These observations of Walter *et al.* (1973) on synchronous *Chlorella* cells have been confirmed by Pinaev *et al.* (1976) on HeLa and mouse mast cells. Immediately after division, HeLa cells have a low partition, which increases with time. The interphase cells, on the other hand, are distributed in three major groups, presumably in accordance with their stage in the cell cycle. Pinaev *et al.* (1976) also examined the CCD of mouse mast cells in culture. Cells obtained from the exponential growth phase (which is up to 55 hours) had high partition characteristics, the earlier the cells taken from the exponential phase the greater was their partition. Partition decreased as the cells approached the stationary phase, at which the partition was found to be at its lowest and the cells markedly heterogeneous. These results are substantially similar to those reported by Gerston and Bosmann (1974) on the partitioning of mouse leukemic cells. However, consistent with the criticism made earlier, they did not observe any alterations in the electrophoretic mobilities of cells at various stages in the cell cycle, which can be explained by the offsetting of any increase in charge consequent on alterations in the size and surface area of the particles. It is beyond reasonable doubt that the alterations in the partition behaviour reflect the occurrence of a sequence of surface events concomitant with the cyclic programme of cell growth and division.

PARTITION BEHAVIOUR OF DIFFERENTIATING CELLS

The cell surface undergoes a number of complex changes in the course of their differentiation, organisation and morphogenesis (pp. 104–131). Partition of cells in aqueous two phase polymer system has afforded a useful method for investigating these surface changes. Walter and Krob (1975) have used rat intestinal epithelial cells as a differentiating system. These cells begin as mitotically active undifferentiated cells at the base of the crypt and differentiate as they move up the villus, and reach the fully differentiated state at the upper two-thirds of the villus in 36–72 hours (Leblond and Meissier, 1958). Walter and Krob (1975) examined the partition behaviour of cell fractions from the villus tip to the crypt base isolated using a separation method described by Weiser (1973). Usually ten fractions are made, of which the early fractions are rich in villus cells while the later ones contain predominantly crypt cells. In Fractions 1 to 10 there occurs a gradient of villus to crypt cells. The villus (differentiated cells) appear to have low partition but the differentiating cells of the later Weiser fractions, show considerably greater partition with peak to the right of the CCD extraction train. If the fractions are pooled and then subjected to partition, a bimodal distribution is obtained. Using alkaline phosphatase assay, the cells with low partition have been identified as the upper villus cells. The cell peak with high partition has comparatively less alkaline phosphatase specific activity and appear therefore to be crypt cells.

Walter and Krob (1975) also injected rats with ^{14}C-methyl thymidine and sacrificed the animals and isolated Weiser fractions of intestinal epithelium at different times (4 to 48 hours) after injection of the radioactive DNA precursor. The counter current distribution of these samples and their radioactivity distribution pattern are given in Fig. 36. The cells are found to be distributed in the usual bimodal pattern. However, high relative specific radioactivity is found to be associated with cells which lie between the two major peaks. At 11 hours, the specific activity appears to shift to the right until 24 hours, when the trend appears to be reversed. At 48 hours, the specific radioactivity seems to increase from right to left of the extraction train. This shift in the distribution of labelled cells at partition has been interpreted to indicate that the actively dividing cells show an intermediate partition, and that in subsequent differentiation an increased partition occurs. This is followed by reduction in partition during maturation. The reasons for these changes in partition are not clear. The increase in partition concomitant with differentiation may suggest an increase in membrane glycoprotein-containing sialic acids. These conclusions are consistent with the observations discussed elsewhere (pp. 138–140) that cells in mitosis have higher contents of fucose-containing glycoproteins, increased sialyl transferase activity, and also higher surface charge. The observations of Elul *et al.* (1975) suggest the possibility that undifferentiated dividing cells have higher net negative surface charge as compared with differentiated cells.

DRUG INTERACTIONS WITH THE CELL SURFACE

The two phase polymer system lends itself to easy adaptation for monitoring the interaction of drugs with the cell surface. It is possible, for instance, to detect alterations in the surface negativity that may have resulted from a direct interaction of the drug with the components of the surface. Surface changes may be produced as a secondary effect of interference by the drug in the metabolic activities of the cell. These changes may be detected not only as alterations in the surface negativity, but also by measuring the partition coefficient of the cells at various pH values, when an isoelectric point is obtained which is the pH at which the

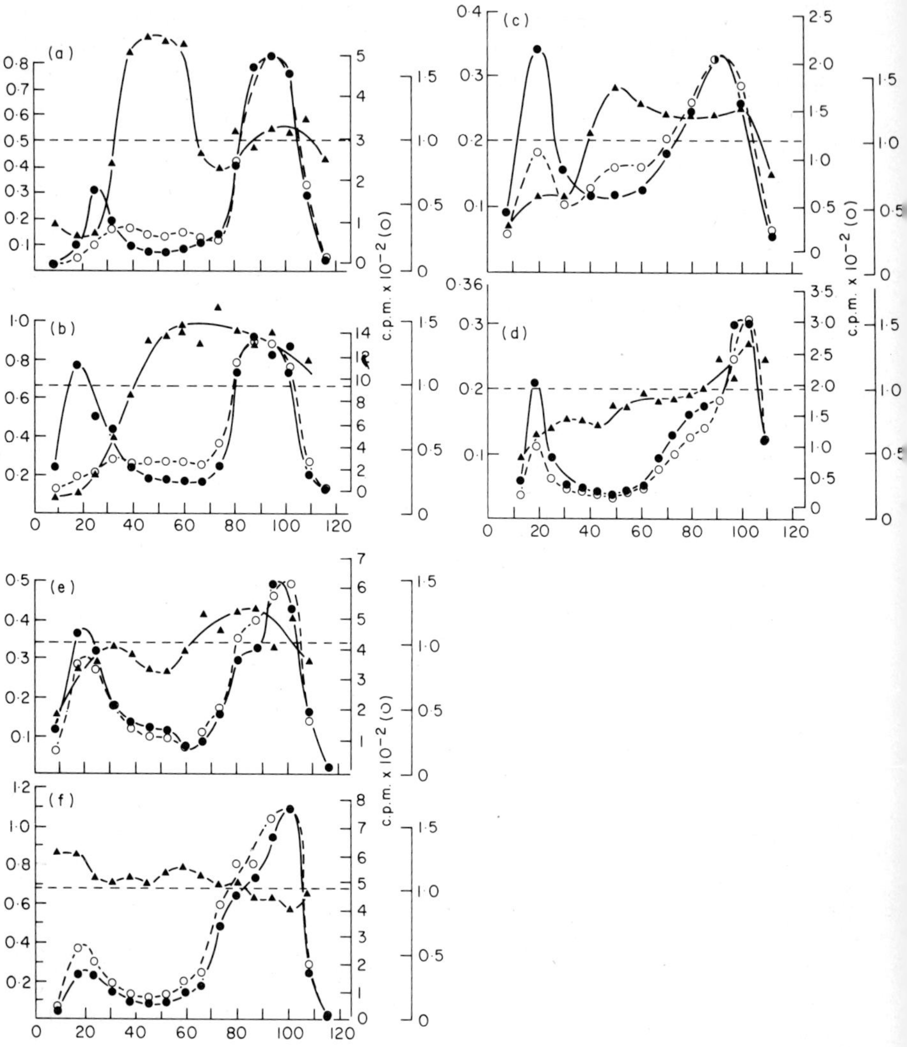

Fig. 36. Counter current distribution and radioactivity distribution of Weiser cell fractions of intestinal epithelium of rat injected with ^{14}C-methyl thymidine. (a), (b), (c), (d) and (e) represent samples fractionated at 4, 11, 24, 30 and 48 hours respectively after injection of the label. Ordinate: left-cell absorbance at 500 nm (●); inner right-radioactivity in c.p.m. (○) another right-relative specific activity (▲).

cells change their partition phase affinity. However, the exact relationship between this isoelectric point and the true isoelectric point is still unclear.

On account of the uncertainties in the interpretation of the theoretical basis of the partition of cells, the technique has not been used, except in preliminary studies, in the investigation of drug interaction with the cell surface. Nonetheless, some of the investigations carried out in the author's laboratory on the effects of dexamethasone on human astrocytoma cells in culture will be discussed here to serve as an illustration of how the two phase aqueous polymer system could be profitably employed to elucidate the possible mechanisms of action of this drug.

Dexamethasone (9α-fluoro-16α-methylprednisolone) is extensively used in the management of patients with tumours of the astrocytic series. Although this drug produces inhibition of cell growth in culture (Mealy *et al.*, 1971; Wilson *et al.*, 1972; Sherbet *et al.*, 1978), the major effect of dexamethasone is thought to be the alleviation of cerebral oedema (Galicich and French, 1961; French and Galicich, 1966) and a consequent lowering of intracranial pressure (Garde, 1965; Rasmussen and Gulati, 1962).

Despite its extensive clinical use, very little work appears to have been done to elucidate the mechanism by which dexamethasone alleviates oedema. It has been suggested that the effect is due to alterations in the membrane structure caused by an intercalation of the steroid with the membrane lipids. This is said to reduce the abnormal permeability associated with the pathological state of the membrane. Abnormal permeability may also result from free radical damage to membrane lipids. There is indirect evidence which suggests that intercalation of steroids reduces free radical initiation and redresses the membrane permeability (Demopoulos *et al.*, 1972).

The intercalation of dexamethasone has been demonstrated by Sherbet, Lakshmi and Haddad (unpublished work) using pure dexamethasone solubilised using dimethylsulphoxide (DMSO) and dexamethasone phosphate (Decadron: Merck, Sharpe and Dohme). The latter molecule bears the phosphate group at position 21. Astrocytoma cultures were treated with dexamethasone (DMSO) or with Decadron for 4 hours, at the end of which their partition behaviour was examined in a Dextran (5%)-PEG (4%) two phase system containing phosphate buffered saline. Dexamethasone solubilised with DMSO had no effect on the partition behaviour of astrocytoma cells. But cells treated with dexamethasone phosphate (Decadron) showed between 30 and 60% increase in the partition coefficient. This difference in the effect itself demonstrates that dexamethasone molecules are becoming intercalated in the membrane. The steroid being amphipathic in nature would orientate itself in such a way as to present the polar phosphate groups to the aqueous phase where they are detected by the two phase system by virtue of an increased negativity conferred on the surface of the cells by the phosphate groups. Although solubilised with the help of DMSO, dexamethasone (DMSO) will have no polar groups introduced into the molecule. Therefore the intercalation of this molecule is not detectable. In other words, the phosphate group in the Decadron molecule acts as a label. Using the two phase polymer system, we have also made some preliminary observations about the intercalation of dexamethasone by non-malignant cells. These suggest the possibility that the intercalation of dexamethasone into the plasma membrane may be greater in the malignant than in the non-malignant cell.

ZONE CHARACTERISED BY PHASE PARTITION METHOD

There are experimental and theoretical limits on the dimensions of the zone or section of the cell membrane that is amenable to investigation by a given method. In Chapter 5 (pp. 234–235) the dimensions of the zones that are likely to be probed by isoelectric and free electrophoretic methods have been discussed. A clear distinction can be made between the isoelectric zone (IEZ) and the electrokinetic zone (EKZ). It was suggested that the isoelectric focusing technique is able to probe the membrane up to 60–70 Å below the surface, while cell electrophoresis is able to characterise ionisable groups that occur in the 10 Å zone from the plane of shear.

Walter *et al.* (1972) found that although partition in two phase polymer system distinguishes 3 types of erythrocytes of beef (pp. 245–247), these have identical electrophoretic mobility. Treatment with NANase, as discussed earlier, releases sialic acid in greater amounts from cells which show higher partition. Nonetheless, the cells treated with enzyme show the same electrophoretic mobility irrespective of their partition behaviour. Walter *et al.* (1972) therefore suggest that the two phase partitioning measures charge-associated surface properties to a greater depth than the free electrophoresis method. This is a distinct possibility. These conclusions would have been unequivocal if Walter and colleagues had shown that the tree types of erythrocyte were of identical size. No data about their size appears to be given and it may have been assumed that they are all of the same size, even after enzymic treatments.

It should be remembered however that the most important feature of Bronsted's partition theory is the exponential relationship between the partition coefficient and particle attributes such as size, which means that for very small differences in size, for instance, significant alterations may be expected in the partition behaviour.

EPILOGUE

The cell surface has indeed reached a state of pre-eminence on account of the wide ranging roles it plays in biological phenomena, some of which I have discussed in the preceding pages. I have treated the subject in a style which has varied somewhat from the conventional, and have tried to visualise the cell surface by some bioelectric and electrokinetic techniques aided by data obtained by other means. Although the growth of literature in the field of cell surface biology has been phenomenal, this approach has brought the task of writing to manageable proportions.

The methods discussed here are of varying degrees of novelty and technical simplicity, and have been chosen either because they have sufficient latent use or have already re-emerged with an added dimension to their application in the field of cell surface biology. Thus, the electrometric titration of cells, despite its drawbacks, can be improved considerably and can serve to provide a certain amount of background information about the cell surface charge. Cell electrophoresis has recently had several new facets added to its application and it has undoubtedly contributed significantly to our understanding of the roles played by the cell surface. The mechanisms involved in the partition of cells in aqueous two phase polymer systems are yet unclear, but the spectrum of biological events towards the understanding of which this relatively simple method could be directed, is very broad indeed.

Finally, the isoelectric equilibrium method may be considered to have entered a new phase of application, namely in the characterisation of subcellular particles, viruses, bacteria and cells. In common with techniques of the past which found new application, at this early stage in the new phase, the scope of application of the method as an analytical tool is yet undefined and not fully understood. Much of my interpretation of the isoelectric characteristics and the quantitations based on the isoelectric data may need to be improved, revised or altered in the future. The effects of isoelectric focusing on the surface of the cells have not been fully investigated. A considerable refinement of the methods, especially concerning the suitability of the media that are used to provide the density gradients, is urgently needed. There are a host of other problems too, which I have alluded to from time to time. However, we have made a beginning, and probably the most difficult thing, in scientific research as in other walks of life, is to make a beginning.

Bibliography

Aaronson, S. A. and Todaro, G. J. (1968). *Science* **162**, 1024.
Abramson, H. A. (1929). *J. Gen. Physiol.* **12**, 711.
Abramson, H. A. (1934). "Electrokinetic Phenomena and Their Application to Biology and Medicine", Chemical Catalog Comp., New York.
Abramson, H. A., Moyer, L. S. and Gorin, M. H. (1942). "Electrophoresis of Proteins and the Chemistry of Cell Surfaces", Reinhold Publishing Corp., New York.
Adam, G. and Adam, G. (1975). *Exp. Cell Res.* **93**, 71.
Agarwal, H. O. (1964). *Mededel. Landbouwhogeschool Wageningen*, **64–65**, 1–53.
Albertsson, P. Å. (1971). "Partition of Cell Particles and Macromolecules", 2nd edn. Wiley–Interscience, New York.
Albertsson, P. Å. (1973). *Biochemistry* **12**, 2525.
Albertsson, P. Å. and Baird, G. D. (1962). *Exp. Cell Res.* **28**, 296.
Albertsson, P. Å. and Nyns, E. J. (1961). *Arkiv. Kemi.* **17**, 197.
Alexander, A. E. and Saggers, L. (1948). *J. Sci. Instr.* **25**, 374.
Allfrey, V. G., Littau, V. C. and Mirsky, A. E. (1965). *Proc. Nat. Acad. Sci. U.S.A.* **49**, 414.
Allison, A. C. and Valentine, R. C. (1960). *Biochem. Biophys. Acta* **40**, 400.
Allison, F. and Lancaster, M. G. (1964). *Ann. N.Y. Acad. Sci.* **116**, 936.
Alwen, J. and Lawn, A. M. (1974). *Exp. Cell Res.* **89**, 197.
Ambrose, E. J. (1965). *In* "Cell Electrophoresis", (E. J. Ambrose, ed.). J. and A. Churchill, London, p. 194.
Ambrose, E. J. (1966). *Progr. Biophys. Mol. Biol.* **16**, 243.
Ambrose, E. J. (1967). *In* "Canadian Cancer Conference Proceedings" (J. F. Morgan *et al.*, eds). Pergamon Press, Oxford, p. 247.
Ambrose, E. J., James, A. M. and Lowick, J. H. B. (1956). *Nature* **177**, 576.
Anderson, C. G. and Mackie, T. J. (1939). *Brit. J. Exp. Path.* **20**, 270.
Andersson, J., Möller, G. and Sjoberg, O. (1972). *Cell Immunol.* **4**, 381.
Andersson, L. C., Nordling, S. and Häyry, P. (1973). *Cell Immunol.* **8**, 235.
Andersson, L. C., Wasastjerna, C. and Gahmberg, C. G. (1976). *Internat. J. Cancer* **17**, 40.
Armstrong, P. B. (1966). *J. Exp. Zool.* **163**, 99.
Armstrong, P. B. and Jones, D. P. (1968). *J. Exp. Zool.* **167**, 275.
Arndt-Jovin, D. J. and Berg, P. (1971). *J. Virol.* **8**, 716.

Aub, J. C., Sanford, K. H. and Cote, M. N. (1965). *Proc. Nat. Acad. Sci. U.S.A.* **54**, 396.

Ault, K. A., Griffith, A. L., Platsoucas, D. and Catsimpoolas, N. (1976). *J. Immunol.* **117**, 1406.

Ave, K., Kawakami, I. and Sameshima, M. (1968). *Develop. Biol.* **17**, 617.

Axelrad, A. A. and McCulloch, E. A. (1958). *Stain Technol.* **33**, 67.

Bahl, O. P. (1969). *J. Biol. Chem.* **244**, 575.

Baird, G. D., Albertsson, P. Å. and Hofsten, B. V. (1961). *Nature* **192**, 236.

Baker, P. C. (1965). *J. Cell Biol.* **24**, 95.

Balduini, C., Balduini, C. L. and Ascari, E. (1974). *Biochem. J.* **140**, 557.

Bancroft, J. B., Hiebert, E., Rees, M. W. and Markham, R. (1968). *Virology* **34**, 224.

Bangham, A. D. (1961). *Proc. Roy. Soc. B.* **155**, 292.

Bangham, A. D. and Pethica, B. A. (1960). *Proc. Roy. Phys. Soc. Edin.* **28**, 43.

Bangham, A. D., Flemens, R., Heard, D. H. and Seaman, G. V. F. (1958a). *Nature* **182**, 642.

Bangham, A. D., Pethica, B. A. and Seaman, G. V. F. (1958b). *Biochem. J.* **69**, 12.

Baronowski, T., Lisowka, E., Morawiecki, A., Romanowska, E. and Strozecka, K. (1959). *Arch. Immunol. Terapie Doswiadczalnej* **7**, 15.

Barrat, N. and Avramaes, S. (1973). *Exp. Cell Res.* **76**, 451.

Bartholomew, J. W. and Umbreit, W. W. (1944). *J. Bacteriol.* **48**, 567.

Bases, R., Frances-Mendez, L. and Mendez, L. (1973). *Exp. Cell Res.* **76**, 441.

Bayer, M. E. and Remsen, C. C. (1970). *J. Bacteriol.* **101**, 204.

Bedford, J. M. (1963). *Nature* **200**, 1178.

Beebe, S. P. (1904). *Amer. J. Physiol.* **12**, 167.

Ben-Bassat, H. and Goldblum, N. (1975). *Proc. Nat. Acad. Sci. U.S.A.* **72**, 1046.

Ben-Bassat, H., Sharon, N. and Sachs, L. (1970). *Virology* **40**, 854.

Ben-Bassat, H., Inbar, M. and Sachs, L. (1971). *J. Memb. Biol.* **6**, 183.

Ben-Bassat, H., Goldblum, N., Manny, N. and Sachs, L. (1974). *Internat. J. Cancer* **14**, 367.

Ben-Bassat, H., Goldblum, T., Mitrani, S., Bentwich, Z. and Goldblum, N. (1976a). *Progr. Med. Virol.* **21**, 177.

Ben-Bassat, H., Goldblum, N., Mitrani, S., Klein, G. and Johansson, B. (1976b). *Internat. J. Cancer* **17**, 448.

Benda, L. (1966). *Acta Virol. Prague* **10**, 376.

Beniams, H. and Gustavson, R. G. (1942). *J. Phys. Chem.* **46**, 1015.

Bennett, D. and Boyse, E. A. (1973). *Nature* **246**, 308.

Bennett, M. V. L. and Trinkaus, J. P. (1970). *J. Cell Biol.* **44**, 592.

Ben-Or, S., Eisenberg, S. and Doljanski, F. (1960). *Nature* **188**, 1200.

Bernard, P. J., Weiss, L. and Ratcliffe, T. (1969). *Exp. Cell Res.* **54**, 293.

Bert, G., Lajolo di Cassano, D., Recco, P. and Mazzei, D. (1970). *Lancet* **1**, 365.

Bert, G., Forrester, A. J. and Davies, A. J. S. (1971). *Nature New Biol.* **234**, 86.

Bey. E. (1965). *In* "Cell Electrophoresis", (E. J. Ambrose, ed.). J. and A. Churchill, London, p. 142.

Bhattacharya, B. C. (1958). *Zeit. Tierforsch. u. Zuchtungsbiol.* **72**, 250.

Bhattacharya, B. C. (1962). *Zeitschrift Wissenshaftliche Zool.* **166**, 203.

Bier, M. and Nord, F. F. (1951). *Arch. Biochem. Biophys.* **33**, 320.
Bisset, K. A. and Hale, C. M. F. (1951). *J. Gen. Microbiol.* **5**, 150.
Blasie, J. K., Dewey, M. M., Blaurock, A. E. and Worthington, C. R. (1965). *J. Mol. Biol.* **14**, 143.
Blazsek, V. A. and Gyergyay, F. (1966). *Experientia* **22**, 380.
Blix, G. (1958). *In* "Chemistry and Biology of Mucopolysaccharides", (G. E. W. Wolstenholme and M. O'Connor, eds). J. and A. Churchill, London, p. 302.
Bloom, H. J. G. (1950a). *Brit. J. Cancer* **4**, 259.
Bloom, H. J. G. (1950b). *Brit. J. Cancer* **4**, 347.
Bloom, H. J. G. (1965). *Brit. J. Cancer* **19**, 228.
Bloom, H. J. G. and Richardson, W. W. (1957). *Brit. J. Cancer* **11**, 359.
Blumfeld, O. O., Gallop, P. M., Howe, C. and Lee, C. T. (1970). *Biochim. Biophys. Acta* **211**, 109.
Bockstahler, L. E. and Kaesberg, P. (1962). *Biophys. J.* **2**, 1.
Boland, A. V. Jr., Kempf, J. E. and Hanson, R. J. (1957). *J. Immunol.* **79**, 416.
Bolund, L., Darzynkiewicz, N. and Ringertz, N. R. (1970). *Exp. Cell Res.* **62**, 76.
Bona, C., Anteunis, A., Robineaux, R. and Halpern, B. (1972). *Clin. Exp. Immunol.* **12**, 377.
Bonner, J. and Huang, R. C. (1963). *J. Mol. Biol.* **6**, 169.
Booher, J., Sensenbrenner, M. and Mandel, P. (1973). *Neurobiology*, **3**, 335.
Borle, A. B. and Loveday, J. (1968). *Cancer Res.* **28**, 2401.
Born, G. V. R. (1965). *Nature* **206**, 1121.
Born, G. V. R. and Cross, M. J. (1964). *J. Physiol.* **170**, 397.
Borysenko, J. Z., Ukena, Th. E. and Karnovsky, M. J. (1977). *Exp. Cell Res.* **107**, 253.
Bosmann, H. B. (1972). *Biochem. Biophys. Res. Commun.* **49**, 1256.
Bosmann, H. B., Hagopian, A. and Eylar, E. H. (1968). *J. Cell Physiol.* **72**, 81.
Brachet, J. (1964). *Nature* **204**, 1218.
Bradley, W. E. C. and Culp, L. A. (1974). *Exp. Cell Res.* **84**, 335.
Branton, D. (1966). *Proc. Nat. Acad. Sci. U.S.A.* **55**, 1048.
Brent, T. P. and Forrester, J. A. (1967). *Nature* **215**, 92.
Bretscher, M. S. (1971a). *Biochem. J.* **122**, 40.
Bretscher, M. S. (1971b). *J. Mol. Biol.* **58**, 775.
Bretscher, M. S. (1971c). *Nature New Biol.* **231**, 229.
Bretscher, M. S. (1971d). *J. Mol. Biol.* **59**, 351.
Bretscher, M. S. (1972). *J. Mol. Biol.* **71**, 523.
Bretscher, M. S. (1974). *In* "The Cell Surface in Development" (A. A. Moscona, ed.). John Wiley and Sons, New York and Chichester, p.17.
Bretton, R., Wicker, R. and Bernhard, W. (1972). *Internat. J. Cancer* **10**, 397.
Brinton, C. C., Buzzell, A. and Lauffer, M. A. (1954). *Biochim. Biophys. Acta* **15**, 533.
Bronsted, J. N. (1931). *Z. Phys. Chem. Series A* 2571 (Bodenstein-Festband).
Brooks, D. E., Seaman, G. V. F. and Walter, H. (1971). *Nature New Biol.* **234**, 61.
Brossmer, R. and Patscheke, H. (1968). *Hoppe-Seyl. Z. Physiol. Chem.* **349**, 1242.
Brown, H. C. and Broom, J. C. (1932). *Brit. J. Exp. Path.* **13**, 337.
Buck, C. A., Glick, M. C. and Warren, L. (1970). *Biochemistry*, **9**, 4567.

Buck, C. A., Glick, M. C. and Warren, L. (1971). *Science* **172**, 169.
Buggs, C. W. and Green, R. G. (1935a). *J. Bacteriol.* **30**, 447.
Buggs, C. W. and Green, R. G. (1935b). *J. Bacteriol.* **30**, 453.
Bull, H. B. (1964). "An Introduction to Physical Biochemistry". F. A. Davis Co., Philadelphia, p. 319.
Burczyk, J. (1975). *Exp. Cell Res.* **90**, 211.
Burczyk, J., Szurman, N. and Zontek, I. (1976). *Studia Biophy.* **56**, 15.
Burger, M. M. (1969). *Proc. Nat. Acad. Sci. U.S.A.* **62**, 994.
Burger, M. M. (1973). *Fed. Proc.* **32**, 91.
Burger, M. M. and Goldberg, A. R. (1967). *Proc. Nat. Acad. Sci. U.S.A.* **57**, 359.
Burger, M. M. and Martin, G. S. (1972). *Nature New Biol.* **237**, 9.
Burger, M. M., Bombik, B. M., Breckenridge, B. L. and Sheppard, J. R. (1972). *Nature* **239**, 161.
Bussel, R. H. and Robinson, W. S. (1973). *J. Virol.* **12**, 321.
Caen, J. P. and Michel, H. (1972). *Nature* **240**, 148.
Carlson, B. M. (1974). *In* "Neoplasia and Cell Differentiation", (G. V. Sherbet, ed.). Karger, A. G. Basel and New York, p. 60.
Carnegie, P. R., Caspary, E. A. and Field, E. J. (1972). *Biochem. J.* **126**, 5P.
Carnegie, P. R., Caspary, E. A. and Field, E. J. (1973a). *Brit. J. Cancer* **28**, Suppl. I, 219.
Carnegie, P. R., Caspary, E. A., Dickinson, J. P. and Field, E. J. (1973b). *Clin. Exp. Immunol.* **14**, 37.
Carr, C. W. (1953). *Arch. Biochem.* **43**, 147.
Carruthers, C. and Suntzeff, V. (1944). *Science* **19**, 245.
Caspary, E. A. (1972). *Proc. Roy. Soc. Med.* **65**, 636.
Caspary, E. A. and Field, E. J. (1971). *Brit. Med. J.* **2**, 613.
Catchpole, H. R., Engel, M. B. and Joseph, N. R. (1961). *Fed. Proc.* **20**, 143.
Catsimpoolas, N., Hjerten, S., Kolin, A. and Porath, J. (1976). *Nature* **259**, 264.
Chapman, D. L. (1913). *Phil. Mag.* **25**, 475.
Chaudhuri, S. and Lieberman, I. (1965). *Biophys. Biochem. Res. Commun.* **20**, 303.
Chiarugi, V. P. and Urbano, P. (1972). *J. Gen. Virol.* **14**, 133.
Chlumecka, V., D'Obrenen, P. and Colter, J. S. (1973) *Canad. J. Biochem.* **51**, 1521.
Choucroun, N. and Plotz, H. (1934). *C.R. Acad. Sci. Paris* **199**, 165.
Clarke, M. (1971). *Biochem. Biophys. Res. commun.* **45**, 1063.
Cline, M. J. and Livingstone, D. C. (1971). *Nature New Biol.* **232**, 155.
Clowes, G. H. A. and Frisbie, W. S. (1905). *Amer. J. Physiol.* **14**, 173.
Coates, A. S. and Carnegie, P. R. (1975). *Clin. Exp. Immunol.* **22**, 16.
Codington, J. F., Sanford, B. H. and Jeanloz, R. W. (1970). *J. Nat. Cancer Inst.* **45**, 637.
Coggin, J. H., Ambrose, K. R. and Anderson, N. G. (1970). *J. Immunol.* **105**, 524.
Cohen, A. (1963). *In* "Mechanisms of Virus Infection", (W. Smith, ed.). Academic Press, London and New York, p. 153.
Cohen, B., Posswillio, D. E. and Woods, D. A. (1971). *Ann. Roy. Coll. Surg.* **48**, 255.

Cohen, L. B., Salzberg, B. M., Davila, H. V., Ross, W. N., Landowne, D., Waggoner, A. S. and Wang, C. H. (1974). *J. Memb. Biol.* **19**, 1.
Cohen, S. S. (1945). *J. Exp. Med.* **82**, 133.
Collard, J. G. and Temmink, J. H. M. (1975). *J. Cell Sci.* **19**, 21.
Collins, M. F. (1966a). *J. Exp. Zool.* **163**, 23.
Collins, M. F. (1966b). *J. Exp. Zool.* **163**, 39.
Collins, M. F. (1974). *In* "Neoplasia and Cell Differentiation", (G. V. Sherbet, ed.). Karger, A. G. Basel and New York, p. 234.
Coman, D. R. (1944). *Cancer Res.* **4**, 625.
Coman, D. R. (1953). *Cancer Res.* **13**, 397.
Comoglio, P. M. and Filogamo, G. (1973). *J. Cell Sci.* **13**, 415.
Cone, C. D. Jr. (1969). *Trans. N.Y. Acad. Sci.* **31**, 404.
Cone, C. D. Jr. (1970). *Oncology* **24**, 538.
Cone, C. D. Jr. (1971). *J. Theoret. Biol.* **30**, 151.
Cook, G. M. W. (1962). *Nature* **195**, 159.
Cook, G. M. W. and Eylar, E. H. (1965). *Biochim. Biophys. Acta* **101**, 57.
Cook, G. M. W. and Jacobson, W. (1968). *Biochem. J.* **107**, 549.
Cook, G. M. W., Heard, D. H. and Seaman, G. V. F. (1960). *Nature* **188**, 1011.
Cook, G. M. W., Heard, D. H. and Seaman, G. V. F. (1961). *Nature* **191**, 44.
Cook, G. M. W., Heard, D. H. and Seaman, G. V. F. (1962). *Exp. Cell Res.* **28**, 27.
Cooper, G. W. and Bedford, J. M. (1971). *Anat. Rec.* **169**, 300.
Corley, R. D. and Joseph, N. R. (1966). *Proc. Soc. Exp. Biol. Med.* **122**, 1171.
Coulter, C. B. (1920–21). *J. Gen. Physiol.* **3**, 309.
Cox, A. J. (1969). *J. Invest. Dermatol.* **53**, 428.
Craig, L. C. (1960). *In* "A Laboratory Manual of Analytical Methods of Protein Chemistry", (P. Alexander and R. J. Block, eds), Vol. 1. Pergamon Press, Oxford, p. 121.
Craig, L. C. and Craig, D. (1956). *In* "Techniques of Organic Chemistry", (A. Weissberger, ed.), Vol. III, Part 1, 2nd edn. Interscience, New York.
Crain, S. M. (1956). *J. Comp. Neurol.* **104**, 285.
Critchley, D. R. (1974). *Cell* **3**, 121.
Critchley, D. R., Wyke, J. A. and Hynes, R. O. (1976). *Biochim. Biophys. Acta* **436**, 335.
Crozier, E. H., Hollinger, M. E., Woodend, B. E. and Robertson, J. H. (1976). *J. Clin. Path.* **29**, 608.
Cunningham, D. D. and Pardee, A. B. (1969). *Proc. Nat. Acad. Sci. U.S.A.* **64**, 1049.
Curtis, A. S. G. (1960). *Amer. Naturalist* **94**, 37.
Curtis, A. S. G. (1961). *Exp. Cell Res.* Suppl. 8, 107.
Curtis, A. S. G. (1962). *Biol. Rev.* **37**, 82.
Curtis, A. S. G. (1963). *Nature* **200**, 1235.
Curtis, A. S. G. (1964). *J. Cell Biol.* **20**, 199,
Curtis, A. S. G. (1966). *Sci. Progr.* **54**, 61.
Curtis, A. S. G. (1967). "The Cell Surface: Its Molecular Role in Morphogenesis". Academic Press, New York and London.
Dale, G. and Latner, A. L. (1968). *Lancet* April, p. 847.

Dale, G. and Latner, A. L. (1969a). *Clin. Chim. Acta* **24**, 61.
Dale, G. and Latner, A. L. (1969b). *Protides Biol. Fluids* **17**, 427.
Dan, K. (1936). *Physiol. Zool.* **9**, 43.
Dan, K. (1947). *Biol. Bull.* **93**, 259.
Danielli, J. F. (1936). *J. Cell. Comp. Physiol.* **7**, 393.
Danielli, J. F. and Davson, H. (1935). *J. Cell. Comp. Physiol.* **5**, 495.
Danon, D. and Marikovsky, Y. (1961). *Comp. Rend.* **253**, 1271.
Dasgupta, S. and Kung-Ho, C. (1971). *Exp. Cell Res.* **64**, 463.
Davies, H. (1970). *Protides Biol. Fluids* **17**, 389.
Davson, H. and Danielli, J. F. (1952). "The Permeability of Natural Membranes", 2nd edn. University Press, Cambridge.
Dawes, E. A. (1972). "Quantitative Problems in Biochemistry". Churchill Livingstone, Edinburgh.
Dean, R. T. and Messner, M. (1975). *J. Chromatog.* **105**, 353.
Debye, P. and Hückel, E. (1923). *Phys. Z.* **24**, 49, 185, 305, 575.
Defendi, V. (1963). *Proc. Soc. Exp. Biol. Med.* **133**, 12.
De Haan, R. L. (1961). *Carnegie Institution Wash. Year Book* **60**, 419.
De Haan, R. L. (1963a). *In* "Biological Organisation at the Cellular and Supercellular Level", (R. J. C. Harris, ed.). Academic Press, New York and London, p. 147.
De Haan, R. L. (1963b). *Acta Embryol. Morph. Exp.* **6**, 26.
De Haan, R. L. (1963c). *Exp. Cell Res.* **29**, 544.
De Haan, R. L. (1965). *In* "Organogenesis", (R. L. De Haan and H. Ursprung, eds.). Holt, Rinehart and Winston, New York, p. 377.
De Haan, R. L. and Gottlieb, S. H. (1968). *J. Gen. Physiol.* **52**, 643.
Dekker, W. A. L., Van der Meer, C. and Scholtens, R. Th. (1942). *J. Microbiol. Serol.* **8**, 53.
De Laat, S. W. and Bluemink, J. G. (1974). *J. Cell Biol.* **60**, 529.
De Laat, S. W., Luchtel, D. and Bluemink, J. G. (1973). *Develop. Biol.* **31**, 163.
De Laat, S. W., Buwalda, R. J. A. and Habets, A. M. M. C. (1974). *Exp. Cell Res.* **89**, 1.
De Laat, S. W., Wouters, W., Da Silva Guarda, M. M. and Da Silva Guarda, M. A. (1975). *Exp. Cell Res.* **91**, 15.
Deman, J. J. and Bruyneel, E. A. (1974). *Exp. Cell Res.* **89**, 206.
Deman, J. J., Bruyneel, E. A. and Mareel, M. M. (1974). *J. Cell Biol.* **60**, 641.
Deman, J. J., Vakaet, L. C. and Bruyneel, E. A. (1976). *J. Memb. Biol.* **26**, 205.
Demopoulos, H. B. *et al.* (1972). *In* "Steroids and Brain Oedema", (H. J. Reuben and K. Schurmann, eds.). Springer, Berlin, p. 29.
Den, H., Schultz, A. M., Basu, M. and Roseman, S. (1971). *J. Biol. Chem.* **246**, 2721.
De Petris, S. (1967). *J. Ultrastr. Res.* **19**, 45.
De Petris, S. (1974). *Nature* **250**, 54.
De Petris, S. and Raff, M. C. (1972). *Eur. J. Immunol.* **2**, 524.
De Petris, S. and Raff, M. C. (1973). *Nature New Biol.* **241**, 257.
De Petris, S. and Raff, M. C. (1974). *Eur. J. Immunol.* **4**, 130.
De Petris, S., Raff, M. C. and Mallucci, L. (1973). *Nature New Biol.* **244**, 275.

De Reuck, A. V. S. and Knight, J. (eds) (1966). "Histones, Their Role in the Transfer of Genetic Information". J. and A. Churchill, London.
Determan, H. (1967). "Gelchromatographie". Springer-Verlag, Berlin and New York.
Derjaguin, B. V. and Landau, L. (1941). *Acta Physicochim. U.R.S.S.* **14**, 633.
Dick, C. (1968). *Experientia* **24**, 356.
Dickinson, J. P. and Caspary, E. A. (1973). *Brit. J. Cancer* **28**, Suppl. I, 224.
Donald, D. Hutchinson, F., McLeod, T. M. and Raffle, E. J. (1974). *J. Immunol. Methods* **6**, 151.
Donner, M. and Wioland, M. (1975). *Folia Biol. (Praha)* **21**, 146.
Douglas, H. W. (1959). *Trans. Faraday Soc.* **55**, 850.
Drews, J. (1976). *In* "Molecular Base of Malignancy", (E. Deutsch, K. Moser, H. Rainer and A. Stacher, eds). Georg-Thime, Stuttgart, p. 236.
Duff, R. and Rapp, F. (1970). *J. Immunol.* **105**, 521.
Dumont, F. (1974a). *Internat. Arch. Allergy* **47**, 110.
Dumont, F. (1974b). *In* "La stimulation blastique des lymphocytes par les mitogènes" (B. Serrou, ed.). Inserm, Paris, p. 47.
Dumont, F. (1975). *Immunology* **28**, 731.
Dumont, F. and Robert, F. (1975). *Folia Biol. (Praha)* **22**, 1.
Dumont, F. and Robert, F. (1977). *Immunology* **33**, 295.
Dumont, J. N. and Yamada, T. (1972). *Develop. Biol.* **29**, 385.
Dungern, E. Van and Hirzfeld, L. (1911). *Z. Immunforsch.* **8**, 526.
Durandy, A. M., Wioland, M., Sabolovic, D. and Griscelli, C. (1975). *Clin. Immunol. Immunopathol.* **4**, 440.
Dworkin, G. A. (1958). *Biofizika* **3**, 610.
Easty, G. C. (1974). *In* "Neoplasia and Cell Differentiation" (G. V. Sherbet, ed.). Karger, A. G. Basel and New York, p. 189.
Economidou, J., Hughes-Jones, N. C. and Gardner, B. (1967). *Vox Sang.* **12**, 321.
Ede, D. A. and Kelly, W. A. (1964a). *J. Embryol. Exp. Morph.* **12**, 161.
Ede, D. A. and Kelly, W. A. (1964b). *J. Embryol. Exp. Morph.* **12**, 339.
Edelman, G. M. and Millette, C. F. (1971). *Proc. Nat. Acad. Sci. U.S.A.* **68**, 2436.
Edelman, G. M., Yahara, I. and Wang, J. L. (1973). *Proc. Nat. Acad. Sci. U.S.A.* **70**, 1442.
Edidin, M. and Weiss, A. (1972). *Proc. Nat. Acad. Sci. U.S.A.* **60**, 2456.
Edsall, J. T. and Wyman, J. (1958). "Biophysical Chemistry", Vol. 1. Academic Press, New York and London, p. 90.
Eggerth, A. H. (1924). *J. Gen. Physiol.* **6**, 63, 587.
Ehrlich, P. (1957). "Gesammelte arbeiten", (F. Himmelweit, ed.), Vol. 3. Springer-Verlag, Berlin.
Eisenberg, S., Ben-Or, S. and Doljanski, F. (1962). *Exp. Cell Res.* **26**, 451.
Elul, R., Brons, J. and Kravitz, K. (1975). *Nature* **258**, 616.
Engel, M. B., Catchpole, H. R. and Joseph, N. R. (1960). *Science* **132**, 669.
Engel, M. B., Pumper, R. W. and Joseph, N. R. (1968). *Proc. Soc. Exp. Biol. Med.* **128**, 990.
Entlicher, G., Tichá, M., Koštiř, J. V. and Kocourek, J. (1969). *Experientia* **25**, 17.

Entlicher, G., Koštiř, J. V. and Kucourek, J. (1971). *Biochim. Biophys. Acta* **236**, 795.
Eriksson, E., Albertsson, P. Å. and Johansson, G. (1976). *Mol. Cell. Biochem.* **10**, 123.
Eylar, E. H., Madoff, M. A., Brody, O. V. and Oncley, J. L. (1962). *J. Biol. Chem.* **237**, 1992.
Faillard, H. (1956). *Z. Physiol. Chem.* **305**, 145.
Fairbanks, G., Steck, T. L. and Wallach, D. F. H. (1971). *Biochemistry* **10**, 2606.
Fawcett, J. S. (1968). *FEBS Letters* **1**, 81.
Fawcett, J. S. (1969). *Protides Biol. Fluids* **17**, 409.
Feldherr, C. M. (1962). *J. Cell Biol.* **14**, 65.
Feldman, M. (1955). *J. Embryol. Exp. Morph.* **3**, 251.
Fernandez, S. M. and Berlin, R. D. (1976). *Nature* **264**, 411.
Fernandez-Moran, H. (1957). *In* "The Metabolism of the Nervous System" (D. Richter, ed.). Pergamon Press, Oxford, p. 1.
Field, E. J. (1973). "The Folksam Symposium". Bonniers, Stockholm, p. 7.
Field, E. J. and Caspary, E. A. (1970). *Lancet* **2**, 1337.
Field, E. J. and Caspary, E. A. (1972). *Brit. J. Cancer* **26**, 164.
Field, E. J., Caspary, E. A. and Carnegie, P. R. (1971). *Nature* **233**, 284.
Field, E. J., Caspary, E. A. and Smith, K. S. (1973a). *Brit. J. Cancer* **28**, Suppl. I, 208.
Field, E. J., Bates, D., Shaw, D. A., Griffin, S. G., Shenton, B. K. and Smith, K. (1973b). *Lancet* **2**, 675.
Field, E. J., Shenton, B. K. and Joyce, G. (1974). *Brit. Med. J.* **1**, 412.
Fike, R. M. and Van Oss, C. J. (1976). *In vitro* **12**, 428.
Findlay, A. (1949). "Practical Physical Chemistry". Longmans Green and Co., Harlow.
Fingl, E., Woodbarg, L. A. and Hecht, H. H. (1952). *J. Pharmacol. Exp. Therapy* **104**, 103.
Fischbach, G. D., Nameroff, M. and Nelson, P. G. (1971). *J. Cell. Physiol.* **78**, 289.
Fischer, H. and Wagner, L. (1954). *Naturwissenschaften* **41**, 533.
Fischer, L. (1969). *In* "Laboratory Techniques in Biochemistry and Molecular Biology" (T. S. Work and E. Work, eds), Vol. 1, Part 2. North-Holland Publishing Co., Amsterdam and London.
Flanagan, S. D. and Barondes, S. H. (1975). *J. Biol. Chem.* **250**, 1484.
Flanagan, S. D., Taylor, P. and Barondes, S. H. (1975). *Nature* **254**, 441.
Forrester, J. A. (1965). *In* "Cell Electrophoresis", (E. J. Ambrose, ed.). J. and A. Churchill, London, p. 115.
Forrester, J. A., Ambrose, E. J. and Stoker, M. G. P. (1964). *Nature* **201**, 945.
Forrester, J. A., Dumonde, D. C. and Ambrose, E. J. (1965). *Immunology* **8**, 37.
Forstner, G. G. (1971). *Biochem. J.* **121**, 781.
Forstner, J. and Manery, J. F. (1971). *Biochem. J.* **125**, 343.
Foster, D. O. and Pardee, A. B. (1969). *J. Biol. Chem.* 244, 2675.
Foulds, L. (1967). See discussion of Hayflick, L. *Nat. Cancer Inst. Monogr.* **26**, 355.
Fraenkel-Conrat, H. (1944). *J. Biol. Chem.* **154**, 227.
Frampton, V. and Hildebrand, E. M. (1944). *J. Bacteriol.* **48**, 537.

Franks, D. (1968). *Develop. Biol.* **43**, 17.
Frasca, J. N., Aurbach, O., Parks, V. R. and Jamieson, J. D. (1968). *Exp. Mol. Path.* **9**, 380.
Fredriksson, S. (1975). Ph.D. Thesis, Univ. Göteborg.
French, L. A. and Galicich, J. H. (1966). *Clin. Neurosurg.* **10**, 212.
Freedman, M. H. and Painter, R. H. (1971). *J. Biol. Chem.* **246**, 4340.
Freund, J. (1925). *Amer. Rev. Tuberculosis* **12**, 131.
Frithiof, L. (1969). *Acta Path. Scand.* Suppl. **200**, 1.
Frye, L. D. and Edidin, M. (1970). *J. Cell. Sci.* **7**, 319.
Fugita, S., Litwin, S. and Hartman, N. (1975). *J. Exp. Med.* **142**, 1416.
Fuhrman, G. F. (1965). *In* "Cell Electrophoresis", (E. J. Ambrose, ed.). J. and A. Churchill, London, p. 92.
Fuhrman, G. F. and Ruhenstroth-Bauer, G. (1965). *In* "Cell Electrophoresis", (E. J. Ambrose, ed.). J. and A. Churchill, London, p. 22.
Fukuda, M. and Osawa, T. (1973). *J. Biol. Chem.* **248**, 5100.
Furchgott, R. L. and Ponder, E. (1941). *J. Gen. Physiol.* **24**, 447.
Furthmayr, H., Tomita, M. and Marchesi, V. T. (1975). *Biochem. Biophys. Res. Commun.* **65**, 113.
Furuhjelm, U., Myllyla, G., Nevanlina, H. R., Nordling, S., Pirkola, A., Gavin, J., Gooch, A., Sanger, R. and Tippett, P. (1969). *Vox Sang.* **17**, 256.
Gahmberg, C. G. and Hakomori, S. (1973). *Proc. Nat. Acad. Sci. U.S.A.* **70**, 3329.
Gahmberg, C. G. and Hakomori, S. (1976). *Biomembranes* **8**, 131.
Gahmberg, C. G., Kiehn, D. and Hakomori, S. (1974). *Nature* **248**, 413.
Gail, M. H. and Boone, C. W. (1972). *Exp. Cell Res.* **73**, 252.
Galicich, J. H. and French, L. A. (1961). *Amer. Practnr. Dig. Treat.* **12**, 169.
Galtsoff, P. S. (1925). *J. Exp. Zool.* **42**, 183.
Garber, B. B. and Moscona, A. A. (1972). *Develop. Biol.* **27**, 235.
Garrido, J. (1975). *Exp. Cell Res.* **94**, 159.
Garrido, J., Burglen, M., Samolyk, D., Wicker, R. and Bernhard, W. (1974). *Cancer Res.* **34**, 230.
Gasic, G. J. and Galanti, N. L. (1966). *Science* 151, 203.
Gaader, A., Jonsen, J., Laland, S., Hellem, A. and Owren, P. A. (1961). *Nature* 192, 531.
Gardas, A. and Koscielak, J. (1971). *Vox Sang.* **20**, 137.
Garde, A. (1965). *Acta Neurol. Scand.* **13**, Suppl., 439.
Gardos, G., Lassen, U. and Pape, L. (1976). *Biochim. Biophys. Acta* **448**, 599.
Garrod, D. R. and Gingell, D. (1970). *J. Cell Sci.* **6**, 277.
Geelen, J. L. M. C., Van Kammen, A. and Verduin, B. J. M. (1972). *Virology* **49**, 205.
Gelin, L. E. (1956). *Acta Chir. Scand.* Suppl. **210**.
Gelin, L. E. and Ingelman, G. (1961). *Acta Chir. Scand.* **122**, 294.
Gent, W. L. G., Gregson, N. A., Gammack, D. B. and Raper, J. H. (1964). *Nature* **204**, 553.
Georgiev, G. P., Ananieva, L. A. and Koslov, Y. U. (1966). *J. Mol. Biol.* **22**, 365.
Gersten, D. M. and Bosmann, H. B. (1974). *Exp. Cell Res.* **88**, 225.
Gersten, D. M. and Bosmann, H. B. (1975). *Exp. Cell Res.* **96**, 215.

Gilbert, D. L. and Ehrenstein, G. (1969). *Biophys. J.* **9**, 447.

Gittens, G. J. and James, A. M. (1963a). *Biochim. Biophys. Acta* **66**, 237.

Gittens, G. J. and James, A. M. (1963b). *Biochim. Biophys. Acta* **66**, 250.

Glaves, D., Harlos, J. P. and Weiss, L. (1977). *Internat. J. Cancer* **19**, 474.

Glick, M. C., Rabinowitz, Z. and Sachs, L. (1974). *J. Virol.* **13**, 967.

Godson, G. N. (1970). *Analy. Biochem.* **35**, 66.

Goldman, D. E. (1943). *J. Gen. Physiol.* **27**, 37.

Goldstein, I. J. and Misaki, A. (1970). *J. Bacteriol.* **103**, 422.

Goldstein, I. J., Hollerman, C. E. and Merrick, J. M. (1965). *Biochim. Biophys. Acta* **97**, 68.

Goldstone, A. H., Urbaniak, S. J. and Irvine, W. J. (1974). *Clin. Exp. Immunol.* **17**, 113.

Gomulkiewicz, J. (1973). *Stud. Biophys.* **35**, 21.

Good, N. E., Winget, G. D., Winter, W., Connolly, T. N., Izawa, S. and Singh, R. M. M. (1966). *Biochemistry* **5**, 467.

Goodenough, D. A. and Revel, J. P. (1970). *J. Cell Biol.* **45**, 272.

Gordon, J. A., Sharon, N. and Lis, H. (1972). *Biochim. Biophys. Acta* **264**, 387.

Gordon, M., Dandekar, P. V. and Bartoszewicz, W. (1974). *J. Reprod. Fert.* **36**, 211.

Gordon, M. J. (1957). *Proc. Nat. Acad. Sci. U.S.A.* **43**, 913.

Gordon, M. J. (1958). *Sci. Amer.* **199**, 87.

Gorin, M. H. (1941). *J. Phys. Chem.* **45**, 371.

Gorter, E. and Grendel, F. (1927). *J. Exp. Med.* **41**, 439.

Gottschalk, A. (1955). *Biochem. J.* **61**, 302.

Gottschalk, A. (1956). *Biochem. Biophys. Acta* **20**, 560.

Gottschalk, A. (1957). *Biochem. Biophys. Acta* **23**, 645.

Gottschalk, A. (1960a). *Nature* **186**, 949.

Gottschalk, A. (1960b). *In* "The Chemistry and Biology of Sialic Acids and Related Substances". Cambridge University Press, Cambridge, p. 98.

Gottschalk, A. and Graham, E. R. B. (1959). *Biochim. Biophys. Acta* **34**, 380.

Gottschalk, A., Murphy, W. H. and Graham, E. R. B. (1962). *Nature* **194**, 1051.

Gouy, G. L. (1910). *J. Physique* **9**, 457.

Gram, I., Pitra, C. and Abel, H. (1975). *Studia Biophys. Berlin* **50**, 223.

Gräper, L. (1929). *Roux Arch. Entwickl. Mech. Org.* **116**, 382.

Grassetti, D. R., Murray, J. F. Jr. and Ruan, H. T. (1969). *Biochem. Pharmacol.* **18**, 603.

Greenberg, J. H. and Jamieson, G. A. (1974). *Biochim. Biophys. Acta* **345**, 231.

Grefrath, S. P. and Reynolds, J. A. (1974). *Proc. Nat. Acad. Sci. U.S.A.* **71**, 3913.

Griffith, A. L., Catsimpoolas, N. and Wortis, H. H. (1975). *Life Sci.* **16**, 1693.

Grimes, W. J. (1970). *Biochemistry* **9**, 5083.

Grottum, K. A. and Solum, N. O. (1969). *Brit. J. Haematol.* **16**, 275.

Habel, K. (1961). *Proc. Soc. Exp. Biol. Med.* **106**, 722.

Hafs, H. D. and Boyd, L. J. (1971). *Amer. Soc. Animal Sci.* p. 85.

Hafs, H. D. and Boyd, L. J. (1974). *J. Animal Sci.* **38**, 603.

Haglund, H. (1970). *Methods in Biochem. Analysis* **19**, 1.

Haglund, H. (1975). *In* "Isoelectric Focusing", (J. P. Arbuthnott and J. A. Beely, eds). Butterworths, London, p. 30.
Hakamori, S. I. and Jeanloz, R. W. (1961). *J. Biol. Chem.* **236**, 2827.
Hakomori, S. and Murakami, W. T. (1968). *Proc. Nat. Acad. Sci. U.S.A.* **59**, 254.
Hamaguchi, H. and Cleve, H. (1972). *Biochim. Biophys. Acta* **278**, 271.
Hamaker, H. C. (1937). *Physica* **4**, 1058.
Hampton, J. R. and Hardisty, R. M. (1967). *Nature* **213**, 400.
Hampton, J. R. and Mitchell, J. R. A. (1966a). *Nature* **209**, 470.
Hampton, J. R. and Mitchell, J. R. A. (1966b). *Nature* **210**, 1000.
Hannig, M. (1948). *Proc. Soc. Exp. Biol. Med.* **68**, 385.
Hannig, K. and Zeiller, K. (1969). *Hoppe-Seyler Z. Physiol. Chem.* **350**, 467.
Hanson, R. J., Kempf, J. E. and Boland, A. V. Jr. (1957). *J. Immunol.* **79**, 422.
Hardisty, R. M., Dormandy, K. M. and Hutton, R. A. (1964). *Brit. J. Haematol.* **10**, 371.
Harris, H. L. and Zalik, S. E. (1974). *J. Cell Physiol.* **83**, 359.
Harris, H. L. and Zalik, S. E. (1977). *Differentiation* **7**, 83.
Hartley, G. S. and Roe, J. W. (1940). *Trans. Faraday Soc.* **36**, 101.
Hartree, E. F. and Srivastava, P. N. (1965). *J. Reprod. Fert.* **9**, 47.
Hartveit, F., Cater D. B. and Mehrishi, J. N. (1968). *Brit. J. Exp. Path.* **49**, 634.
Hatanaka, M. (1974). *Biochim. Biophys. Acta* **355**, 77.
Hato, M., Ueda, T., Kurihara, K. and Kobatake, Y. (1976). *Biochim. Biophys. Acta* **426**, 73.
Hausman, S. J. and Bosma, M. J. (1975). *J. Exp. Med.* **142**, 998.
Haydon, D. A. (1961). *Biochim. Biophys. Acta* **50**, 450.
Haydon, D. A. and Seaman, G. V. F. (1967). *Arch. Biochem. Biophys.* **122**, 126.
Häyry, P. and Defendi, V. (1968). *Virology* **36**, 317.
Häyry, P. and Defendi, V. (1969). *Transpl. Proc.* **1**, 119.
Häyry, P. and Defendi, V. (1970). *Virology* **41**, 22.
Häyry, P., Andersson, L. C., Gahmberg, C., Roberts, P., Ranki, A. and Nordling, S. (1975). *Israel J. Med. Sci.* **11**, 1299.
Heard, D. H. and Seaman, G. V. F. (1960). *J. Gen. Physiol.* **43**, 635.
Heard, D. H. and Seaman, G. V. F. (1961). *Biochim. Biophys. Acta* **53**, 366.
Heard, D. H., Seaman, G. V. F. and Simon-Reuss, I. (1961). *Nature* **190**, 1009.
Heidelberger, M. (1927). *Physiol. Rev.* **7**, 107.
Helmholz, H. (1879). *Wied. Ann.* **7**, 337.
Henry, D. C. (1931). *Proc. Roy. Soc. A.* **133**, 106.
Henry, D. C. (1948). *Trans. Faraday Soc.* **44**, 1021.
Herbst, C. (1900). *Roux Arch. Entwickl. Mech. Org.* **9**, 424.
Herschman, H. R. and Helinski, D. R. (1967). *J. Biol. Chem.* **242**, 5360.
Hickie, R. A. and Kalant, H. (1967). *Cancer Res.* **27**, 1053.
Hill, B. J., Baxby, D. and Douglas, H. W. (1922). *J. Gen. Virol.* **16**, 39.
Hill, M. J., James, A. M. and Maxted, W. R. (1963a). *Biochim. Biophys. Acta* **66**, 264.
Hill, M. J., James, A. M. and Maxted, W. R. (1963b). *Biochim. Biophys. Acta* **75**, 402.
Hirsch, J. G. (1958). *J. Exp. Med.* **108**, 925.

Hirst, G. (1941). *Science* **94**, 22.
Hirst, G. K. (1942a). *J. Exp. Med.* **75**, 49.
Hirst, G. K. (1942b). *J. Exp. Med.* **76**, 195.
Hirst, G. K. (1948). *J. Exp. Med.* **87**, 301.
Hochberg, E. and Becker, Y. (1968). *J. Gen. Virol.* **2**, 231.
Hodgkin, A. L. and Katz, B. (1949). *J. Physiol. (Lond.)* **108**, 37.
Hodgman, C. D., Weast, R. C. and Selby, S. M. (1958). "Handbook of Chemistry and Physics". Chemical Rubber Publishing Co., Cleveland, Ohio, p. 1745.
Hoffman, D. R., Grossberg, A. L. and Pressman, D. (1972). *J. Immunol.* **108**, 18.
Hofschneider, P. H. and Martin, H. H. (1968). *J. Gen. Microbiol.* **51**, 23.
Holland, J. J. (1962). *Virology* **16**, 163.
Holtfreter, J. (1938). *Roux Arch. Entwickl. Mech. Org.* **138**, 522.
Holtfreter, J. (1939). *Arch. Exptl. Zellforsch.* **23**, 169.
Holfreter, J. (1943a). *J. Exp. Zool.* **94**, 261.
Holtfreter, J. (1943b). *J. Exp. Zool.* **93**, 251.
Holtfreter, J. (1944a). *J. Exp. Zool.* **95**, 171.
Holtfreter, J. (1944b). *Rev. Canad. Biol.* **3**, 220.
Horvath, J. (1944). *Strahlentherapie* **75**, 119.
Houwink, A. L. and Van Iterson, W. (1950). *Biochim. Biophys Acta* **5**, 10.
Howard, A. and Virella, G. (1969). *Protides Biol. Fluids* **17**, 449.
Howe, C. and Morgan, C. (1969). *J. Virol.* **3**, 70.
Howitt, F. O. (1934). *Biochem. J.* **28**, 1165.
Huang, R. C. and Bonner, J. (1962). *Proc. Nat. Acad. Sci. U.S.A.* **48**, 1276.
Hubbard, A. L. and Cohn, Z. (1972). *J. Cell Biol.* **55**, 390.
Hückel, E. (1924). *Phys. Z.* **25**, 204.
Huggins, J. W., Chesnut, R. W., Durham, N. N. and Carraway, K. L. (1976). *Biochim. Biophys. Acta* **426**, 630.
Humphrey, T. (1963). *Develop. Biol.* **8**, 27.
Hynes, R. O. (1973). *Proc. Nat. Acad. Sci. U.S.A.* **70**, 3170.
Hynes, R. O. (1974). *Cell* **1**, 147.
Hynes, R. O. and Wyke, J. A. (1975). *Virology* **64**, 492.
Ikeda, S. and Suzuki, S. (1912). U.S. Pat. 1 015 891.
Inbar, M. and Sachs, L. (1969a). *Proc. Nat. Acad. Sci. U.S.A.* **63**, 1418.
Inbar, M. and Sachs, L. (1969b). *Nature* **223**, 710.
Inbar, M. and Sachs, L. (1973). *FEBS Letters* **32**, 124.
Inbar, M., Ben-Bassat, H. and Sachs, L. (1971a). *Proc. Nat. Acad. Sci. U.S.A.* **68**, 2748.
Inbar, M. Ben-Bassat, H. and Sachs, L. (1971b). *J. Memb. Biol.* **6**, 195.
Inbar, M., Ben-Bassat, H. and Sachs, L. (1972a). *Nature New Biol.* **236**, 3.
Inbar, M., Ben-Bassat, H. and Sachs, L. (1972b). *Nature New Biol.* **236**, 16.
Inbar, M., Ben-Bassat, H. and Sachs, L. (1972c). *Internat. J. Cancer* **9**, 143.
Inbar, M., Ben-Bassat, H. and Sachs, L. (1973a). *Internat. J. Cancer* **12**, 93.
Inbar, M., Huet, C., Oseroff, A. R., Ben-Bassat, H. and Sachs, L. (1973b). *Biochim. Biophys. Acta* **311**, 594.
Irlin, I. S. (1967). *Virology* **32**, 725.
Isselbacher, K. J. (1972). *Proc. Nat. Acad. Sci. U.S.A.* **69**, 585.

Iwai, K. (1969) (ed.). "Histones and Gene Function". Research Group on Histones and Gene Function in Japan.
Jackson, L. J. and Seaman, G. V. F. (1972). *Biochemistry* **11**, 44.
Jacobson, C. F., Léonis, J. Linderstrøm-Lang, K. and Ottesen, M. (1957). *In* "Methods of Biochemical Analysis", (D. Glick, ed.), Vol. IV. Interscience, New York, p. 171.
Jamakosmanović, A. and Loewenstein, W. R. (1968). *J. Cell Biol.* **38**, 556.
James, A. M. (1957). *Progr. Biophys. Biophysical Chem.* **8**, 95.
James, A. M. (1965). *In* "Cell Electrophoresis", (E. J. Ambrose, ed.) J. and A. Churchill, London, p. 154.
James, A. M. and Barry, P. J. (1954). *Biochim. Biophys. Acta* **15**, 186.
Jandl, J. H. (1961). "9th Internat. Congress Pediatrics". University Toronto Press, Toronto, p. 41.
Janossy, G. and Greaves, M. F. (1972). *Clin. Exp. Immunol.* **10**, 525.
Jenkinson, E. J. and Searle, R. F. (1977). *Exp. Cell Res.* **106**, 386.
Ji, T. H. and Nicolson, G. L. (1974). *Proc. Nat. Acad. Sci. U.S.A.* **71**, 2212.
Joffe, E. W. and Mudd, S. (1935). *J. Gen. Physiol.* **18**, 599.
Johansson, G. (1970a). *Biochim. Biophys. Acta* **221**, 387.
Johansson, G. (1970b). *Biochim. Biophys. Acta* **222**, 381.
Johnson, R. G. and Sheridan, J. D. (1971). *Science* **174**, 717.
Joklik, W. K. and Darnell, J. E. (1961). *Virology* **13**, 439.
Jones, B. M. and Kemp, R. B. (1970). *Exp. Cell Res.* **63**, 301.
Jones, L. (1931). *Proc. Soc. Exp. Biol. Med.* **28**, 883.
Jonsson, M., Pettersson, S. and Rilbe, H. (1973). *Analy. Biochem.* **51**, 557.
Joseph, N. R., Catchpole, H. R., Laskin, D. M. and Engel, M. B. (1959). *Arch. Biochem. Biophys.* **84**, 224.
Just, W. W., Leon-V., J. O. and Werner, G. (1975). *In* "Progress in Isoelectric Focusing and Isotachophoresis", (P. G. Righetti, ed.). ASP Biol. Med. Press B. V., Amsterdam, p. 265.
Joseph, N. R. and Engel, M. B. (1959). *Arch. Biochem. Biophys.* **85**, 209.
Kabat, E. A. (1956). "Blood Group Substances". Academic Press, New York and London.
Kalb, A. J. and Levitzki, A. (1968). *Biochem. J.* **109**, 669.
Kanno, Y. and Loewenstein, W. R. (1963). *Exp. Cell Res.* **31**, 149.
Kanno, Y. and Matsui, Y. (1968). *Nature* **218**, 775.
Karpas, A., Sandler, R. M. and Thorburn, R. J. (1977a). *Brit. J. Cancer* **36**, 177.
Karpas, A., Hayhoe, F. G. J., Greenberger, J. S., Barker, C. R., Crawley, J. C., Lowenthal, K. M. and Moloney, W. C. (1977b), *Leukemia Res.* **1**, 35.
Katchalsky, A., Danon, D., Nevo, A. and De Vries, A. (1959). *Biochim. Biophys. Acta* **33**, 120.
Kates, J. R., Winterton, R. and Schlesinger, K. (1971). *Nature* **229**, 345.
Kathan, R. H. and Adamany, A. (1967). *J. Biol. Chem.* **242**, 1716.
Kathan, R. H., Winzler, R. J. and Johnson, C. A. (1961). *J. Exp. Med.* **113**, 37.
Kellenberger, E. and Ryter, A. (1958). *J. Biophys. Biochem. Cytol.* **4**, 323.
Kemp, R. B. (1968). *Nature* **218**, 1255.
Kemp, R. B. (1969). *Cytobios* **1**, 187.

Kemp, R. B. and Jones, B. M. (1970). *Exp. Cell Res.* **63**, 293.
Keski-Oja, J., Vaheri, A. and Ruoslahti, E. (1976). *Internat. J. Cancer* **17**, 261.
Khan, M. A., Litt, M., Kwart, H. and Rosenlund, R. (1973). *Tex. Reports Biol. Med.* **31**, 665.
Kidson, C. S. and Kirby, K. S. (1964). *Nature* **203**, 599.
Kidson, C. S. and Kirby, K. S. (1965). *Cancer Res.* **25**, 472.
Kleinschuster, S. J. and Moscona, A. A. (1972). *Exp. Cell Res.* **70**, 397.
Klenk, E. (1958). *In* "Chemistry and Biology of Mucopolysaccharides", (G. E. W. Wolstenholme and M. O'Connor, eds). J. and A. Churchill, London, p. 296.
Klenk, E. and Lempfrid, H. (1957). *Hoppe-Seyler Z. Physiol. Chem.* **307**, 278.
Klenk, E. and Uhlenbruck, G. (1958). *Hoppe-Seyler Z. Physiol. Chem.* **311**, 227.
Klenk, E. and Uhlenbruck, G. (1960). *Z. Physiol. Chem.* **319**, 151.
Kletzien, R. F. and Perdue, J. F. (1974a). *J. Biol. Chem.* **249**, 3883.
Kletzien, R. F. and Perdue, J. F. (1974b). *J. Biol. Chem.* **249**, 3375.
Koch, H. J. A. and Backx, J. (1969). *Sci. Tools* **16**, 44.
Koch, M. A. and Sabin, A. B. (1963). *Proc. Soc. Exp. Biol. Med.* **113**, 4.
Kojima, K. and Maekawa, A. (1970). *Cancer Res.* **30**, 2858.
Kojima, K. and Yamagata, Y. (1971). *Exp. Cell Res.* **67**, 142.
Kojima, K., Takeichi, N., Kobayashi, H. and Maekawa, A. (1970). *Nagoya Med. J.* **16**, 7.
Kolin, A. (1954). *J. Chem. Phys.* **22**, 1628.
Kolin, A. (1955a). *J. Chem. Phys.* **23**, 407.
Kolin, A. (1955b). *Proc. Nat. Acad. Sci. U.S.A.* **41**, 101.
Kolin, A. (1958). *Methods in Biochem. Analy.* **6**, 259.
Kolodny, G. M. (1973). *J. Cell Physiol.* **81**, 233.
Koscielak, J. and Zakrezewski, K. (1960). *Nature* **182**, 516.
Kourilsky, F. M., Silvestre, C., Neauport-Sautes, C., Loosfelt, Y. and Dausset, J. (1972). *Eur. J. Immunol.* **2**, 249.
Kozawa, S. (1914). *Biochem. Z.* **60**, 146.
Kraemer, P. M. (1966). *J. Cell Physiol.* **67**, 23.
Kraemer, P. M., Tobey, R. A. and Van Dilla, M. A. (1973). *J. Cell Physiol.* **81**, 305.
Kram, R. and Tomkins, G. M. (1973). *Proc. Nat. Acad. Sci. U.S.A.* **70**, 1659.
Krzanowski, M. (1970). *J. Reprod. Fert.* **23**, 11.
Kruyt, H. R. (1948). "Colloid Science", Vol. 2. Elsevier, Amsterdam, p.
Kuhns, W. J., Weinbaum, G., Turner, R. S. and Burger, M. M. (1973). *In* "Humoral Control of Growth and Differentiation" (J. LoBue, J. and A. S. Gordon, eds), Vol. II. Academic Press, New York and London, p. 59.
Kuroda, Y. (1973). *Gann* **64**, 555.
Kuroda, Y. (1974). *Exp. Cell Res.* **84**, 351.
Kuusela, P., Ruoslahti, E. and Vaheri, A. (1975). *Biochim. Biophys. Acta* **379**, 295.
Lakshmi, M. S. and Sherbet, G. V. (1974). *In* "Neoplasia and Cell Differentiation" (G. V. Sherbet, ed.). Karger, A. G. Basel and New York, p. 380.
Lamb, H. (1888). *Phil. Mag.* **25**, 52.
Landsteiner, K. (1900). *Zentr. Bakteriol Parasitenk.* **27**, 357.
Landsteiner, K. (1901). *Wein Klin. Wochschr.* **14**, 1132.
Landsteiner, K. and Levine, P. (1927). *Proc. Soc. Exp. Biol. Med.* **24**, 941.

Landsteiner, K. and Levine, P. (1928). *J. Exp. Med.* **48**, 731.
Landsteiner, K. and Wiener, A. S. (1940). *Proc. Soc. Exp. Biol. Med.* **43**, 223.
Laskov, R. and Scharff, M. D. (1970). *J. Exp. Med.* **131**, 515.
Lassen, U., Pape, L. and Vestergaard-Bogind, B. (1976). *J. Memb. Biol.* **26**, 51.
Latner, A. L. and Longstaff, E. (1969). *Nature* **224**, 71.
Latner, A. L. and Longstaff, E. (1971). *Brit. J. Cancer* **25**, 280.
Latner, A. L. and Turner, G. A. (1974). *J. Cell Sci.* **14**, 203.
Latner, A. L., Longstaff, E. and Turner, G. A. (1973). *Brit. J. Cancer* **27**, 218.
Latner, A. L., Longstaff, E. and Turner, G. A. (1974). *Eur. J. Cancer* **10**, 601.
Laurence, D. J. R. and Neville, A. M. (1972). *Brit. J. Cancer* **26**, 335.
Leaback, D. H. and Rutter, A. C. (1968). *Biochem. Biophys. Res. Commun.* **32**, 447.
Leaback, D. H., Rutter, A. C. and Walker, P. G. (1969). *Protides Biol. Fluids* **17**, 423.
Leblond, C. P. and Meissier, B. (1958). *Anat. Rec.* **132**, 247.
Lerche, D., Klaus, H. and Kunter, U. (1976). *Studia Biophys.* **56**, 21.
Lewis, G. N. and Randall, M. (1921). *J. Amer. Chem. Soc.* **43**, 1112.
Lefford, F. (1972). *Exp. Cell Res.* **70**, 242.
LeVine, D., Kaplan, M. J. and Greenway, P. J. (1972). *Biochem. J.* **129**, 847.
Levinson, C., Smith, T. C. and Mikiten, T. M. (1972). *J. Cell. Physiol.* **80**, 149.
Lewis, A. M. Jr., Levin, M. J., Wiese, W. H., Crumpacker, C. S. and Henry, H. (1969). *Proc. Nat. Acad. Sci. U.S.A.* **63**, 1128.
Lewis, E. A., St. Cyr, M. J. and Smith, F. (1967). *Carbohydrate Res.* **5**, 194.
Lewkonia, R. M., Kerr, E. J. L. and Irvine, W. J. (1974). *Brit. J. Cancer* **30**, 532.
Lhérisson, A. M., Meyer, G. and Bonneau, H. (1967). *Bull. Cancer* **54**, 419.
Li, N. C., Tang, P. and Mathur, R. (1961). *J. Phys. Chem.* **65**, 1074.
Lichtman, M. A. A. and Weed, R. I. (1970). *Blood* **35**, 12.
Lilien, J. E. (1968). *Develop. Biol.* **17**, 657.
Lilien, J. E. (1969). *In* "Current Topics in Developmental Biology", (A. A. Moscona and A. Monroy, eds). Academic Press, New York and London, p. 169.
Lilien, J. E. and Moscona, A. A. (1967). *Science* **157**, 70.
Lin, P. S., Schmidtullrich, R. and Wallach, D. F. H. (1977). *Proc. Nat. Acad. Sci. U.S.A.* **74**, 2495.
Lindahl, D. (1958). *Nature* **181**, 784.
Linderstrøm-Lang, K. and Nielson, S. O. (1959). *In* "Electrophoresis", (M. Bier, ed.). Vol. II. Academic Press, London and New York, p. 35).
Lipman, K. M., Dodelson, R. and Hays, R. M. (1966). *J. Gen. Physiol.* **49**, 501.
Lis, H. and Sharon, N. (1973). *Ann. Rev. Biochem.* **42**, 541.
Lis, H., Sela, B. A., Sachs, L. and Sharon, N. (1970). *Biochim. Biophys. Acta* **211**, 582.
Lisowska, E. (1969). *Eur. J. Biochem.* **10**, 574.
Litt, M., Khan, M. A., Kwart, H. and Rosenlund, M. L. (1976). *Tex. Reports Biol. Med.* **34**, 151.
Lodge, O. (1886). *Brit. Association Reports*, p. 319.
Loewenstein, W. R. (1966). *Ann. N.Y. Acad. Sci.* **137**, 441.
Loewenstein, W. R. and Kanno, Y. (1966). *Nature* **209**, 1248.

Loewenstein, W. R. and Kanno, Y. (1967). *J. Cell Biol.* **33**, 225.

Loewenstein, W. R. and Penn, R. D. (1967). *J. Cell Biol.* **33**, 235.

Longton, R. W., Cole, J. S. and Grays, R. (1974). *Differentiation* **2**, 43.

Loor, F. (1973). *Exp. Cell Res.* **82**, 415.

Loor, F., Forni, L. and Pernis, B. (1972). *Eur. J. Immunol.* **2**, 203.

Loos, J. A. and Roos, D. (1974). *Exp. Cell Res.* **86**, 331.

Lowick, J. H. B. and James, A. M. (1955). *Biochim. Biophys. Acta* **17**, 424.

Lowick, J. H. B. and James, A. M. (1957). *Biochem. J.* **65**, 431.

Lowick, J. H. B., Purdom, L., James, A. M. and Ambrose, E. J. (1961). *J. Roy. Micros. Soc.* **80**, 47.

Lucy, J. A. and Glauert, A. M. (1964). *J. Mol. Biol.* **8**, 727.

Lukiewicz, S. and Korohoda, W. (1965). *In* "Cell Electrophoresis", (E. J. Ambrose, ed.). J. and A. Churchill, London, p. 26.

Lundberg, A. (1955). *Exp. Cell Res.* **9**, 393.

Luner, S. J., Sturgeon, P., Szklarck, D. and McQuiston, D. T. (1975). *Vox Sang.* **29**, 440.

McClay, D. R. and Moscona, A. A. (1974). *Exp. Cell Res.* **87**, 438.

McClelland, D. A. and Bridges, J. M. (1973). *Brit. J. Cancer* **27**, 114.

McClelland, L. and Hare, R. (1941). *Canad. J. Publ. Health* **32**, 530.

McCutcheon, M., Coman, D. R. and Fontaine, B. M. (1948). *Cancer* **1**, 460.

McDermott, J. R., Caspary, E. A. and Dickinson, J. P. (1974). *Clin. Exp. Immunol.* **17**, 103.

McDonald, T. F., Sachs, H. G., Orr, C. W. M. and Ebert, J. D. (1972a). *Develop. Biol.* **28**, 290.

McDonald, T. F., Sachs, H. G., Orr, C. W. M. and Ebert, J. D. (1972b). *Exp. Cell Res.* **74**, 201.

McNutt, N. S., Culp, L. A. and Black, P. H. (1971). *J. Cell Biol.* **50**, 691.

MacMurdo, H. L. and Zalik, S. E. (1970). *Experientia* **26**, 406.

MacMurdo-Harris, H. and Zalik, S. E. (1971). *Develop. Biol.* **24**, 335.

Machowka, W. W. and Schegaloff, S. B. (1935). *Roux Arch. Entwickl. Mech. Org.* **133**, 694.

Macpherson, I. and Stoker, M. (1962). *Virology* **16**, 147.

Maizels, M., Remington, M. and Truscoe, R. (1958). *J. Physiol.* **140**, 48.

Majhi, S., Medda, J. and Bose, A. (1967). *Naturwissenschaften* **54**, 208.

Majerus, P. W. and Brodie, G. N. (1972). *J. Biol. Chem.* **247**, 4253.

Mäkela, O., Miettinen, T. and Pesda, R. (1960). *Vox Sang.* **5**, 492.

Mallucci, L. (1971). *Nature New Biol.* **233**, 241.

Mallucci, L., Poste, G. and Wells, V. (1972). *Nature New Biol.* **235**, 222.

Malmgren, R. A., Takemoto, K. K. and Carney, P. G. (1968). *J. Natl. Cancer Inst.* **40**, 263.

Mandel, B. (1967). *Virology* **31**, 248.

Mandel, B. (1971). *Virology* **44**, 554.

Marchesi, V. T. and Andrews, E. P. (1971). *Science* **174**, 1247.

Marchesi, V. T., Steers, E. Jr., Tillack, T. W. and Marchesi, S. L. (1969). *In* "The Red Cell Membrane, Structure and Function", (G. A. Jamieson and T. J. Greenwalt, eds). Lippincott, Philadelphia, p. 117.

Marchesi, V. T., Tillack, T. W., Jackson, R. L., Segrest, J. P. and Scott, R. E. (1972). *Proc. Nat. Acad. Sci. U.S.A.* **69**, 1445.

Margoliash, E., Schenk, J. R., Hargie, M. P., Burokas, S., Richter, W. R., Barlow, G. H. and Moscona, A. A. (1965). *Biochem. Biophys. Res. Commun.* **20**, 383.

Marikovsky, Y. and Danon, D. (1969). *J. Cell Biol.* **43**, 1.

Marikovsky, Y., Lotan, R., Lis, H., Sharon, N. and Danon, D. (1976). *Exp. Cell Res.* **99**, 453.

Màrquez, E. D. (1976). *Exp. Cell Res.* **101**, 425.

Martin, H. H. and Frank, H. (1962). *Z. Naturforsch.* **178**, 190.

Martinez-Palomo, A. (1970). *Internat. Rev. Cytol.* **29**, 29.

Martinez-Palomo, A. and Braislovsky, C. (1968). *Virology* **34**, 379.

Martinez-Palomo, A., Wicker, R. and Bernhard, W. (1972). *Internat. J. Cancer* **9**, 676.

Martz, E., Phillips, H. M. and Steinberg, M. S. (1974). *J. Cell Biol.* **16**, 401.

Maslow, D. E. (1970). *Exp. Cell Res.* **61**, 266.

Maslow, D. E. and Weiss, L. (1972). *Exp. Cell Res.* **71**, 204.

Mattson, S. (1933). *J. Phys. Chem.* **37**, 223.

Mayhew, E. (1966). *J. Gen. Physiol.* **49**, 717.

Mayhew, E. and O'Grady, E. A. (1965). *Nature* **207**, 86.

Mayhew, E. and Weiss, L. (1968). *Exp. Cell Res.* **50**, 441.

Mazia, D. and Ruby, A. (1968). *Proc. Nat. Acad. Sci. U.S.A.* **61**, 1005.

Mealy, J., Chen, T. T. and Schanz, G. P. (1971). *J. Neurosurg.* **34**, 324.

Meezan, E., Wu, H. C., Black, P. H. and Robbins, P. W. (1969). *Biochemistry* **8**, 2518.

Mehrishi, J. N. (1969). *Eur. J. Cancer* **5**, 427.

Mehrishi, J. N. (1970a). *Eur. J. Cancer* **6**, 127.

Mehrishi, J. N. (1970b). *Nature* **226**, 452.

Mehrishi, J. N. (1970c). *Nature* **228**, 364.

Mehrishi, J. N. (1972). *Progr. Biophys. Mol. Biol.* **25**, 3.

Mehrishi, J. N. (1973). *Bibl. Anat.* **11**, 260.

Mehrishi, J. N. and Grasetti, D. R. (1969). *Nature* **224**, 563.

Mehrishi, J. N. and Thomson, A. E. R. (1968). *Nature* **219**, 1080.

Mehrishi, J. N. and Zeiller, K. (1974). *Brit. Med. J.* **1**, 360.

Mercer, E. H. (1965). *In* "Organogenesis", (R. L. De Haan and H. Ursprung, eds). Holt, Rinehart and Winston, New York, p. 29.

Mester, L., Szabados, L., Born, G. V. R. and Michal, F. (1972). *Nature New Biol.* **236**, 213.

Miller, G. L. and Stanley, W. M. (1942). *J. Biol. Chem.* **146**, 311.

Minden, P. and Farr, R. S. (1973). *In* "Handbook of Experimental Immunology", (D. M. Weir, ed.), Vol. I. Blackwell Scientific Publications, Oxford, p. 15.1.

Minor, A. H. and Burnett, L. (1952). *Blood* **7**, 393.

Minor, A. H. and Burnett, L. (1953). *N.Y. State J. Med.* **53**, 547.

Mintz, B. and Sachs, L. (1975). *Proc. Nat. Acad. Sci. U.S.A.* **72**, 2428.

Mirsky, A. E. and Osawa, S. (1961). *In* "The Cell", (J. Brachet and A. E. Mirsky, eds), Vol. 2. Academic Press, New York and London, p. 693.
Mitchell, J. R. A. and Sharp, A. A. (1964). *Brit. J. Haematol.* **10**, 78.
Mitchison, J. M. (1971). "The Biology of the Cell Cycle". Cambridge University Press, Cambridge.
Miyamoto, K. and Gilden, R. V. (1971). *J. Virol.* **7**, 395.
Mollinson, P. L. (1972). "Blood Transfusion". Blackwell Scientific Publications, Oxford.
Mooney, M. (1924). *Phys. Rev.* **23**, 396.
Moore, H. D. M. and Hibbit, K. G. (1975). *J. Reprod. Fert.* **44**, 329.
Mora, P. T., Brady, R. O., Bradley, R. M. and McFarland, V. W. (1969). *Proc. Nat. Acad. Sci. U.S.A.* **63**, 1290.
Morawiecki, A. (1964). *Biochim. Biophys. Acta* **83**, 339.
Morell, A., Skvaril, F., Steinberg, A. G., Loghem, V. E. and Terry, W. D. (1972). *J. Immunol.* **108**, 195.
Morell, A. Skvaril, F. and Rufener, J. L. (1973). *Vox Sang.* **24**, 323.
Morgan, C., Rose, H. M. and Mednis, B. (1968). *J. Virol.* **2**, 507.
Morgan, W. T. J. and Van Heynigen, R. (1944). *Brit. J. Exp. Pathol.* **25**, 5.
Mori, Y., Akedo, H. and Tanigaki, Y. (1973). *Exp. Cell Res.* **78**, 360.
Morrill, G. A. and Watson, D. E. (1966). *J. Cell Physiol.* **67**, 85.
Morrill, G. A., Kostellow, A. B. and Murphy, J. B. (1971). *Exp. Cell Res.* **66**, 289.
Morton, J. A. (1962). *Brit. J. Haematol.* **8**, 134.
Moscona, A. A. (1961a). *Exp. Cell Res.* **22**, 455.
Moscona, A. A. (1961b). *Nature* **190**, 408.
Moscona, A. A. (1962). *J. Cell Comp. Physiol.* **60**, Suppl. 1, 65.
Moscona, A. A. (1963a). *Nature* **199**, 379.
Moscona, A. A. (1963b). *Proc. Nat. Acad. Sci. U.S.A.* **49**, 742.
Moscona, A. A. (1968). *Develop. Biol.* **18**, 250.
Moscona, A. A. (1971). *Science* **171**, 905.
Moscona, A. A. and Moscona, H. (1952). *J. Anat.* **86**, 287.
Moyer, L. S. (1936a). *J. Bacteriol.* **32**, 433.
Moyer, L. S. (1936b). *J. Bacteriol.* **31**, 531.
Moyer, L. S. (1940). *Cold Spring Harbor Symp.* **8**, 78.
Mudd, S. (1933). *Cold Spring Harbor Symp.* **1**, 71.
Murray, R. G. E., Steed, P. and Elson, H. E. (1965). *Canad. J. Microbiol.* **11**, 547.
Myllyla, G., Furuhjelm, U., Nordling, S., Pirkola, A., Tippett, P., Gavin, J. and Sanger, R. (1971). *Vox Sang.* **20**, 7.
Nachman, R. L. and Ferris, B. (1972). *J. Biol. Chem.* **247**, 4475.
Neihof, R. and Schuldiner, S. (1960). *Nature* **185**, 526.
Nelson, P. G., Peacock, J. H., Amano, T. and Minna, J. (1971). *J. Cell Physiol.* **77**, 337.
Neri, A., Robertson, M., Connolly, D. T. and Oppenheimer, S. B. (1975). *Nature* **258**, 342.
Nermut, M. V. and Murray, R. G. E. (1967). *J. Bacteriol.* **93**, 1949.
Netter, H. (1925). *Arch. Ges. Physiol.* **208**, 16.
Nevo, A., De Vries, A. and Katchalsky, A. (1955). *Biochim. Biophys. Acta*, **17**, 536.

Niblett, C. L. and Semancik, J. S. (1969). *Virology* **38**, 685.
Nicolson, G. L. (1971). *Nature New Biol.* **233**, 244.
Nicolson, G. L. (1972a). *Nature New Biol.* **239**, 193.
Nicolson, G. L. (1972b) *In* "Membrane Research", (C. F. Fox, ed.). Academic Press, New York and London, p. 53.
Nicolson, G. L. (1973). *Nature New Biol.* **243**, 218.
Nicolson, G. L. (1974a). *Internat. Rev. Cytol.* **39**, 89.
Nicolson, G. L. (1974b). *In* "Control of Proliferation of Animal Cells". Cold Spring Harbor Laboratories, p. 251.
Nicolson, G. L. (1975). *In* "Cellular Membranes and Tumour Cell Behaviour", 25th Annual Symposium on Fundamental Cancer Research, M. D. Anderson Hospital and Tumour Institute. Williams and Hopkins, Baltimore, p. 81.
Nicolson, G. L. (1976a). *Biochim. Biophys. Acta* **457**, 57.
Nicolson, G. L. (1976b). *Biochim. Biophys. Acta* **458**, 1.
Nicolson, G. L. and Blaustein, J. (1972). *Biochim. Biophys. Acta* **266**, 543.
Nicolson, G. L. and Painter, R. G. (1973). *J. Cell Biol.* **59**, 395.
Nicolson, G. L. and Yanagimachi, R. (1972). *Science*, **177**, 276.
Nicolson, G. L., Lacorbiere, M. and Yanagimachi, R. (1972). *Proc. Soc. Exp. Biol. Med.* **141**, 661.
Nicolson, G. L., Smith, J. R. and Poste, G. (1976). *J. Cell Biol.* **68**, 395.
Nilsson, O., Lindqvist, I. and Ronquist, G. (1973). *Exp. Cell Res.* **83**, 421.
Nilsson, P., Wadström, T. and Vesterberg, O. (1970). *Biochim. Biophys. Acta* **221**, 146.
Noonan, K. D. and Burger, M. M. (1973a). *J. Cell Biol.* **59**, 134.
Noonan, K. D. and Burger, M. M. (1973b). *J. Biol. Chem.* **248**, 4286.
Noonan, K. D., Ranger, H. C., Basilico, C. and Burger, M. M. (1973). *Proc. Nat. Acad. Sci. U.S.A.* **70**, 347.
Nordling, S., Sanger, R., Gavin, J., Furuhjelm, U., Myllyla, G. and Metaxas, M. N. (1969). *Vox Sang.* **17**, 300.
Nordling, S., Andersson, L. and Häyry, P. (1972a). *Eur. J. Immunol.* **2**, 405.
Nordling, S., Andersson, L. C. and Häyry, P. (1972b). *Science* **178**, 1001.
Northrop, J. H. (1921–22). *J. Gen. Physiol.* **4**, 629.
Northrop, J. H. and Kunitz, M. (1925). *J. Gen. Physiol.* **7**, 729.
Nurden, A. T. and Caen, J. P. (1974). *Brit. J. Haematol.* **28**, 253.
Nurden, A. T. and Caen, J. P. (1975). *Nature* **255**, 720.
O'Brien, J. S. (1967). *J. Theoret. Biol.* **15**, 307.
Odland, G. and Ross, R. (1968). *J. Cell Biol.* **39**, 135.
Ogura, M. (1963). *J. Ultrastr. Res.* **8**, 251.
Ohkuma, S. and Ikemoto, S. (1965). *Proc. Japan. Acad.* **41**, 482.
Ohkuma, S., Ikemoto, S., Miyauchi, C. and Furuhata, T. (1966). *Proc. Japan. Acad.* **42**, 275.
Olsen, R. L., Norquist, R. and Everet, M. A. (1968). *Cancer Res.* **28**, 2078.
Osterman, L. (1970). *Sci. Tools* **17**, 31.
Overbeek, J. T. G. (1952a). *In* "Colloid Science", (H. R. Kruyt, ed.), Vol. 1. Elsevier, Amsterdam, p. 194.

Overbeek, J. T. G. (1952b). *In* "Colloid Science", (H. R. Kruyt, ed.), Vol. 1. Elsevier, Amsterdam, p. 245.
Overton, E. (1895). *Vierteljahresschr. Naturforsch. Ges. Zurich* **40**, 159.
Ozanne, B. and Sambrook, J. (1971). *Nature New Biol.* **232**, 156.
Palay, S. L. (1960). *J. Biophys. Biochem. Cytol.* **7**, 391.
Pankhurst, K. G. A. (1968). *In* "Surface Phenomena in Chemistry and Biology", (J. F. Danielli, K. G. A. Pankhurst and A. C. Riddiford, eds). Pergamon Press, Oxford, p. 100.
Pape, L., Kirstensen, B. I. and Bengtson, O. (1975). *Biochim. Biophys. Acta* **406**, 516.
Parish, J. H. and Kirby, K. S. (1966). *Biochim. Biophys. Acta* **114**, 198.
Parker, C. W. (1973). *In* "Handbook of Experimental Immunology" (D. M. Weir, ed.), Vol. 1. Blackwell, Oxford, p. 14, 1.
Pasteels, J. (1935). *C.R. Soc. Biol.* **120**, 1362.
Pastuschenko, V. and Donath, E. (1976a). *Studia Biophys.* **56**, 7.
Pastuschenko, V. and Donath, E. (1976b). *Studia Biophys.* **56**, 9.
Patinkin, D., Schlesinger, M. and Doljanski, F. (1970a). *Cancer Res.* **30**, 489.
Patinkin, D., Zaritsky, A. and Doljanski, F. (1970b). *Cancer Res.* **30**, 498.
Paul, J. and Gilmour, R. S. (1966). *J. Mol. Biol.* **16**, 242.
Paul, J. and Gilmour, R. S. (1968). *J. Mol. Biol.* **34**, 305.
Pauling, L. Itano, H. A., Singer, S. J. and Wells, I. C. (1949). *Science* **110**, 543.
Pedlow, J. L. and Lisse, M. W. (1936). *J. Bacteriol.* **31**, 235.
Perdue, J. F. (1973). *J. Cell biol.* **58**, 265.
Perry, M. M. and Waddington, C. H. (1966). *J. Embryol. Exp. Morph.* **15**, 317.
Pessac, B. and Defendi, V. (1972). *Nature New Biol.* **238**, 13.
Pessac-Pejsachowicz, B. and Alliot-Mayet, F. (1968). *C.R. Acad. Sci. Paris* **266**, 1809.
Pethica, B. A. (1961). *Exp. Cell Res.* Suppl. 8, 123.
Phillips, D. and Morrison, M. (1971). *Biochemistry* **10**, 1766.
Phillips, D. M. P. (1962). *Progr. Biophys. Mol. Biol.* **12**, 211.
Phillips, P. G., Furmanski, P. and Lubin, M. (1974). *Exp. Cell Res.* **86**, 301.
Phondke, G. P. and Sundaram, K. (1971). *Immunol.* **21**, 1.
Picton, H. and Linder, S. E. (1892). *J. Chem. Soc.* **61**, 148.
Pinaev, G., Hoorn, B. and Albertsson, P. Å. (1976). *Exp. Cell Res.* **98**, 127.
Pinto Da Silva, J. and Branton, S. D. (1970). *J. Cell Biol.* **45**, 598.
Pinto Da Silva, J. and Douglas, S. D. (1970). *Nature* **232**, 194.
Pinto, Da Silva, P. and Nicolson, G. L. (1974). *Biochim. Biophys. Acta* **363**, 311.
Plummer, D. T. and James, A. M. (1961). *Biochim. Biophys. Acta* **53**, 453.
Ponder, E. (1951). *Blood* **6**, 350.
Pondman, K. V. and Mastenboek, G. G. A. (1954). *Vox Sang.* **4**, 98.
Pope, J. H. and Rowe, W. P. (1964). *J. Exp. Med.* **120**, 121.
Poste, G. (1970). *Exp. Cell Res.* **77**, 264.
Poste, G. (1971). *Exp. Cell Res.* **65**, 359.
Poste, G. (1973). *Exp. Cell Res.* **77**, 264.
Poste, G. and Moss, C. (1972). *In* "Progress in Surface Science", (S. G. Davidson, ed.), Vol. 2. Pergamon Press, Oxford, p. 139.

Poste, G. and Reeve, P. (1972). *J. Gen. Virol.* **16**, 21.

Poste, G., Greenham, L. W., Mallucci, L., Reeve, P. and Alexander, D. J. (1973). *Exp. Cell. Res.* **78**, 303.

Poste, G., Papahadjopoulos, D., Jacobson, K. and Vail, W. J. (1975a). *Biochim. Biophys. Acta* **394**, 520.

Poste, G., Papahadjoupoulos, D. and Nicolson, G. L. (1975b). *Proc. Nat. Acad. Sci. U.S.A.* **72**, 4430.

Poste, G., Papahadjoupoulos, D., Jacobson, K. and Vail, W. J. (1975c). *Nature* **253**, 552.

Prankerd, T. A. J. (1958). *J. Physiol. London* **143**, 325.

Preece, A. W. and Light, P. A. (1974). *Clin. Exp. Immunol.* **18**, 543.

Price, P. G. (1970). *J. Memb. Biol.* **2**, 300.

Pritchard, J. A. V., Moore, J. L., Sutherland, W. H. and Joslin, C. A. F. (1972). *Lancet* **2**, 627.

Pritchard, J. A. V., Moore, J. L., Sutherland, W. H. and Joslin, C. A. F. (1973a). *Brit. J. Cancer* **27**, 1.

Pritchard, J. A. V., Moore, J. L., Sutherland, W. H. and Joslin, C. A. F. (1973b). *Brit. J. Cancer* **28** Suppl. I, 229.

Pritchard, J. A. V., Moore, J. L., Sutherland, W. H. and Joslin, C. A. F. (1976). *Brit. J. Cancer* **34**, 1.

Puck, T. T. and Tolmach, L. J. (1954). *Arch. Biochem. Biophys.* **51**, 229.

Purdom, L., Ambrose, E. J. and Klein, G. (1958). *Nature* **181**, 1586.

Quincke, G. (1959). *Pogg. Ann.* **107**, 1.

Rabinovitch, M. and De Stefano, M. J. (1974). *Exp. Cell Res.* **88**, 153.

Rabinovitch, M. and De Stefano, M. J. (1975). *J. Cell. Physiol.* **85**, 189.

Rabinovitch, M. and De Stefano, M. J. (1976). *J. Exp. Med.* **143**, 290.

Rambourg, A. (1972). *Internat. Rev. Cytol.* **31**, 57.

Rao, K. V. (1977). *Ind. J. Exp. Biol.* **15**, 552.

Rapp, F., Butel, J. S. and Melnick, J. L. (1964). *Proc. Soc. Exp. Biol. Med.* **116**, 1131.

Rasmussen, T. and Gulati, D. R. (1962). *J. Neurosurg.* **19**, 535.

Ratishauser, U. and Sachs, L. (1974). *Proc. Nat. Acad. Sci. U.S.A.* **71**, 2456.

Rawles, M. E. (1948). *Physiol. Rev.* **28**, 383.

Ray, P. K. and Chatterjee, S. (1975). *Experientia* **31**, 1075.

Redmann, K. (1967). *Biophysik* **4**, 92.

Redmann, K. (1971). *Acta Biol. Med. Germ.* **27**, 55.

Redmann, K., Müller, V., St. Tanneberger and Kalkoff, W. (1972). *Acta Biol. Med. Germ.* **28**, 853.

Redmann, K., Jenssen, H. L. and Köhler, H. J. (1974). *Exp. Cell Res.* **87**, 281.

Reed, G. B. and Gardiner, B. G. (1932). *Canad. J. Res.* **6**, 622.

Reitherman, R., Flanagan, S. D. and Barondes, S. H. (1973). *Biochim. Biophys. Acta* **227**, 193.

Repacholi, M. H., Woodcock, J. P., Newman, D. L. and Taylor, K. J. W. (1971). *Phys. Med. Biol.* **16**, 221, 227.

Reuss, F. F. (1809). *Memoires de la Société Imperiales de Naturalistes de Moskou* **2**, 327.

Revel, J. P. (1972). *In* "Comparative Molecular Biology of Extracellular Mattrices", (M. C. Slavkin, ed.). Academic Press, New York and London, p. 77.
Rice, R. H. and Horst, J. (1972). *Virology* **49**, 602.
Rilbe, H. and Petterson, S. (1975). *In* "Isoelectric Focusing", (J. P. Arbuthnott and J. A. Beely, eds). Butterworth, London, p. 44.
Riley, P. A. and Seal, P. (1968). *Nature* **220**, 922.
Riley, R. F. and Coleman, M. K. (1968). *J. Lab. Clin. Med.* **72**, 714.
Robertson, J. D. (1959). *Biochem. Soc. Symp.* **16**, 3.
Robertson, J. D. (1963). *J. Cell Biol.* **19**, 201.
Robertson, M., Neri, A. and Oppenheimer, S. B. (1975). *Science* **189**, 639.
Rogers, H. J. and Perkins, H. R. (1968). "Cell Walls and Cell Membranes". E. and F. N. Spon Ltd, London.
Romanowska, E. (1961). *Nature* **191**, 1408.
Roos, E. and Temmink, H. M. (1975). *Exp. Cell Res.* **94**, 140.
Rosenberg, S. B. and Charalampous, F. C. (1977). *Arch. Biochem. Biophys.* **181**, 117.
Rosenblith, J. Z., Ukena, T. E., Yin, H. H., Berlin, R. D. and Karnovsky, M. J. (1973). *Proc. Nat. Acad. Sci. U.S.A.* **70**, 1625.
Ross, R. and Odland, G. (1968). *J. Cell Biol.* **39**, 152.
Ross, W. N., Salzberg, B. M., Cohen, L. B., Grinwald, A., Davila, H. V., Waggoner, A. S. and Wang, C. H. (1977). *J. Memb. Biol.*, **33**, 141.
Rossomando, E. F., Steffek, A. J., Mujwid, D. K. and Alexander, S. (1974). *Exp. Cell Res.* **85**, 73.
Rothman, S., Adelson, E., Schewebel, A. and Langdell, R. D. (1957). *Vox Sang.* **2**, 104.
Rottmann, W. L., Walther, B. T., Hellerqvist, C. G., Umbreit, J. and Roseman, S. (1974), *J. Biol. Chem.* **249**, 373.
Rowe, D. S., Hug, K., Forni, L. and Pernis, B. (1973). *J. Exp. Med.* **138**, 965.
Rowlatt, C., Wicker, R. and Berhhard, W. (1973). *Internat. J. Cancer* **11**, 314.
Rubin, H. and Franklin, R. M. (1957). *Virology* **3**, 84.
Rueff, F., Fuhrmann, G. F. and Ruhenstroth-Bauer, G. (1963). *Münch. Med. Wschr.* **105**, 1242.
Ruhenstroth-Bauer, G. (1965). *In* "Cell Electrophoresis", (E. J. Ambrose, ed.). J. and A. Churchill, London, p. 66.
Ruhenstroth-Bauer, G. and Lucke-Huhle, C. (1968). *J. Cell Biol.* **37**, 196.
Ruhenstroth-Bauer, G., Fuhrmann, G. F., Granzer, E., Kuelber, W. and Rueff, F. (1962). *Naturwissenschaften* **49**, 363.
Rutgers, A. J., Facq, L. and Van der Minne, J. L. (1950). *Nature* **166**, 100.
Ryan, G. B., Unanue, E. R. and Karnovsky, M. J. (1974a). *Nature* **250**, 56.
Ryan, G. B., Borysenko, J. Z. and Karnovsky, M. J. (1974b). *J. Cell Biol.* **62**, 351.
Sabin, A. B. and Koch, M. A. (1964). *Proc. Nat. Acad. Sci. U.S.A.* **52**, 1131.
Sabolovic, N., Sabolovic, D., Dumont, F. and Siest, G. (1974). *Biomedicine* **21**, 86.
Sachs, H. G., Stambrook, P. J. and Ebert, J. D. (1974). *Exp. Cell Res.* **83**, 362.
Sachtleben, P. (1965). *In* "Cell Electrophoresis", (E. J. Ambrose, ed.). J. and A. Churchill, London, p. 100.

Sachtleben, P. and Straub, E. (1959). *Z. ge. Exp. Med.* **131**, 493.
Sachtleben, P., Fuhrman, G. F., Straub, E. and Ruhenstroth-Bauer, G. (1961). *Klin. Wschr.* **39**, 839.
Sachtleben, P., Gsell, R. and Mehrishi, J. N. (1973). *Vox Sang.* **25**, 519.
Sakai, I. (1967). *Nagoya Med. J.* **13**, 51.
Salton, M. R. J. (1957). *Bacteriol. Rev.* **21**, 82.
Sanders, E. J. and Zalik, S. E. (1972). *J. Cell Physiol.* **79**, 235.
Saric, S. P. and Schofield, R. K. (1946). *Proc. Roy. Soc. London Series A* **185**, 431.
Sato, C. and Kojima, K. (1971). *Exp. Cell Res.* **69**, 435.
Sato, C. and Kojima, K. (1974). *Radiation Res.* **60**, 506.
Sato, C. and Takasawa-Nishizawa, K. (1974). *Exp. Cell Res.* **89**, 121.
Sato, C., Kojima, K., Onozawa, M. and Matsuzawa, T. (1972). *Internat. J. Radiat. Biol.* **22**, 479.
Sato, C., Kojima, K., Nishizawa, K., Shimizu, S. and Inoue, M. (1976). *Biochim. Biophys. Acta* **448**, 379.
Schilling, E. (1971). *Amer. Soc. Animal Sci.* p. 76.
Schmidt-Ullrich, S., Wallach, D. F. H. and Hendricks, J. (1976). *Biochim. Biophys. Acta* **443**, 587.
Schollenberger, C. J. (1927). *Science* **65**, 552.
Schollenberger, C. J. and Driebelbis, F. R. (1930). *Soil Sci.* **30**, 161.
Schröder, V. N. (1934). *Animal Breeding Abstracts* **3**, 166.
Schröder, V. N. (1942). *Animal Breeding Abstracts* **10**, 252.
Schubert, M. (1964). *Biophys. J.* **4**, 119.
Schubert, J. C. F., Walther, F., Holzberg, E. and Zeiller, K. (1972). *Klin. Wschr.*
Seal, P., Riley, P. A. and Inman, D. R. (1969). *J. Invest. Dermatol.* **52**, 259.
Seaman, G. V. F. (1958). "Microelectrophoresis of Red Blood Cells", Thesis, Cambridge University.
Seaman, G. V. F. (1965). *In* "Cell Electrophoresis", (E. J. Ambrose, ed.). J. and A. Churchill, London, p. 4.
Seaman, G. V. F. (1967). *Thrombos. Diath. Haemorrh.* Suppl. 26, 53.
Seaman, G. V. F. and Cook, G. M. W. (1965). *In* "Cell Electrophoresis", (E. J. Ambrose, ed.). J. and A. Churchill, London, p. 48.
Seaman, G. V. F. and Heard, D. H. (1960). *J. Gen. Physiol.* **44**, 251.
Seaman, G. V. F. and Uhlenbruck, G. (1962). *Biochim. Biophys. Acta* **64**, 570.
Seaman, G. V. F. and Vassar, P. S. (1966). *Arch. Biochem. Biophys.* **117**, 10.
Seeds, N. W., Gilman, A. G., Amano, T. and Nirenberg, M. W. (1970). *Proc. Nat. Acad. Sci. U.S.A.* **66**, 160.
Segrest, J. P., Kahne, I., Jackson, R. L. and Marchesi, V. T. (1973). *Arch. Biochem. Biophys.* **155**, 167.
Sela, B. A., Lis, H., Sharon, N. and Sachs, L. (1970). *J. Memb. Biol.* **3**, 267.
Sevinç, A. (1968). *J. Reprod. Fert.* **16**, 7.
Shanbhag, V. P. and Johansson, G. (1974). *Biochim. Biophys. Acta* **61**, 1141.
Shaw, A., Ettin, G. and McPherson, T. A. (1976). *Brit. J. Cancer* **34**, 7.
Shea, L. and Ginsburg, V. (1968). *In* "Biological Properties of the Mammalian Surface Membrane", (A. Manson, ed.). Wistar Institute Press, Philadelphia, p. 67.

Sheinin, R. and Onodera, K. (1972). *Biochim. Biophys. Acta* **274**, 49.
Shek, P. N., Chou, C. T., Dubiski, S. and Cinnader, B. (1976). *Immunology* **31**, 129.
Sherbet, G. V. (1966a). *J. Embryol. Exp. Morph.* **16**, 159.
Sherbet, G. V. (1966b). *In* "Histones, Their Role in the Transfer of Genetic Information", (A. V. S. De Reuck and J. Knight, eds.), Ciba Foundation Study Group 24. J. and A. Churchill, London, p. 81.
Sherbet, G. V. (1966c). *Progr. Biophys. Mol. Biol.* **16**, 89.
Sherbet, G. V. (1970). *Advan. Cancer Res.* **13**, 97.
Sherbet, G. V. and Lakshmi, M. S. (1967a). *Internat. Rev. Cytol.* **22**, 147.
Sherbet, G. V. and Lakshmi, M. S. (1967b). *Nature* **215**, 1089.
Sherbet, G. V. and Lakshmi, M. S. (1968). *Brit. Empire Cancer Res. Campaign Annual Report,* p. 57.
Sherbet, G. V. and Lakshmi, M. S. (1969a). *Brit. Empire Cancer Res. Campaign Annual Report* p. 92.
Sherbet, G. V. and Lakshmi, M. S. (1969b). *Experientia* **25**, 481.
Sherbet, G. V. and Lakshmi, M. S. (1970). *Oncology* **24**, 58.
Sherbet, G. V. and Lakshmi, M. S. (1973). *Biochim. Biophys. Acta* **298**, 50.
Sherbet, G. V. and Lakshmi, M. S. (1974a). *Oncology* **29**, 335.
Sherbet, G. V. and Lakshmi, M. S. (1974b). *J. Nat. Cancer Inst.* **52**, 681.
Sherbet, G. V. and Lakshmi, M. S. (1974c). *J. Nat. Cancer Inst.* **52**, 687.
Sherbet, G. V. and Lakshmi, M. S. (1975). *In* "Isoelectric Focusing", (J. P. Arbuthnott and J. A. Beely, eds). Butterworth, London, p. 338.
Sherbet, G. V. and Lakshmi, M. S. (1976). *In "Molecular Base of Malignancy"*, (E. Deutsch, K. Moser, H. Rainer and A. Stacher, eds). Georg Thieme, Stuttgart, p. 5.
Sherbet, G. V. and Lakshmi, M. S. (1978). *Exp. Cell Biol.* **46**, 82.
Sherbet, G. V., Lakshmi, M. S. and Morris, H. P. (1970). *J. Nat. Cancer Inst.* **45**, 419.
Sherbet, G. V., Lakshmi, M. S. and Rao, K. V. (1972). *Exp. Cell Res.* **70**, 113.
Sherbet, G. V., Lakshmi, M. S. and Coakham, H. B. (1974). *Internat. Res. Commun. System* **2**, 1485.
Sherbet, G. V., Lakshmi, M. S. and Haddad, S. K. (1977). *J. Neurosurg.* **47**, 864.
Shibley, G. S. (1924). *J. Exp. Med.* **40**, 453.
Shibley, G. S. (1926). *J. Exp. Med.* **44**, 667.
Shibley, G. S. (1929). *J. Exp. Med.* **50**, 825.
Shoham, J. and Sachs, L. (1972). *Proc. Nat. Acad. Sci. U.S.A.* **69**, 2479.
Shoham, J. and Sachs, L. (1974). *Exp. Cell Res.* **85**, 8.
Shoham, J., Inbar, M. and Sachs, L. (1970). *Nature* **227**, 1244.
Silver, D. J., Babcock, J. and Bruening, G. (1976). *Virology* **71**, 560.
Silverstein, S. C. and Dales, S. (1968). *J. Cell Biol.* **36**, 197.
Simon-Reuss, I., Cook, G. M. W., Seaman, G. V. F. and Heard, D. H., *Cancer Res.* **24**, 2038.
Sims, P. J., Waggoner, A. S., Wang, C. H. and Hoffman, J. F. (1974). *Biochemistry* **13**, 3315.
Singer, S. J. (1959). *Nature* **183**, 1523.

Singer, S. J. (1971). *In* "Structure and Function of Biological Membranes", (L. I. Rothfield, ed.). Academic Press, New York and London, p. 145.
Singer, S. J. (1972). *Ann. N.Y. Acad. Sci.* **195**, 16.
Singer, S. J. (1974). *Ann. Rev. Biochem.* **43**, 805.
Singer, S. J. and Nicolson, G. L. (1972). *Science* **175**, 720.
Sitz, T. O., Kent, A. B., Hopkins, H. A. and Schmidt, R. R. (1970). *Science* **168**, 1231.
Sjögren, H. O. and Hellström, I. (1967). *In* "Subviral Carcinogenesis", First Internat. Symp. on Tumour Viruses. Nagoya, p. 207.
Sjögren, H. O., Hellström, J. and Klein, G. (1961). *Exp. Cell Res.* **23**, 204.
Sjorstrand, F. S. (1963a). *Nature* **199**, 1262.
Sjorstrand, F. S. (1963b). *J. Ultrastr. Res.* **8**, 517.
Sjorstrand, F. S. (1963c). *J. Ultrastr. Res.* **9**, 340.
Smith, B. A., Ware, B. A. and Weiner, R. S. (1976). *Proc. Nat. Acad. Sci. U.S.A.* **73**, 2388.
Smith, H. S., Turner, S., Leong, J. A. and Rigby, P. W. J. (1976). *J. Virol.* **19**, 146.
Smith, M. E. and Lisse, M. W. (1936). *J. Phys. Chem.* **40**, 399.
Smith, T. C. (1976). *J. Cell Physiol.* **87**, 47.
Smith, T. C. and Adams, R. (1977). *J. Memb. Biol.* **35**, 57.
Smith, T. C., Mikiten, T. M. and Levinson, C. (1972). *J. Cell Physiol.* **79**, 117.
Smoluchowski, M. V. (1921). *In* "Handbuch der Elektrizitat und des Magnetismus", (V. Graetz, ed.), Vol. 2. Barth, Liepzig, p. 366.
So, L. L. and Goldstein, I. J. (1968). *Biochim. Biophys. Acta* **165**, 398.
So, L. L. and Goldstein, I. J. (1969). *Carbohydrate Res.* **10**, 231.
Spemann, H. and Mangold, H. (1924). *Roux Arch. Entwickl. Mech. Org.* **100**, 599.
Spiro, R. G. (1969). *New Eng. J. Med.* **281**, 991.
Spratt, N. T. (1946). *J. Exp. Zool.* **103**, 259.
Springer, G. F. (1964). *In* "Colloquium der Gesselschaft für Physiologische Chemie". Mosbach, Berlin, p. 90.
Springer, G. F., Nagai, Y. and Tegtmeyer, H. (1966). *Biochemistry* **5**, 3254.
Stalder, K. and Springer, G. F. (1960). *Fed. Proc.* **19**, 70.
Stambaugh, R. and Buckley, J. (1971a). *J. Reprod. Fert.* **25**, 275.
Stambaugh, R. and Buckley, J. (1971b). *Amer. Soc. Animal Sci.* p. 59.
Stearns, T. W. and Roepke, M. H. (1941). *J. Bacteriol.* **42**, 411.
Stein, G. (1975). *Z. Immunitatforsch.* **150**, 68.
Stein, G. (1976). *In* "Techniques of Separation and Characterisation of Human Lymphocytes", (D. Sabolovic and B. Serrou, eds), Technological Colloquiums and Seminars, Vol. 57.
Steinberg, M. S. (1958). *Amer. Naturalist* **92**, 65.
Steinberg, M. S. (1962a). *Proc. Nat. Acad. Sci. U.S.A.* **48**, 1577.
Steinberg, M. S. (1962b). *Science* **137**, 762.
Steinberg, M. S. (1962c). *Proc. Nat. Acad. Sci. U.S.A.* **48**, 1769.
Steinberg, M. S. (1962d). *In* "Biological Interactions in Normal and Neoplastic Growth". Henry Ford Hosp. Symp, p. 127.
Steinberg, M. S. (1964). *In* "Cellular Membranes in Development" (M. Locke, ed.). Academic Press, New York and London, p. 321.

Steinberg, M. S. (1970). *J. Exp. Zool.* **173**, 395.
Steinberg, M. S. and Gepner, I. A. (1973). *Nature New Biol.* **241**, 249.
Steinberg, M. S., Armstrong, P. B. and Granger, R. E. (1973). *J. Memb. Biol.* **13**, 97.
Steinhardt, R. A. and Mazia, D. (1973). *Nature* **241**, 400.
Steinhardt, R. A., Lundin, L. and Mazia, D. (1971). *Proc. Nat. Acad. Sci. U.S.A.* **68**, 2426.
Stern, O. (1924). *Z. Elektrochem.* **30**, 508.
Stone, J. D. (1947). *Aust. J. Exp. Biol. Med. Sci.* **25**, 137.
Stone, K. R., Smith, R. E. and Joklik, W. K. (1974). *Virology* **58**, 86.
Straub, E. (1965). *In* "Cell Electrophoresis", (E. J. Ambrose, ed.). J. and A. Churchill, London, p. 125.
Straub, E. and Sachtleben, E. (1963). *Z. Naturforsch* **186/187**, 530.
Street, M. (1956). *Aust. J. Chem.* **9**, 333.
Sturgeon, P., Kolin, A., Kwak, K. S. and Luner, S. J. (1972). *Haematologia* **6**, 93.
Sturgeon, P., Luner, S. J. and McQuiston, D. T. (1973a). *Vox. Sang.* **25**, 498.
Sturgeon, P., McQuiston, D. T., Taswell, H. F. and Allan, C. (1973b). *Vox Sang.* **25**, 481.
Subjeck, J. R., Weiss, L. and Warren, L. (1977). *J. Cell. Physiol.* **91**, 329.
Sugár, J. (1968). *Eur. J. Cancer* **4**, 33.
Sundaram, K., Phondke, G. P. and Ambrose, E. J. (1967a). *Immunology* **12**, 21.
Sundaram, K., Phondke, G. P. and Sundaresan, P. (1967b). *Immunology* **13**, 433.
Suntzeff, V. and Carruthers, C. (1944). *J. Biol. Chem.* **153**, 521.
Suzuki, S., Kojima, K. and Utsumi, K. R. (1970). *Biochim. Biophys. Acta* **222**, 240.
Svennerholm, L. (1956). *Acta Soc. Med. Upsalien* **61**, 75.
Svensson, H. (1960). *Lab. Manual Analy. Methods in Protein Chemistry including Peptides* **1**, 195.
Svensson, H. (1961). *Acta Chem. Scand.* **15**, 325.
Svensson, H. (1962). *Arch. Biochem. Biophys.* Suppl. 1, 132.
Swank, R. L., Fellman, J. H. and Hissen, W. W. (1963). *Circulation Res.* **13**, 392.
Takeichi, M. (1971). *Exp. Cell Res.* **68**, 88.
Takemoto, K. K. and Fabish, P. (1964). *Proc. Soc. Exp. Biol. Med.* **116**, 140.
Tanford, C. (1961). "Physical Chemistry of Macromolecules". John Wiley and Sons, New York and Chichester.
Tanford, C. (1962). *Advan. Protein Chem.* **17**, 69.
Tanner, M. J. A. and Boxer, D. H. (1972). *Biochem. J.* **129**, 333.
Tarin, D. (1968). *Internat. J. Cancer* **3**, 734.
Talbot, P. (1975). In "Isoelectric Focusing" (J. P. Arbuthnott and J. A. Beely, eds). Butterworth, London, p. 270.
Takahashi, M. and Ito, S. (1968). *Zool. Mag.* **77**, 307.
Taylor, R. B., Duffus, W. P. H., Raff, M. C. and De Petris, S. (1971). *Nature New Biol.* **233**, 225.
Templeton, J. G. and Milne, G. R. (1975). In "Isoelectric Focusing" (J. P. Arbuthnott and J. A. Beely, eds). Butterworth, London, p. 313.
Tevethia, S. S., Katz, M. and Rapp, F. (1965). *Proc. Soc. Exp. Biol. Med.* **119**, 896.
Terayama, H. (1952). *J. Polymer Sci.* **8**, 243.

Terayama, H. (1954). *Arch. Biochem. Biophys.* **50**, 55.
Terayama, H. (1962). *Exp. Cell Res.* **28**, 113.
Theuvenet, A. P. R. and Borst-Pauwels, G. W. F. H. (1976). *Bioelectrochem. Bioenerget.* **3**, 230.
Thomas, D. B. and Winzler, R. J. (1971). *Biochem. J.* **124**, 55.
Thompson, R. L. (1931a). *Amer. J. Hyg.* **14**, 235.
Thompson, R. L. (1931b). *Amer. J. Hyg.* **14**, 244.
Thompson, R. L. (1932). *Amer. J. Hyg.* **15**, 712.
Thomson, A. E. R. and Mehrishi, J. N. (1969). *Eur. J. Cancer* **5**, 195.
Tillack, T. and Marchesi, V. T. (1970). *J. Cell. Biol.* **45**, 649.
Tillack, T. W., Scott, R. E. and Machesi, V. T. (1972). *J. Exp. Med.* **135**, 1209.
Tiselius, A. (1930). "The Moving Boundary Method of Studying the Electrophoresis of Proteins". Thesis, Uppsala University.
Tittsler, R. P. and Lisse, N. W. (1928). *J. Bacteriol.* **15**, 105.
Tjernberg, B. and Zajicek, J. (1965). *Acta Cytol.* **9**, 197.
Todd, C. (1927). *Brit. J. Exp. Path.* **8**, 369.
Tomasz, A. (1971). *Nature* **234**, 389.
Tomita, M. and Marchesi, V. T. (1976). *Proc. Nat. Acad. Sci. U.S.A.* **72**, 2964.
Townes, P. L. and Holtfreter, J. (1955). *J. Exp. Zool.* **128**, 53.
Trinkaus, J. P. (1969). "Cells Into Organs". Prentice-Hall, Englewood Cliffs, N.J., U.S.A.
Trowbridge, I. S. and Hilborn, D. A. (1974). *Nature* **250**, 304.
Tupper, J. T. (1972). *Develop. Biol.* **29**, 273.
Tupper, J. T. (1973). *Develop. Biol.* **32**, 140.
Tupper, J. T. and Powers, R. D. (1973). *J. Exp. Zool.* **184**, 354.
Turner, R. S. and Burger, M. M. (1973). *Nature* **244**, 509.
Uhlenbrook, G. (1964). *Vox Sang.* **21**, 338.
Ukena, T. E., Borysenko, J. Z., Karnovsky, M. J. and Berlin, R. D. (1974). *J. Cell Biol.* **61**, 70.
Unanue, R. R., Perkins, W. D. and Karnovsky, M. J. (1972a). *J. Immunol.* **108**, 569.
Unanue, E. R., Perkins, W. D. and Karnovsky, M. J. (1972b). *J. Exp. Med.* **136**, 885.
Unanue, E. R., Karnovsky, M. J. and Engers, H. D. (1973). *J. Exp. Med.* **137**, 675.
Unger, F. M. (1976). *In* "Molecular Base of Malignancy", (E. Deutsch, K. Moser, H. Rainer and A. Stacher, eds). Georg Thieme, Stuttgart, p. 245.
Vaheri, A. and Ruoslahti, E. (1974). *Internat. J. Cancer* **13**, 579.
Vaheri, A., Ruoslahti, E., Westermark, B. and Pontén, J. (1976). *J. Exp. Med.* **143**, 64.
Vaidya, R. A., Glass, R. H., Dandekar, P. and Johnson, K. (1971). *J. Reprod. Fert.* **24**, 299.
Valentine, R. C. and Allison, A. C. (1959). *Biochim. Biophys. Acta* **34**, 10.
Valmet, E. (1968). *Sci. Tools* **15**, 8.
Valmet, E. (1969). *Protides Biol. Fluids* **17**, 401.
Van Beek, W. P., Smets, L. A. and Emmelot, P. (1973). *Cancer Res.* **33**, 2913.
Van Beek, W. P., Smets, L. A. and Emmelot, P. (1975). *Nature* **253**, 457.

Vanderkooi, G. and Green, D. E. (1970). *Proc. Nat. Acad. Sci. U.S.A.* **66**, 615.
Vanderkooi, G. (1972). *Ann. N.Y. Acad. Sci.* **195**, 6.
Van Gool, A. P. and Nanniga, N. (1971). *J. Bacteriol.* **108**, 474.
Van Oss, C. J., Fike, R. M., Good, R. J. and Reinig, J. M. (1974). *Analy. Biochem.* **60**, 242.
Vassar, P. F. (1963a). *Lab. Invest.* **12**, 1072.
Vassar, P. F. (1963b). *Nature* **197**, 1215.
Verwey, E. J. W. and Overbeek, J. Th. G. (1948). "Theory of the Stability of Lyophobic Colloids". Elsevier, Amsterdam.
Vesterberg, O. (1969a). *Sci. Tools* **16**, 24.
Vesterberg, O. (1969b). *Acta Chem. Scand.* **23**, 2653.
Vesterberg, O. and Svensson, H. (1966). *Acta Chem. Scand.* **20**, 820.
Vogt, P. K. (1967). *Virology* **33**, 175.
Vold, M. J. (1961). *J. Colloid Sci.* **16**, 1.
Vlodavsky, I. and Sachs, L. (1975). *Exp. Cell Res.* **93**, 111.
Vlodavsky, V., Inbar, M. and Sachs, L. (1973). *Proc. Nat. Acad. Sci. U.S.A.* **70**, 1780.
Vorbrodt, A. and Koprowski, H. (1969). *J. Nat. Cancer Inst.* **43**, 1241.
Waddington, C. H. (1932). *Phil. Trans. Roy. Soc. B.* **221**, 179.
Waddington, C. H. (1940). "Organisers and Genes". Cambridge University Press, Cambridge.
Wadström, T. (1970). *Biochem. J.* **120**, 745.
Wadström, T. and Hisatsune, K. (1970). *Biochem. J.* (120), 725.
Wadström, T., Möllby, R., Jeansson, S. and Wretlind, B. (1973). *Sci. Tools* **21**, 2.
Walter, H. (1975). *In* "Methods in Cell Biology" (D. M. Prescott, ed.), Vol. 9. Academic Press, New York and London, p. 25.
Walter, H. and Albertsson, P. Å. (1966). *Biochem. Biophys. Acta* **25**, 670.
Walter, H. and Coyle, R. P. (1968). *Biochim. Biophys. Acta* **165**, 540.
Walter, H. and Krob, E. J. (1975). *Exp. Cell Res.* **91**, 6.
Walter, H. and Krob, E. J. (1976). *FEBS Letters* **61**, 290.
Walter, H. and Selby, F. W. (1966). *Biochim. Biophys. Acta* **112**, 146.
Walter, H. and Selby, F. W. (1967). *Biochim. Biophys. Acta* **148**, 517.
Walter, H., Winge, R. and Selby, F. W. (1965). *Biochim. Biophys. Acta* **109**, 293.
Walter, H., Selby, F. W. and Garza, R. (1967). *Biochim. Biophys. Acta* **136**, 148.
Walter, H., Garza, R. and Coyle, R. P. (1968). *Biochim. Biophys. Acta* **156**, 409.
Walter, H., Krob, E. J. and Ascher, G. S. (1969a). *Exp. Cell Res.* **55**, 279.
Walter, H., Krob, E. J., Garza, R. and Ascher, G. S. (1969b). *Exp. Cell Res.* **55**, 57.
Walter, H., Miller, A., Krob, E. J. and Ascher, G. S. (1971). *Exp. Cell Res.* **69**, 416.
Walter, H., Tung, R., Jackson, L. J. and Seaman, G. V. F. (1972). *Biochem. Biophys. Res. Commun.* **48**, 565.
Walter, H., Eriksson, G., Taube, O. and Albertsson, P. Å. (1973). *Exp. Cell Res.* **77**, 361.
Walter, H., Krob, E. J. and Brookes, D. E. (1976a). *Biochemistry* **15**, 2959.
Walter, H., Krob, E. J. and Tung, R. (1976b). *Exp. Cell Res.* **102**, 14.
Ward, P. D. and Ambrose, E. J. (1969). *J. Cell Sci.* **4**, 289.
Warren, D. and Thorne, H. V. (1969). *J. Virol.* **4**, 380.

Warren, L. (1959). *J. Biol. Chem.* **234**, 1971.
Warren, L., Fuhrer, J. P. and Buck, C. A. (1972a). *Proc. Nat. Acad. Sci. U.S.A.* **69**, 1838.
Warren, L., Critchley, D. and Macpherson, I. (1972b). *Nature* **235**, 275.
Watkins, W. M. (1966). *Science* **152**, 172.
Watkins, W. M., Koscielak, J. and Morgan, W. T. J. (1964). *Proc. Internat. Congr. Blood Transfusion*, p. 213.
Watterson, R. L. (1942). *Physiol. Zool.* **15**, 234.
Weber, J. (1973). *J. Cell Physiol.* **81**, 49.
Weed, R. I., Lacelle, P. and Merrill, W. (1969). *J. Clin. Invest.* **48**, 795.
Weidel, W., Frank, H. and Martin, H. H. (1960). *J. Gen. Microbiol.* **22**, 158.
Weinbaum, G. and Burger, M. M. (1973). *Nature* **244**, 510.
Weinstein, R. S. (1969). *In* "Red Cell Membrane, Structure and Function", (G. A. Jamieson and T. J. Grenwalt, eds). Lippincott, Philadelphia, p. 36.
Weiser, M. M. (1973a). *J. Biol. Chem.* **248**, 2536.
Weiser, M. M. (1973b). *J. Biol. Chem.* **248**, 2542.
Weiss, L. (1960). *J. Cell Biol.* **26**, 733.
Weiss, L. (1961a). *Exp. Cell Res.* Suppl. 8, 141.
Weiss, L. (1961b). *Nature* **191**, 1108.
Weiss, L. (1963). *Exp. Cell Res.* **30**, 509.
Weiss, L. (1965). *J. Cell Biol.* **26**, 735.
Weiss, L. (1967). "The Cell Periphery, Metastasis and Other Contact Phenomena". North-Holland Publishing Co., Amsterdam.
Weiss, L. (1968). *Exp. Cell Res.* **53**, 603.
Weiss, L. (1972). *In* "The Chemistry of Biosurfaces", (M. L. Hair, ed.). Marcel Dekker Inc., New York, p. 377.
Weiss, L. (1974). *Exp. Cell Res.* **86**, 223.
Weiss, L. and Haushka, T. S. (1970). *Internat. J. Cancer* **6**, 270.
Weiss, L. and Horoszewicz, J. S. (1971). *Internat. J. Cancer* **7**, 149.
Weiss, L. and Lachman, P. J. (1964). *Exp. Cell Res.* **36**, 86.
Weiss, L. and Levison, C. (1969). *J. Cell Physiol.* **73**, 31.
Weiss, L. and Mayhew, E. *J. Cell Physiol.* **68**, 345.
Weiss, L. and Mayhew, E. (1967). *J. Cell Physiol.* **69**, 281.
Weiss, L. and Neiders, M. (1971). *J. Periodont Res.* **6**, 28.
Weiss, L. and Sinks, L. F. (1970). *Cancer Res.* **30**, 90.
Weiss, L., Subjeck, J. R. and Poste, G. (1975). *Internat. J. Cancer* **16**, 914.
Weiss, L., Subjeck, J. R. and Glaves, D. (1976a). *Exp. Cell Res.* **100**, 172.
Weiss, L., Glaves, D. and Subjeck, J. R. (1976b). *Exp. Cell Res.* **102**, 104.
Weiss, P. (1947). *Yale J. Biol. Med.* **19**, 235.
Weiss, P. (1958). *Internat. Rev. Cytol.* **7**, 391.
Weiss, P. (1961). *Exp. Cell Res.* **8**, 260.
Weiss, P. and Scott, B. I. H. (1963). *Proc. Nat. Acad. Sci. U.S.A.* **50**, 330.
Weiss, P. and Taylor, A. C. (1960). *Proc. Nat. Acad. Sci. U.S.A.* **46**, 1177.
Weston, J. A. (1963). *Develop. Biol.* **6**, 279.
Weston, J. A. (1970). *Advan. Morphogen.* **8**, 41.
Weston, J. A. and Butler, S. L. (1966). *Develop. Biol.* **14**, 246.

Wetzel, R. (1929). *Roux Arch. Entwickl. Mech. Org.* **119**, 118.

Whittemore, N. B., Thrabold, N. C., Reed, C. F. and Weed, R. I. (1969). *Vox Sang.* **17**, 289.

Wickus, C. G. and Robbins, P. W. (1973). *Nature New Biol.* **32**, 80.

Wiig, J. N. (1974). *Scand. J. Immunol.* **3**, 357.

Wiig, J. N., Maehle, B. O. (1975). *Exp. Cell Res.* **93**, 31.

Wiig, J. N. and Thunold, S. (1973). *Clin. Exp. Immunol.* **15**, 497.

Wildy, P. and Ridley, M. (1958). *Nature* **182**, 1801.

Wilkins, D. J., Ottewill, R. H. and Bangham, A. D. (1962). *J. Theoret. Biol.* **2**, 165.

Williams, N., Kraft, N. and Shortman, K. (1972). *Immunology* **22**, 885.

Wilson, C. B., Barker, M., Hoshino, T., Oliver, A. and Downie, R. (1972). *In* "Steroids and Brain Oedema", (H. J. Reuben and K. Schurmann, eds). Springer, Berlin, p. 95.

Wilson, H. V. (1907). *J. Exp. Zool.* **5**, 245.

Winkler, K. C. and Bungenberg De Jong, H. G. (1940–41). *Arch. Neerl. Physiol.* **25**, 431.

Winzler, R. J. (1969). *In* "Red Cell Membrane, Structure and Function", (G. A. Jamieson and T. J. Greenwalt, eds). Lippincott, Philadelphia, p. 157.

Winzler, R. J. (1970). *Internat. Rev. Cytol.* **29**, 77.

Wioland, M., Sabolovic, D. and Burg, C. (1972). *Nature New Biol.* **237**, 274.

Woeber, K. (1949). *Strahlentherapie* **79**, 563.

Woeber, K. (1959). *Internat. J. Physiol. Med.* **4**, 10.

Woo, J. and Cater, D. B. (1972). *Biochem. J.* **128**, 1273.

Woods, D. A. and Smith, C. J. (1969a). *J. Invest. Dermatol.* **52**, 259.

Woods, D. A. and Smith, C. J. (1969b). *Exp. Mol. Pathol.* **10**, 107.

Woodward, D. J. (1968). *J. Gen. Physiol.* **52**, 509.

Wrigley, C. W. (1968a). *J. Chromatog.* **36**, 362.

Wrigley, C. W. (1968b). *Proc. Aust. Biochem. Soc.* p. 36.

Wrigley, C. W. (1968c). *Sci. Tools* **15**, 17.

Wright, A. and Kanegasaki, S. (1971). *Physiol. Rev.* **51**, 748.

Wu, G. J. and Bruening, G. (1971). *Virology* **46**, 596.

Wu, H. C., Meezan, E., Black, P. H. and Robbins, P. W. (1969). *Biochemistry* **8**, 2509.

Yahara, I. and Edelman, G. M. (1972). *Proc. Nat. Acad. Sci. U.S.A.* **69**, 608.

Yahara, I. and Edelman, G. M. (1973a). *Exp. Cell Res.* **81**, 143.

Yahara, I. and Edelman, G. M. (1973b). *Nature* **236**, 152.

Yamada, T. (1967). *In* "Current Topics in Developmental Biology", (A. Monroy and A. A. Moscona, eds), Vol. 2. Academic Press, New York and London, p. 283.

Yamada, T. (1972). *In* "Cell Differentiation" (R. Harris, P. Allin and D. Viza, eds.). Munksgaard, Copenhagen, p. 56.

Yamada, T. and Yamada, M. (1973). *Nature* **244**, 297.

Yamakawa, T. and Suzuki, S. (1952). *J. Biochem.* (*Tokyo*) **39**, 393.

Yamakawa, I., Irie, R. and Inanago, M. (1960). *J. Biochem.* **48**, 490.

Yamamoto, K., Omato, S., Ohnishi, T. and Terayama, H. (1973). *Cancer Res.* **33**, 567.

Yaoi, Y. and Kanaseki, T. (1972). *Nature New Biol.* **237**, 283.
Yin, H. H., Ukena, T. E. and Berlin, R. D. (1972). *Science* **178**, 867.
Young, N. M., Leon, M. A., Takahashi, T., Howard, I. K. and Sage, H. J. (1971). *J. Biol. Chem.* **246**, 1596.
Zalik, S. E. and Cook, G. M. W. (1976). *Biochim. Biophys. Acta* **419**, 119.
Zalik, S. E. and Scott, V. (1972). *J. Cell Biol.* **55**, 134.
Zalik, S. E. and Scott, V. (1973). *Nature New Biol.* **244**, 212.
Zalik, S. E., Sanders, E. J. and Tilley, C. (1972). *J. Cell. Physiol.* **79**, 225.
Zbinden, G., Mehrishi, J. N. and Tomlin, S. (1970). *Throm. Diath. Haemorrh.* **23**, 261.
Zeiller, K. and Pascher, G. (1973). *Eur. J. Immunol.* **3**, 614.
Zeiller, K., Holzberg, E., Pascher, G. and Hannig, K. (1972). *Hoppe-Seyler Z. Physiol. Chem.* **353**, 105.
Zeiller, K., Pascher, G. and Hannig, K. (1976). *Immunology* **31**, 863.
Zhdanov, V. M. and Bukrinskaya, A. G. (1962). *Acta Virol. Engl. Ed.* (*Praha*) **6**, 97.
Zucker, M. B. (1964). *Thromb. Diath. Haemorrh.* **13** Suppl., 301.
Zucker, M. B. and Levine, R. U. (1963). *Throm. Diath. Haemorrh.* **10**, 1.

Subject Index